D1116479

Principles of Package Development

Second Edition

FW
FISCHER i
WORTMANN
KOPRIVNICA

Principles of Package Development

Second Edition

Roger C. Griffin, Jr.

Michigan State University
East Lansing, Michigan

Stanley Sacharow

Princeton Packaging Group
Milltown, New Jersey

Aaron L. Brody

Container Corporation of America
Oaks, Pennsylvania

AVI PUBLISHING COMPANY, INC.
Westport, Connecticut

Frontispiece: Four methods of packaging liquids. *Roman amphora ca. 100 BC*, courtesy of John Harvey and Sons Ltd. *Northern European Beer Cask*, courtesy of Walter Landor Associates, Museum of Packaging Antiquities. *Onion-shaped wine bottle ca. 1670–1680,* courtesy of John Harvey and Sons Ltd. *Yugoslavian alcoholic spirits bottle ca. 1920*, courtesy of Ante Rodin, Zagreb.

THE AVI PUBLISHING COMPANY, INC.
250 Post Road East
P.O. Box 831
Westport, Connecticut 06881

Library of Congress Cataloging in Publication data:

Griffin, Roger C.
Principles of package development.

Includes bibliographies and index.
1. Packaging. I. Sacharow, Stanley. II. Brody, Aaron L. III. Title.
TS195.G75 1985 658.5′64 84-24624

ISBN 0–87055–465–4

Printed in the United States of America
ABCD 4321098765

To my family, friends, and associates

To Beverly Lynn, Scott Hunter, and Brian Evan—without whom all this would have been a dream

To Carolyn and to Stephen Russell, Glen Alan, and Robyn Todd—my gift to them and to all

Contents

Preface

Since the first edition of "Principles of Packaging Development" was published, the packaging industry has undergone many profound changes. These have included the virtual elimination of cellophane and its replacement with oriented polypropylene as a carton overwrap, fluid milk in blow-molded HDPE bottles, PET beverage bottles, cookie bags and cartons lined with polyolefin coextrusions instead of waxed glassine, and bread in reclosable polyolefin and coextruded film bags. New phrases have also worked their way into the lexicon of the practicing packaging technologist, such as "child resistance" and "tamper-evident." This most popular text on packaging demanded updating.

How these phrases and ideas have affected the industry in the 1980s and how they will probably alter its course in the future are treated. New concepts of packaging system planning and forecasting techniques are intruding into package management, and new chapters will introduce them to the reader.

The years have added a certain degree of maturity to the packaging industry. Not only have the original authors broadened their perspectives and changed professional responsibilities, we have also included a third co-author, Dr. Aaron L. Brody, whose experience in the industry, academic background, and erudite insights into the very nature of packaging have added an unparalled degree of depth to this book. We would like to thank David L. May of Reynolds Metals Company for revising and adding his valuable experience to the chapter on testing. In addition, we would like to thank all those people that have helped give this book a "new view."

Roger C. Griffin, Jr.
Stanley Sacharow
Aaron L. Brody

1

Origins of Packaging Development

Packaging, which by the mid-twentieth century dominated the consciousness of the American consumer, had entered the lives of the Americans unheralded and unchronicled for the very reasons which made it distinctive. The use of packaging was a parable of the unnoticed, multiplex, anonymous sources of innovation.
Daniel J. Boorstin (1973).

Packaging development begins with man's earliest beginnings. Ancient artifacts give us some idea of when certain packages were first used. It is not known who invented the very earliest forms, but it is rather obvious the packages were created to make transportation easier. For example, it is easier to drink liquids from a cup, than one's "cupped" hand—the cup transports the liquid to the mouth. Similarly, a man cannot carry large quantities of liquids in his hands for any distance. This transportation need was solved by using hollow shells and gourds, hollowed-out logs, animal skins and bladders, and, later, pottery and glass or metallic vessels. Likewise, a man cannot carry a large quantity of small objects, such as grain, in his hands. This transportation need was met by the invention of sacks and woven baskets. Beasts of burden helped carry numerous or large containers. Wheeled vehicles helped the animals carry even more. Baskets and leather "wine skins" date back before the earliest of written history. The in-

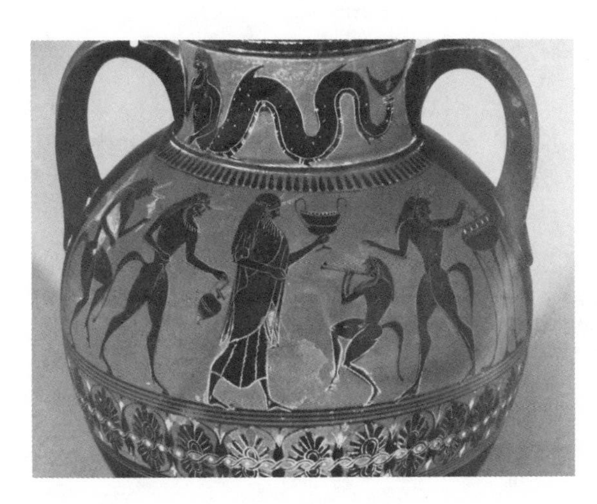

FIG. 1.1. Greek amphora made about 540 BC of clear, red-brown clay.
Courtesy of Christie, Manson, and Woods.

vention of woven cloth and clay pottery took place before 8000 BC, and crude glassware appeared not long after that date. Metals date from the end of the Stone Age. Wooden chests, kegs, barrels, and boxes were known in early Greek and Roman days.

As man progressed beyond the stage of a hunter and gatherer and became a farmer, he stopped wandering and settled in communities, which grew from villages to towns and cities. Within these cities jobs became specialized; that is, one man became a brick mason, another man a potter, another a weaver, and so on. As production of goods and services increased, trade became common between communities and nations, and this led to greater and greater transportation distances. Land caravans and sea routes were established for world trade. New lands were explored and settled to expand trade and find new products and resources. For transportation over longer distances it was desirable that the package prevent spillage and contamination. Closed containers such as barrels and container closures such as bottle plug seals were invented to accomplish this.

The use of packaging and overpackaging to protect the contents was a logical evolution. An example of this can be seen in the ancient Egyptian custom of packaging mummies.

The mummy's head was covered with an ornate mask over linen wrapping. The lungs, stomach, liver, and intestines were mummified separately and entombed in canopic jars. The wrapped mummy was then placed in an anthropoid wooden coffin and enclosed in several outer coffins before placement in a stone or wooden sarcophagus.

For centuries few changes were made in basic package forms. Thus about 1200 AD the principal package forms were as follows.

Material	*Package Form and Use*
Leather	Wrappings, bags, bottles
Cloth	Wrappings, sacks
Wood	Barrels, boxes, kegs, chests
Grass or split wood	Baskets, matting
Stone	Small pots or jars
Earthenware	Pots, jars, urns, ewers, bowls, etc.
Metal	Pots, bowls, cups, etc.
Glass	Jars, bottles, cups, bowls, etc.

EARLY HISTORY

Within the next few hundred years the industrial revolution occurred, which substituted machinery for man and beast power, there-by greatly increasing productivity. This increased the need for packaging and encouraged development of more convenient forms, particularly the tin can, the paper bag, and the paperboard box or carton.

The manufacture of paper and the art of printing had been developed in Europe during the Renaissance. However, the earliest paper was made in China from mulberry bark about 200 BC. The Arabs learned the art when a Chinese army attacked Samarkand in 751 AD and some paper makers were captured. Papers were first made from flax fiber and later from old linen rags. Although some say the Crusaders brought paper home with them from the Holy Land (not unlikely), it is certain that the Arabs introduced it to Sicily and thence to Italy and South Germany, and the Moors introduced it to Spain in the twelfth century, from whence it spread through France, West Germany, the Netherlands, Belgium, and England. Earliest records of English paper making date to about 1310 AD. The first paper making in America was in 1690 in Germantown, Pennsylvania. Cardboard was also a Chinese invention of about the sixteenth century. The first paper-making machinery in which fibers were laid down on a moving wire cloth was used in 1799 in England. The use of wood pulp in paper making was introduced in 1867.

Decoration and ornamentation predate history. Ancient artifacts show that primitive man learned to paint, carve, and sculpt. For many thousands of years all such work was applied directly to the object to be decorated. No one knows when the first experimentation with indirect

FIG. 1.2. A sealed wine bottle, *ca.* 1774; with parallel sides marked "In Williams 1774."

Courtesy of John Harvey and Sons Ltd.

methods occurred. Whoever first applied a color or ink to one surface and then pressed it against another to transfer the color was the first printer. We have evidence that the first printing on paper from carved wooden blocks took place in China about 868 AD. Also in China, at about 1401 AD, individual wooden blocks for printing of characters were used. The art of printing on paper from wooden blocks undoubtedly existed in Europe prior to Gutenberg's use of movable type in 1454 and probably was predated by block printing on textiles, but the earliest surviving printed package was a wrapper used by a German papermaker, Andreas Bernhart, in the 1550's. During the next hundred years the use of printed paper wrappers spread to other products, from patent medicines and dentifrices to tobacco and various foods. In the mid-1700s, engraved copper or steel plates were being used in place of wood blocks for printed labels. In 1798, the principles of lithography were discovered by Senefelder in Bavaria. This, together with the advent of cheaper machine-made paper, gave graphic arts a strong impetus. In the 1830s, printed color was introduced on matchbox labels. In 1875, Robert Barclay in England invented offset lithography. He applied slow-drying inks to a glazed cardboard cylinder, from which they were transferred to the desired surface. Later the cardboard was replaced with rubber-covered canvas, and this was replaced by a rubber roll.

FIG. 1.3. Delft pottery jar for snuff; Delft, Holland, *ca.* 1739–1763. *Courtesy of U.S. Tobacco Museum.*

Flexography was also born in England about 1890 in an attempt to keep up with the speed of paper bag–making machines. Here the inks were fast-drying "aniline" types and were applied to resilient rubber blocks or plates, which were used for the actual printing.

During the nineteenth and twentieth centuries, photographic techniques were invented whereby printing plates could be prepared and chemically etched. Process color separation by photography lowered costs of plate preparation and reduced the number of colored inks required. By the turn of the twentieth century, four-color lithography was commonplace and six colors were possible.

Modern high-speed rotary presses with their automatic controls all come from these earlier beginnings.

THE INDUSTRIAL REVOLUTION

Prior to the eighteenth century, most articles of manufacture were the product of hand labor. The word "manufacture" itself comes from the Latin *manu* (by hand) and *factus* (made). Only simple machines powered by muscle, wind, gravity, or water were available. The foot treadle and crankshaft turned the spinning wheel, the potter's wheel, the grindstone, and the carpenter's lathe. Dogs were trained to run on treadmills to turn roasts on spits. Horses or oxen were harnessed to windlasses to move great loads. Windmills pumped water; waterwheels turned gristmills or powered sawmills.

All this was changed by the coming of the Machine Age, which began in the latter half of the seventeenth century and within 200 years had completely revolutionized industry and the civilization dependent on it. Newcomb and Watt brought the steam engine to practical reality in the 1700's, while Lavoisier, Dalton, Priestley, and others were building the foundations of modern chemistry and Galvani, Volta, and Ohm were investigating the mysteries of electricity.

Mass production, standardization of parts, and power-driven machinery were combined successfully in 1700 by Polhem in Sweden. Similar manufacture was carried on in France in 1762. In 1785, a French gunsmith, LeBlanc, showed an automated gun making process to the American ambassador, Thomas Jefferson, who spoke and wrote about it to his contemporaries. Eli Whitney pioneered the automated procedure in the United States. In 1803, after years of tooling up, his gun factory in Hamden, Connecticut, began producing muskets at three times the rate skilled gunsmiths could make them by hand. Before long, mass production came to be known as "the American System."

The continual search for more efficient methods of production led to the invention of more and more labor-saving machines and even of machines for making machines—the machine tools, which today are regarded as the lifeblood of an industrial nation. The new machines created a need for new materials of construction. Wooden parts could

FIG. 1.4. Various figural bottles of salt-glazed stone were used as flasks, *ca.* 1840–1850.
Courtesy of Philips International.

not stand great strain or long wear. Brass was expensive. Cast iron was hard but brittle. Wrought iron was tough but too soft. The only available steels were made from wrought iron and were expensive. In 1856, Henry Bessemer answered the need. His process for converting pig iron into steel not only made the metal plentiful and cheap, but also created medium-carbon steels heretofore unknown. These mild steels were just right for long-wearing machine parts.

The textile industry was one of the first to convert from hand labor and is a good example of how swiftly progress followed. In rapid succession from 1733 came Kay's flying shuttle, Hargreave's spinning mule, and Cartwright's power loom. These machines could spin and weave better-quality and stronger thread much faster than hand labor. They could process pure cotton fiber instead of mixtures. They made cotton "king" of the agricultural world.

FIG. 1.5. Round tin of Baker's breakfast cocoa.
Courtesy of Walter Landor and Assoc.

FIG. 1.6. Late nineteenth century printed paperboard carton for dessert pudding tablets.
Courtesy of Walter Landor and Assoc.

During the 1800s, new machinery, new processes, and new scientific discoveries poured forth at a staggering rate. It would be a formidable task to name them all; thus here are but a few:

Prime power: Steam engines, internal combustion engines, the dynamo, the electric motor

Communications and graphic arts: Telegraphy, photography, lithography, rotogravure printing, the telephone, the typewriter

Transportation: Steamships, steam railroads, the automobile, the airplane, the bicycle, the balloon

Medicine: Bacteriology, antiseptic surgery, anesthesia, chemical therapy, immunization, sanitation

Food, clothing, agriculture: Food processing, seed planters, cultivators, harvesters, sewing machines, power spinning, power weaving

Primary industry: Steel, petroleum, coal, coke, coal-tar products, soap, rubber, pulp and paper, plastics

THE BIRTH OF PACKAGING DEVELOPMENT

By the end of the nineteenth century, the Industrial Revolution had created a high level of productivity and mass transportation means for moving products to the consumer. Consumers were now able to pick and choose what they would buy and what they would not. As the factory produced goods, it also created wealth and higher wages. The consumer acquired an increasing level of buying power.

FIG. 1.7. Bitters bottle, 1890.

FIG. 1.8. Early lithographed tinplate can.
Courtesy of Walter Landor and Assoc.

In this buyer's market the consumer began to demand more for his money. First of all he wanted safety. He did not want to be poisoned by the product as a result of adulteration or contamination. Secondly, he wanted quality. The product had to be well-made and its quality protected against deterioration. These demands led to legislation, unit packaging, and brand identification and advertising. Legislation came into being with respect to sanitation and purity in food, drug, and cosmetics manufacture; truthfulness in labeling; and safety in the manufacture and transportation of goods in commerce. Unit packaging came into being because it gave greater protection to the product. Brand identification made it possible for the consumer to select products of high quality, this knowledge coming from previous experience or as a result of persuasion through advertising.

Until comparatively modern times, about 1930–1940, packaging was considered pretty much a necessary evil. The consumer demands mentioned above had to be met, but this was an added cost, and businessmen were principally concerned with holding this cost to a minimum.

With the advent of a more affluent society, new factors entered the picture. As people worked shorter hours, they had more leisure time and began to demand more conveniences in both food purchasing and preparation. Supermarkets were invented, where the customer no longer relied on a sales clerk in selecting a purchase. Now the package had to provide convenience and sell the product. Easy-opening devices, easy reclosures, smaller unit portions, prepared foods, and easy-dispensing packages are all examples of convenience items. Use of the graphic arts, brand symbols, unique package shapes, and see-through wrapping are all examples of package selling devices.

The relatively new profession of packaging development came into being in the last several decades to accomplish the aims of the present-day packaging "industry." It is a multidisciplined profession requiring

FIG. 1.9. A superb selection of perfume bottles, *ca.* 1900–1930.
Courtesy of Phillips.

a knowledge of packaging materials; packaging materials' converting; package fabrication, filling, and closing; packaging machinery; package testing; product properties; shipping, storage, and handling procedures; packaging economics; commercial art and design; marketing and advertising; and legal regulations. The cross-influences of each of these knowledge areas will be discussed in the ensuing chapters.

NATIONAL AND INTERNATIONAL RECOGNITION

That packaging has been recognized as an essential function by industry throughout the world is reflected in the formation of various industrial and technical societies relating to its various aspects.

In 1952, packaging leaders in several European countries formed a packaging alliance on the continent that united the various individual packaging associations into the European Packaging Federation. The

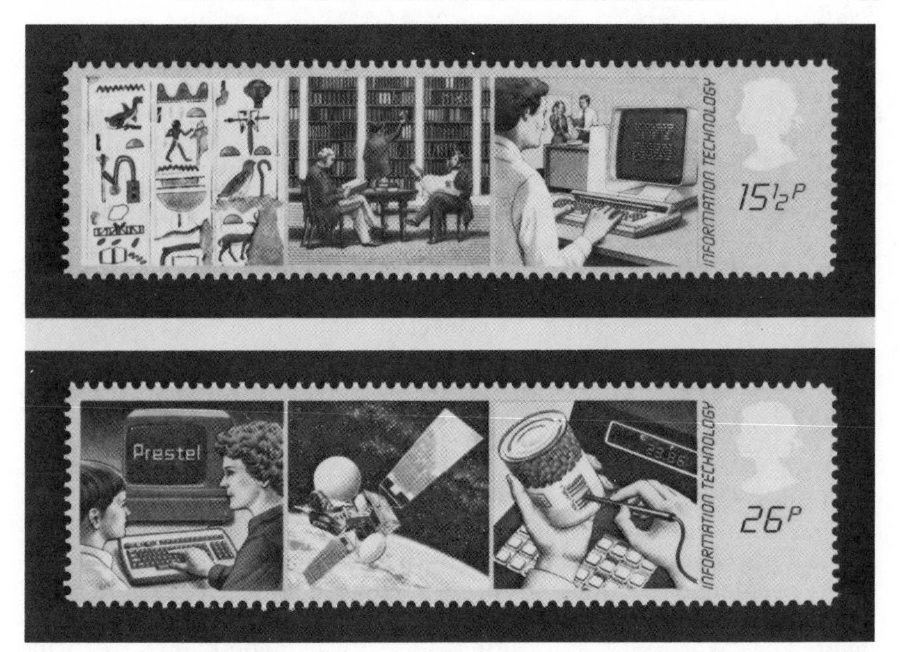

FIG. 1.10. British postage stamp, 1982. These British postage stamps were issued to commemorate information technology. Note the use of the universal product code (UPC) on the stamp in the lower right.
Courtesy of The Post Office (U.K.).

success of this organization encouraged its members to appoint a study group in 1966 to prepare proposals for a World Packaging Organization. Also in 1966, the Asian Packaging Federation was formed and immediately announced interest in participating in a world organization. On September 6, 1968, the Inaugural General Assembly of the World Packaging Organization was held at the Tokyo International Conference at the time of the Tokyo World's Fair. Representatives of 21 nations from the European Federation and of 7 nations from the Asian Federation attended, along with observers from Australia, Mexico, Canada, and the United States. The Inaugural General Assembly approved the following Declaration of Intent:

> The main objectives of the World Packaging Organization are to provide a forum for expansion of knowledge on packaging; to promote the development of packaging; to develop skills and expertise in packaging; and in general, to create conditions for conservation, preservation, and distribution of world food production; to preserve and protect quality and effectiveness of medical and hygiene supplies, and to contribute to the development of worldwide trade.

Under the approved statutes, voting membership in the World Packaging Organization is limited to continental packaging federations. Therefore, The Packaging Association of Canada and the

Packaging Institute (United States), being the sole fully integrated packaging organizations of their respective countries, were approached to determine whether they would join in forming a North American Packaging Federation and then, as such, become a member of the world body. This was agreed, and on April 15, 1969, the North American Packaging Federation was launched in Chicago.

Before the North American Packaging Federation had been started, the World Packaging Organization Administrative Council was actively initiating programs of global interest. The project of "World Hunger, As It Is Affected by Packaging," was assigned to the India Institute of Packaging, and the project on "Education" was assigned to the British Institute of Packaging. At the first the council meeting attended by North American Packaging Federation representatives, held in London in April 1969. The Package Institute (United States) was requested to accept a project on "Solid Waste Disposal."

Realizing that discarded packaging constitutes 13 percent of the total solid waste problem, and that literally millions of tons of valuable raw materials are lost through such wastes each year, The Packaging Institute accepted this assignment with enthusiasm.

The problem of "Uniform International Food and Drug Regulations" was referred to the German Packaging Institute for review and comment. The "World Star Packaging" award was made known, with the first competition taking place at World Pack in Utrecht, Holland, in April 1970.

The global approach of the packaging industry to global problems shared by all mankind truly reflects that the science and art of packaging technology have grown to full maturity and that packaging is no longer regarded as a necessary evil.

The packaging industry likewise depends on the ingenuity and skill of package development scientists, whose continuing efforts ensure a never-ending supply of new and better packages for the products of world commerce and industry.

BIBLIOGRAPHY

ANON. 1970A. Industrial packaging and the contract packer. Packag. Rev. *91*, 17–19, 22–23.

ANON. 1970B. Will Industry Sell Recycling. Mod. Packag. *43*, 46–49.

ANON. 1970C. Component protection. Envirón. Eng. *43*, 19.

ANON. 1971A. Guide to Foodpack. Packag. News *18*(2).

ANON. 1971B. Packaging and Container. Army Test and Evaluation Command. Rep. No. MTP–10–2–211, p. 15.

ANON. 1972. Trade prospects and development—Industrialists' view on packaging. Packag. Rev. *92*(10).

ANON. 1974. Helping the retailer reduce in store losses Mod. Packag. *27* (1), 36–37.

ANON. 1975A. Packaging studies. Can. Packag. *28* (6), 7.

ANON. 1975B. Trends in Packaging. Spec. Rep. No. 18. New York State Inst. Food Technol. Proc. Ninth Annu. Symp.

ANON. 1976A. Container, review and outlook. Containers Packag. *28* (4), 3–11.

ANON. 1976B. The packaging story in Europe PIEC News, 1–11.
ANON. 1976C. Turning point. Mod. Plast. Int. *6* (1), 8–9.
ANON. 1977. Flexibles future. Mod. Packag. *50* (7), 82–83.
ANON. 1978A. Packaging and package printing 1975–1977. Bibliography No.B8114. Rochester Institute of Technology, Graphic Arts Research Center. Rochester, New York.
ANON. 1978B. New applications in packaging technology. J. Soc. Environ. Eng. *17* (3), 25–27.
ANON. 1978-79. Containers and packaging review and outlook. Containers Packag. *31* (2), 3–18.
ANON. 1980. Packaging trends and challenges. Can. Packag. *33* (2), 19–22.
ANON. 1981 The Package's Role in Product Success, Food Eng. *53* (1), 79.
BANKS, W. H. 1976. (Editor). Research and development of relevance to packaging in British Universities and Colleges of Technology. Pira Bibliography No. 829.
BOORSTIN, D. J. 1973. The Americans: The Democratic Experience. Random House, New York.
BRODATSCH, E. 1981. Development of corrugators in the 80's. Paperboard Packag. *66* (11).
BUREAU, W. H. 1981. Paper: Its far-reading users. Graph. Arts Monogr. *53* (3), 198–199.
COVELL, P. 1977. Packaging in the U.S.: A sign of things to come. Packag. Rev. *97* (2).
DAVIS, A. 1967. Package and Print. Clarkson N. Potter, New York.
DAY, F. T. 1970. Into the seventies with glass. Flavor Ind. *1* (3), 179–181.
DEAN, J. M. 1970. Packaging in the seventies. Packaging *41* (488).
FRIEDMAN, W. F. and KIPNESS, J. J. 1960. Industrial Packaging. John Wiley & Sons, New York.
HANLON, J. F. 1981. Elements of packaging: approaches, scope, function, technology, Part I. Packag. Technol. *11* (6), 29–32.
HANLON, J. F. 1982. Elements of packaging: approaches, scope, function, technology, Part II. Packag. Technol. *2* (1), 4–10.
HECHT, M. 1970. Trends in the carton industry. Graphic Arts Mon. *42* (3), 44–45.
HECKMAN, J. H. 1971. Role of FDA in packaging. Plast. Technol. *17* (1), 21–22.
KANE, J. N. 1950. Famous First Facts. H. W. Wilson Co, New York.
KAUFMAN, M. 1963. The First Century of Plastics. The Plastics Institute London.
KELSEY, R. J. 1978. Packaging in Today's Society. St. Regis Paper Co., New York.
KRAUS, M. 1977. Glass in packaging today and tomorrow. Verpack Rundsch *28* (2), (German)
LIGHT, A. The primary role of packaging. Marketing, April, pp. 31–32.
LOPEZ, A. 1969. A Complete Course in Canning, 9th Edition. Canning Trade, Baltimore, Maryland.
MARDER, H. 1976. Packages, man your battle station. Mod. Packag. *29* (3), 51–52.
MERMELSTEIN, N. H. 1976. An overview of the retort pouch in the U.S. Food Technol. *30* (2), 28–29, 32, 34–37.
PAINE, F. A. (Editor). 1962. Fundamentals of Packaging. Blackie and Son, Ltd., London.
PAINE, F. A. 1975. What packaging means to the quality of life. Packag. Technol. *21* (13), 1–4, 16.
PATTERSON, W., and SEARLE, G. (Editor) 1973. Packaging in Britain. A policy for containment. Friends of the Earth Ltd., London.
SACHAROW, S. 1968. Packaging materials. *In* Encyclopedia of Polymer Science and Technology, Vol. 9, H. F. Mark and N.G. Gaylord (Editors). John Wiley & Sons, New York.
SACHAROW, S., and GRIFFIN R. C. 1980. Principles of Food Packaging 2nd Edition. AVI Publishing Co., Westport CT.
SELIN, J. 1977. Packaging in the third world. Packag. News, Spring quarter, p. 5.
SEYMOUR, R. B. 1975. Packaging: The American angle. Plast. Rubber Wkly. No. 567, 16–17.

SLADE, R. C. 1976. Information and innovation are key words for packaging user. Packag. Technol. *22* (141), 11–13, 16.

SHARMAN, H. 1977. The scientific approach to packaging. Marketing, October, 77, 79–80, 82.

SMITH, A. 1977A. A history of tinplate. Tin Int. *50* (3), 85–87.

SMITH, A. 1977B. A history of tinplate. Tin Int. *50* (9), 325–326, 344.

SMITH, A. 1978. A history of tinplate. Tin Int. *51* (6), 212–215.

TAKATSUKI, K. and YAYASHIBARA, I. 1973. Quality control of packaging materials. Jpn. Plast. Age *11* (12), 23–28.

U.S. Food and Drug Administration. 1971. FDA and the Packaging Industry. FDA (Food Drug Adm) Pap. *5* (8), 11–14.

WHITE, D. 1977. What do you want from packaging? Design, No. 338, 44–49.

2

Package Development: An Overview

The greater the similarity between products, the less part reason plays in brand selection.

David Ogilvy (1973)

INTERRELATIONSHIPS

The ultimate design of a package is a choice which represents the distillation of a multitude of lesser decisions, each relating to a specific package or product requirement as defined by management, markets, sales, manufacturing, or research and development. Each of these groups approaches the subject from a different viewpoint, yet makes an important contribution to the whole.

Management's Role

Packaging has both immediate and long-range influences on sales and profits. Immediate influences are those that bear on impulse or first-purchase buying. Long-range influences are those that convince the purchaser to rebuy the product because of a convenience feature or a quality feature. In today's highly competitive market, an enlightened management must be aware of the role that packaging plays in profit making and must be prepared to support an imaginative packaging development program. The packaging development program is usually the responsibility of a packaging executive, who may be called a director, manager, or coordinator. In some companies the packaging development program is handled by a committee reporting to a corporate executive.

Regardless of the organizational arrangement, the packaging development department has a number of important responsibilities:

1. To be aware of existing and new packaging materials and have a general knowledge of their properties and costs.
2. To be aware of marketing and technological developments that create new packaging requirements and make old ones obsolete.

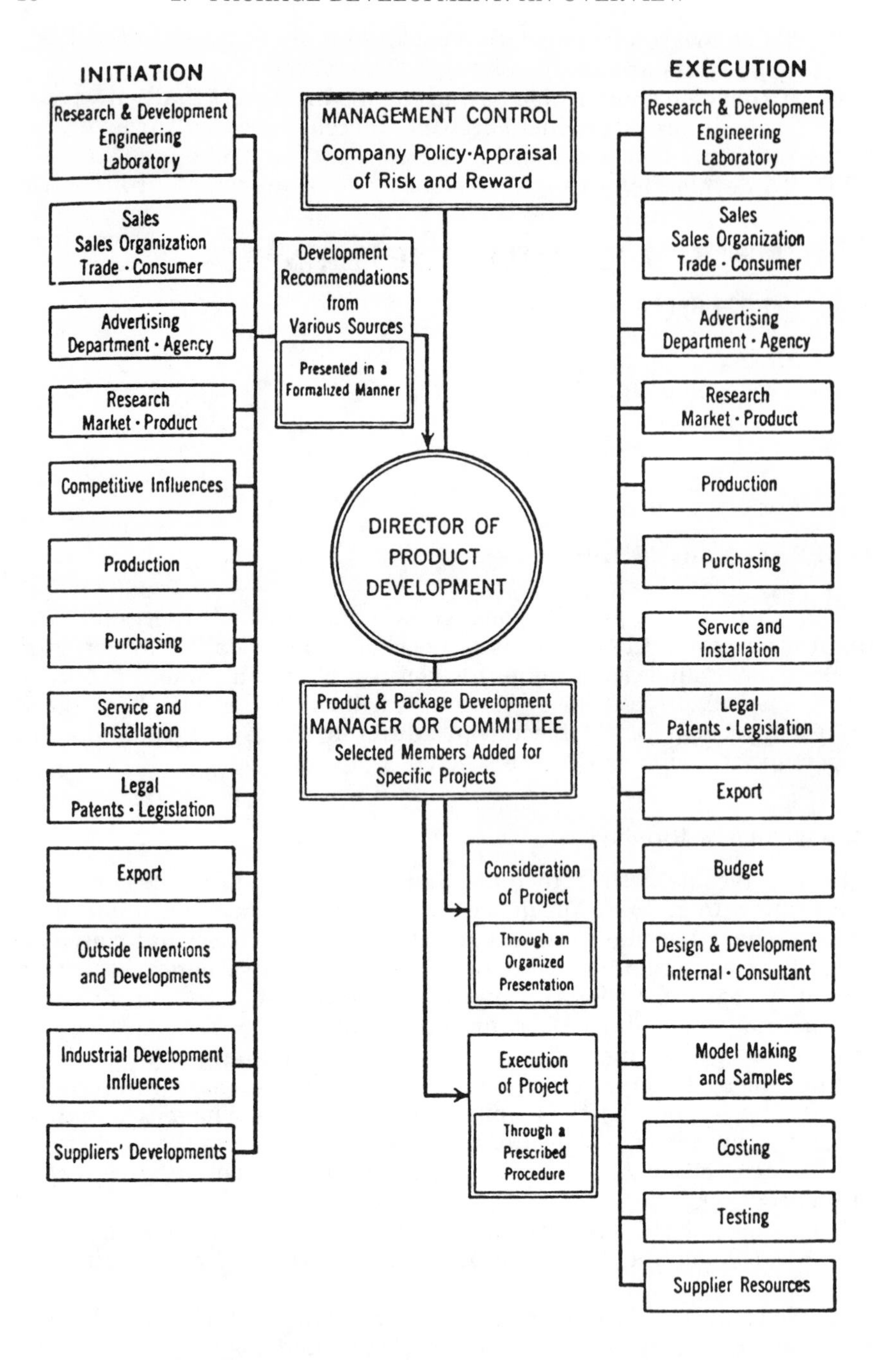

FIG. 2.1 Coordination of the project

3. To be aware of competitive packaging, its composition and economics, its advantages and disadvantages.
4. To develop short-range packaging modifications that will reduce costs, improve product shelf-life, increase product turnover, improve product acceptance, or help introduce new products.
5. To develop long-range packaging programs which promote the corporate image, coordinate a line of products, reduce waste, extend market areas, or help the company expand into new markets.
6. To develop new packaging innovations.
7. To be aware of available packaging machinery and its capabilities.
8. To develop complete packaging systems which integrate package, product, and equipment needs.

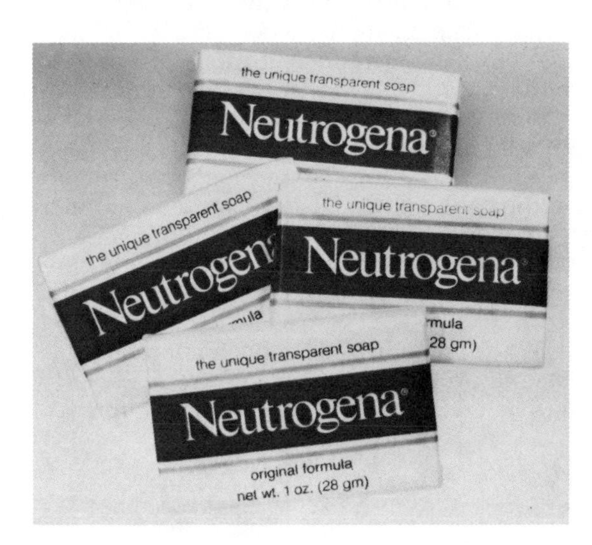

FIG. 2.2 Neutrogena soap has changed from a transparent overwrap to a new opaque polypropylene wrap. This is a superb example of increasing both the package's moisture vapor transmission rate and its sales appeal.
Courtesy of Hercules, Inc.

The amount of effort expended in a development project in packaging depends on the type of development required. There are several types, which may be categorized as follows.

1. Modification of an existing package for an existing product in order to improve sales through (a) improved aesthetics of package design; (b) improved price due to lower cost materials, improved manufacturing efficiencies, or improved performance in handling, distribution, and storage; (c) improved product quality due to improved package protection; (d) improved product utility due to improved package performance and convenience features.

2. Expansion of a product line through use of a well-tried and proven package that has been used for (a) similar products or (b) radically different products, but can be adapted to the proposed use.
3. Development of a new package concept for a proven product to improve sales through (a) revised customer interest, (b) improved product quality, and (c) improved product utility.
4. Development of a new package concept for a new and untried product (see Fig. 2.3).

The development project selected also must be considered in terms of its overall cost versus the potential return in the form of improved sales and in terms of the time required to bring the final result to the marketplace. Too long a time spent in development can render the product obsolete before it is completed, or a competitor may get there first. In modifying an existing package, management must consider not only whether the change can be accomplished quickly, but also whether it should be accomplished abruptly or in gradual stages. An abrupt change may lose some customers due to loss of consumer identification with the product, but it is justified where there has been a major improvement in the product, where there has been a major change in marketing or distribution, or where there has been a major change in corporate ownership or in the desired corporate image. Major, abrupt changes are usually introduced when a product's sales are leveling off, it is beginning to lose its share of the market, and profits are disappearing.

Minor, gradual changes are ordinarily introduced when a product is enjoying a good, strong sales and profit level. Changes are made only to reduce costs, improve quality, add convenience, or to keep up with competition.

DEVELOPMENTAL PATHS[1]

There are two paths relating to the development of a new product/package system. The main path is the development of the total system from concept to marketplace, and the secondary path is the development of the package itself as an integral part of the whole system. Which path the packaging development department follows depends on the type of company involved. Primary producers of products that need to be packaged follow the main path. They may also use the secondary path, or they may call on package materials suppliers and converters to follow the secondary path and to develop a suitable package for their product. In order to identify the packaging development function, it is necessary to define both of these pathways.

[1] In an excellent publication Melis (1979) summarizes many of the creative steps inherent in proper package development.

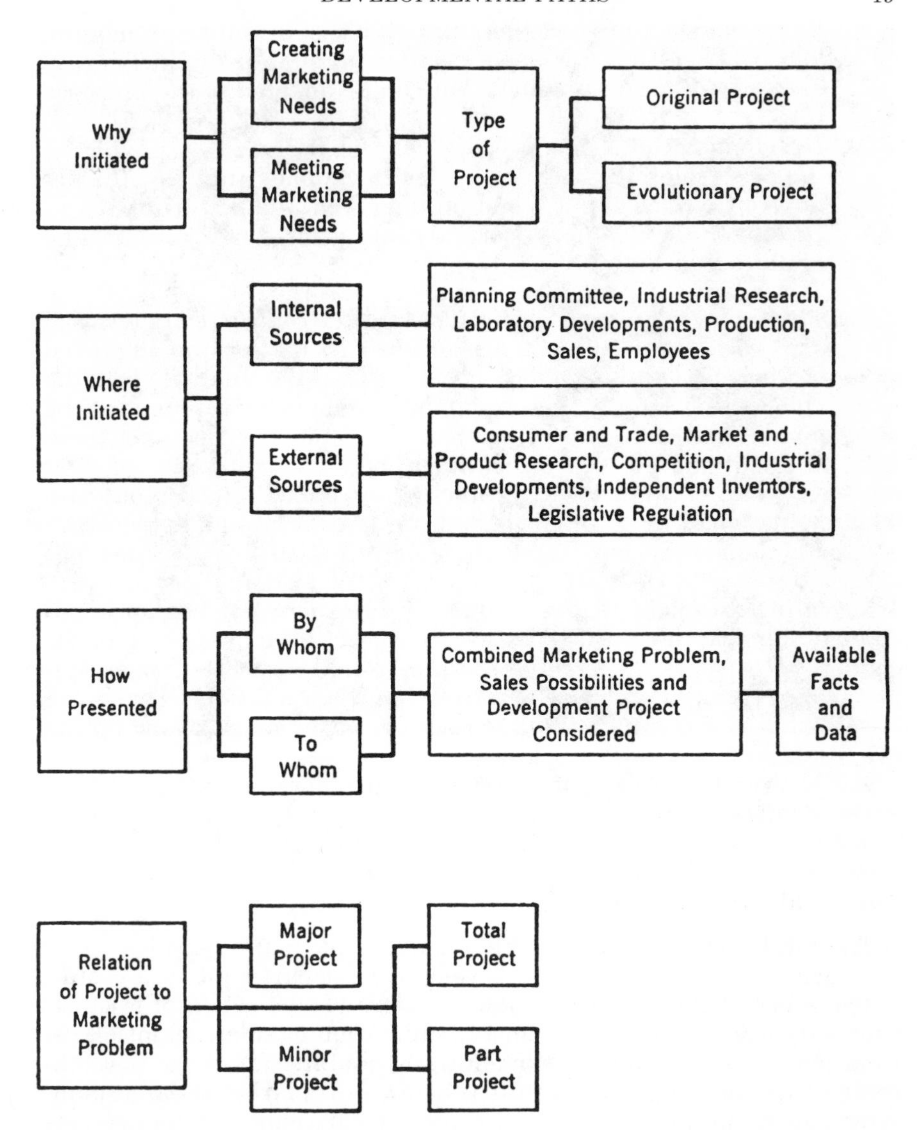

FIG. 2.3 Initiation of the project

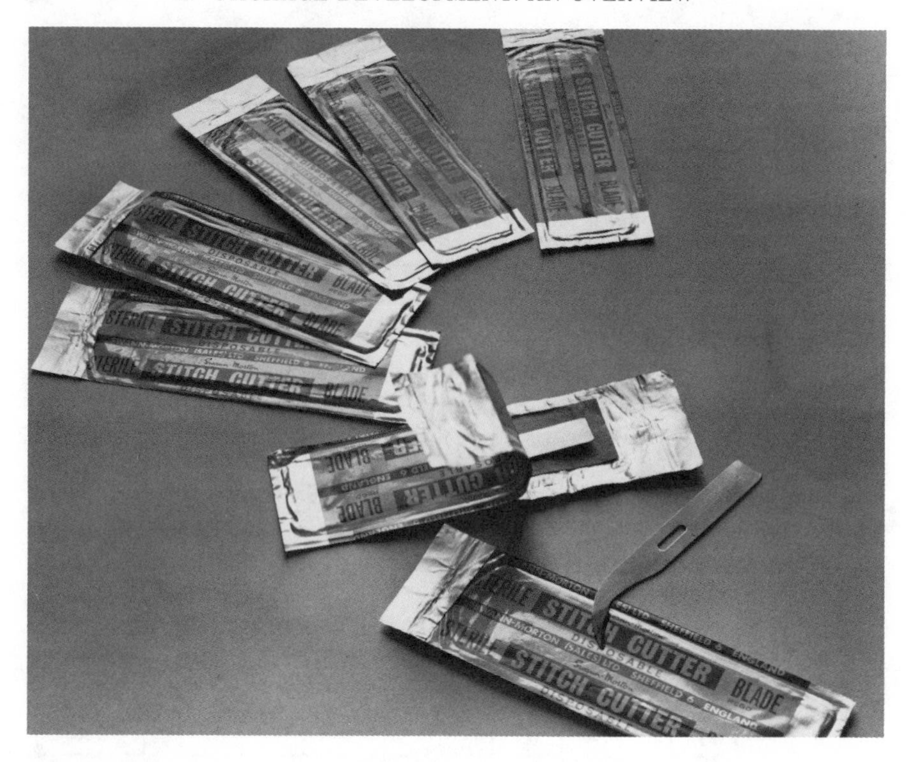

FIG. 2.4 Peel-seal foil package for surgical blades.
Courtesy of Aluminum Foils Ltd.

The Total System Path

The total system path involves management, markets, sales, manufacturing, and the packaging development department in a coordinated effort. It includes the following steps: (1) definition of the goal (initial reason for the development, estimation of sales potential, estimation of acceptable developmental cost and final package cost); (2) the package development path (see below); (3) market testing (planning, execution, and analysis of results); (4) decision whether to proceed, modify and retest, or to drop further effort; and (5) full production (planning and execution of gear-up to full production together with national sales and advertising programs) (see Fig. 2.5).

The Package Development Path

The package development path involves packaging management and in-house, supplier, or contracted package development personnel. It comprises the following steps: (1) definition of the product properties

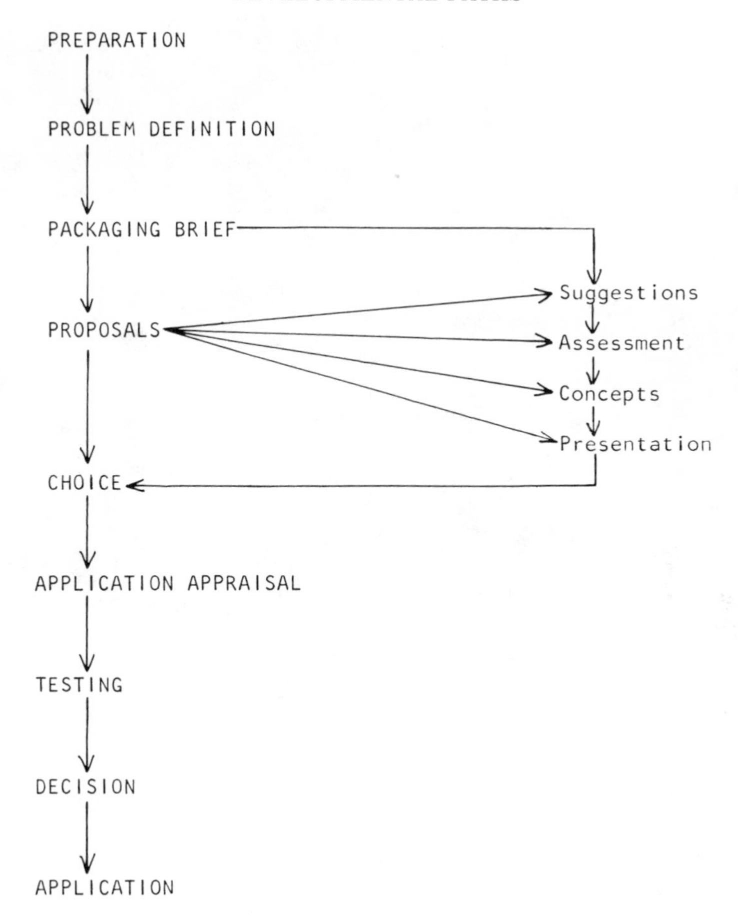

FIG. 2.5 Packaging development flow sheet

as they relate to package technical requirements; (2) definition of package technical and functional requirements; (3) definition of package styling and design requirements; (4) identification of legal or other restrictions; (5) selection of possible package designs and materials; (6) estimation of the probable cost of development; (7) deciding whether to proceed; (8) package preparation and testing for technical performance, consumer preference, and economic feasibility; and (9) deciding whether to proceed to market testing.

Should the decision to go to a test market be affirmative, the packaging development personnel will be involved in the final steps of the total system path.

Initial Reason for the Development. The various reasons for changing a package have been described previously. The initial impetus for

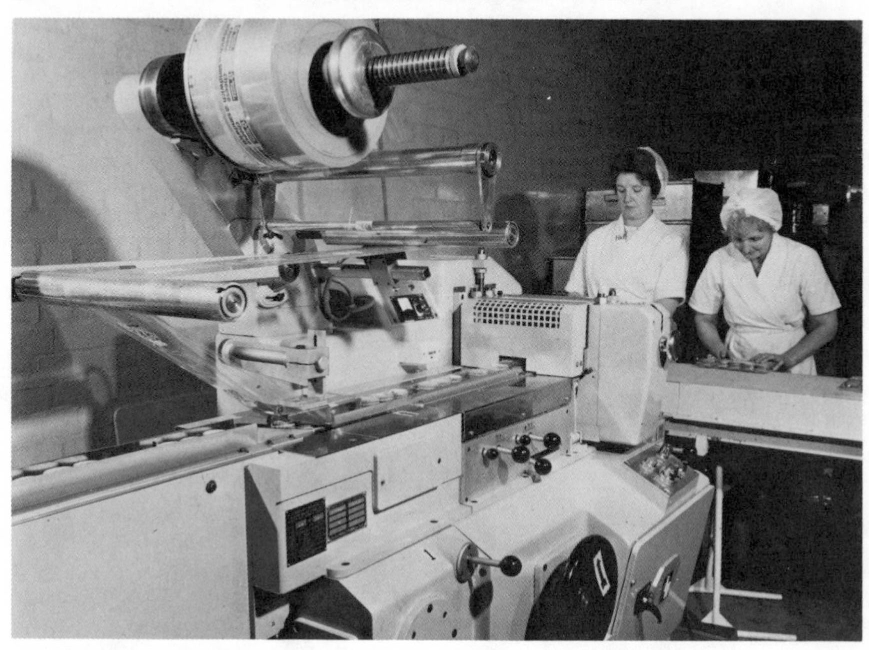

FIG. 2.6 Packaging biscuits in the Huntley and Palmers factory at Huyton, Lancs., United Kingdom, using polypropylene film.
Courtesy of Imperial Chemical Industries, Ltd.

the development may originate anywhere. A new technical development may produce a more efficient process, a new source of supply may develop for a cheaper or higher-performance material, or advertising may want to create a "new image."

Economic Parameters. Market research should enter the picture at this point to establish basic economic criteria. What is the estimated sales potential for the proposed product? Does the selling price include a sufficient profit margin to recover the developmental cost in a reasonable time? Will the consumer pay a premium? What special features will the package have to have to command such a premium? Artists' sketches or mock-up models of proposed packages can be included in this investigation. If careful market research is not done at this point but is deferred until the test market stage, a great deal of development money may be spent only to find that the package meets all requirements except the economic ones. Profit vanishes and costs may never be recovered. The developmental path may be followed through step 6 above at little cost, but after step 7 (the decision to proceed) the developmental cost will become significantly larger. It cannot be emphasized too strongly that step 1 of the total system path and steps 1 through 7 of the developmental path are critical and interrelated.

Definition of Product Properties Affecting Package Technical Properties. In order to decide what properties a package must have, it is first necessary to know the properties of the product it is to contain.

General Category. First it is necessary to consider the general category of the product. Is it a food, a drug, a cosmetic, a beverage, a textile, a machine, a chemical, a hardware part, a toy, glassware, an instrument, a piece of furniture, etc.? This may seem obvious, but each major classification may have special rules or regulations for packaging or for shipping and handling. It is best to know the ground rules right from the beginning.

Physical Form. The general physical form of the product will have great bearing on the type of packaging to be used. It is therefore necessary to know, for example, whether it is (1) a massive bulk or a small unit; (2) a solid, liquid, or gas or a combination, such as a form or an emulsion; (3) massive, chunky, granular, or powdery, if a solid, or watery and thin, or thick and viscous, if a liquid; and (4) soft and light or hard and dense.

Special Properties. It is necessary to note any special properties of the product that will require special features in the package. Is the product sensitive to temperature? Must it be protected against extreme heat or cold? Will it be marketed frozen? Will the entrance of moisture or solvent or evaporation of the product make it unsalable? Is the product fragile, requiring special cushioning against shock or special protection against crushing? Is it susceptible to attack by insects, fungi, mildew, rust, or bacteria? How long must the package protect? If the product normally deteriorates in three months, it is foolish to provide a package that gives protection for two years.

Hazards. It is particularly important to be aware of possible hazards inherent in the product. If it is poisonous, corrosive, flammable, sharp, explosive, or radioactive, extra precautions are needed and special regulations apply. If it has a very strong odor, it might be offensive to people, or it could affect other products stored near it.

Definition of Package Technical and Functional Requirements. The functional requirements for a new package must be defined exactly and completely if the development of the package is to be accomplished with economy and dispatch. All too often a package development nears completion and then the packaging scientist is informed, "This is not what we want! We forgot to tell you that another property is needed!" Therefore, it is vital to the success of packaging development to be complete in fact gathering so as to define the target accurately. Since the number of questions that need to be answered are many, a checklist is of extreme help in keeping track of and properly organizing the answers.

From the data supplied by markets and the information gathered about properties of the product itself, the package engineer begins to list the properties the package must have and at the same time begins to eliminate certain materials and packages from consideration. In

FIG. 2.7 A Transwrap 330 vertical machine produces flat-bottom or pillow bags at speeds of up to 80 per minute.
Courtesy of Bosch Packaging Machinery.

addition, unless a radical departure from the conventional approach is desired, experience can be drawn on for guidance. For this reason it is extremely desirable to know what packages have been or are being used for similar products. A study of the advantages or disadvantages of these packages will help in making a selection. Conventional packages will reveal the current market unit sizes or counts, existing prices, and whether product visibility is expected and will give some evidence on the shelf-life that may be anticipated.

In considering the type of inner and outer packaging needed for a product, it is necessary to determine the nature of the handling, storage, and distribution cycle from the point of manufacture to the point of consumption of the product. The less handling a package receives, the less likelihood there is of damage due to shock, vibration, dropping, crushing, and the like. The area of distribution is a factor in this. A product sold within a radius of, say, 50 to 100 miles will require less shipping protection than one that is distributed nationally or is exported. Industrial products are often shipped more directly than are consumer products—that is, from plant to warehouse to warehouse to plant. Consumer products may be shipped from plant to warehouse, to regional distributor warehouse, to area distributor warehouse, to local wholesaler, and to retail market and may thus change hands more times and be exposed to more handling damage.

The effect of climate in a given market area must be considered. Sensitive products shipped to or through humid tropical climates require extra protection against moisture and heat. High altitudes can cause pressured packages to explode. Thus the climate (or environment) within vehicles of transportation and within warehouses is equal in importance to the type of packaging chosen.

The pallet-stacking or warehouse stacking procedures will influence the strength of materials used in shipping containers; some delicate products may require revised procedures.

Package Functional Requirements Related to Package or Product Manufacture. There will be a number of package limitations imposed on the packaging engineer by the very nature of the package manufacturing process. For example, certain methods of forming aluminum cans or containers will not permit the depth to diameter ratio to exceed specified limits. Other limitations may be imposed by the manufacturing line for the product itself. Filling speeds may require a wider mouth. Production-line speeds may dictate what type of heat-sealing compound or labeling adhesive must be used. Thus the package development engineer should be fully cognizant of all phases of package and product manufacturing that can affect the package's functional requirements.

Definition of Package Marketing Requirements

Styling and Design. Marketing, advertising, and sales people will study the market and propose the styling and design requirements for the package. The package development engineer must be aware of these proposals from the start, as they may put limitations on the technical and functional properties of the package. An extremely-complex-process gravure print design will demand a higher-quality paper or paperboard than a simple one- or two-color line printing. An unusual shape may need heavier wall thicknesses or deeper score lines.

Shelf display requirements may dictate different package orientation in the shipper or a different package shape. If the package is to be displayed by hanging from a hook, a different design is needed than if it will be in a box.

Utility Features. Conventional packaging will reveal what utility features may be expected by the consumer in the marketplace. These include such items as easy-opening devices, reclosure features, pilferage protection, and, of increasing importance today, ease of disposability versus reusability. Markets will define what utility features are required.

Identification of Legal Restrictions. The packaging development engineer must be fully aware of all legal and other regulatory restrictions that may influence his choice of a packaging material or design. One of the most obvious limitations is the possible infringement of an existing patent. Usually suppliers are fully cognizant of patent status on their products and will call attention to the need for obtaining licenses for their use. The lack of such notice, however, does not absolve the user from infringement.

The Food Additives Amendment (1958) to the Food, Drug, and Cosmetics Act of 1938 puts severe restrictions on the use of packaging materials that might with normal or abnormal handling contaminate the product and be consumed. All such packaging materials must be given approval. Methods for obtaining this approval are discussed in Chapter 9.

The Federal Trade Commission Act of 1914 relates to provisions for requiring safety in transportation of goods, whereas the Hazardous Substance Labeling Act of 1960 and the Fair Packaging and Labeling Act of 1966 set strict standards on the information required on labels, on what can and cannot be said on packages, and on the proliferation of package sizes.

Failure to be cognizant of statutory limitations can prove disastrously costly in the development of a proposed package.

The engineer must also be aware of other nonstatutory regulations that may affect his choice of a package. Industry-accepted standards may not include a dimension or size desired. Some religions (particularly the Jewish faith) have strict rules applying to packages that are used for foods. Ethnic groups may have prejudices against certain colors or shapes.

Although there are not yet statutory limitations pertaining to package disposability, the present worldwide concern over ecology and our contaminated land, water, and atmosphere could well lead to future regulations. The packaging scientist must now also consider these factors in selecting materials for a new package system.

Selection of Possible Package Designs. The packaging development engineer, working together with styling and design people, now prepares a list of possible package designs which he considers to be technically feasible. The list will include specifications on materials, methods of manufacture, and estimated costs. This list will be subjected to a preliminary screening, and the preferred few possibilities will then be rendered into artists' sketches or actual package mockups so that some idea of consumer preference can be obtained by market research.

Estimation of Developmental Costs. After proceeding to this point, the packaging engineer should have a good grasp on the magnitude of the problem he faces. For each proposed package design he should be able to estimate the developmental cost needed to bring it to the marketplace and the probability of success. He should also be able to point out whether new or modified production equipment might be required.

Decision Whether to Proceed. It now becomes a management decision whether to proceed with the development. If all the foregoing has been done, management will be able to make an intelligent decision with minimal risk of failure.

Evaluation Testing of Proposed Package Designs. The packaging engineer then must obtain a number of packages for each design concept so that evaluation tests can be conducted. He may purchase the packages or he may make them by hand, by pilot machinery, or in a regular manufacturing line. He may choose several alternative materials or fabrication methods for each concept. During this procedure he eliminates only those items which prove impossible to manufacture or too costly. When the packages are in hand, he subjects them to product compatibility tests, design fulfillment tests, and shipping and abuse tests.

Product compatibility tests determine whether the package tends to adversely affect the product and whether the product tends to adversely affect the package. Design fulfillment tests determine whether the package meets the design and performance criteria. Will it hold the desired quantity, will it dispense, does it provide critical protection needs, does it provide minimum shelf-life requirements? Shipping and abuse tests determine whether the package will survive the normal handling to be expected in passing through distribution channels. Abuse testing determines the margin of safety built into the package. How much longer shelf-life than the minimum can be expected? How many times can the package be dropped?

Finally, it may be desirable to take the best design concepts and submit them to a consumer placement test. This helps gauge the consumer's reaction to the design and the likelihood of its being a commercial success. Unexpected consumer dislikes may eliminate some concepts or require some design modifications.

The package development engineer must be familiar with the costs of materials and of final package forms so that in selecting a package design he will not choose one that is excessively expensive for the product need. He walks a fine line between inadequate packaging, which causes claims and loss of sales, and overadequate packaging,

FIG. 2.8 Vibration table used to test shipping ability of packages.
Courtesy of Gaynes Engineering Co.

which gives unnecessary protection at too high a cost. In the beginning it is best to err on the latter side, as process refinements or new technological developments frequently permit the package costs to be reduced without loss of performance in the market place.

It is also hard to judge how much the consumer will pay for the added convenience provided by a new product-package combination. In today's affluent society, consumers seem willing to pay a large premium for a product that is easier to use and saves time in preparation, use, or after-use disposal. Frequently this justifies a more expensive package choice. One needs only to compare the costs of aerosol can dispensers for paints, cosmetics, and household products with the costs of the less convenient competing packages to see this convenience premium. Or one can note the tremendous increase in the number of prepackaged, prepared meals that reduce the homemaker's time in the kitchen.

With highly competitive commodity items, the packaging engineer must maintain a constant vigil for new materials or new ideas that can result in lower packaging costs or better product protection.

Decision to Proceed to Market Test. Management again must make a decision. Now one or two package designs have been thoroughly laboratory tested. Limited production runs have indicated that the package can be made and used. Shipping tests and shelf-life tests have indicated that it will function. Estimated costs are reasonable. All that remains is to find out whether the consumer will buy the product and at what price. Management must decide whether the cost of a market test is justified.

The decision is one of the most critical management decisions in the development of a product, and the package design will be a major factor in whether the test market will succeed or fail. Up to this point all that has been risked has been development costs. Beyond this point not only are larger expenses for materials and services risked, but the entire company image is exposed to the adverse reactions that may occur in the marketplace should the package fail to do its intended job. A successful package development may or may not help a product attain marketing success, but a package failure almost certainly will damage the chances of a good product attaining success. For this reason it may be prudent to retreat in the face of uncertainties and undertake further development work rather than risk failure in the test market. It may be necessary to redefine marketing or technical parameters, to find other design concepts, and to make further evaluation tests on previous concepts to eliminate some of the uncertainties.

Market Test. The planning, execution, and analysis of results of a market test are primarily the responsibility of marketing personnel. However, other departments will be heavily involved. These include market research personnel, who plan the scope of the test and analyze the results; advertising personnel, who plan the type and amount of advertising to be used; sales personnel, who help select target areas and enlist field cooperation; manufacturing personnel, who produce the product for the test market; engineering personnel, who provide machine modifications if required; purchasing personnel, who locate adequate sources of raw materials; and packaging development personnel, who must provide the specifications and technical advice and be prepared to cope with unexpected minor problems.

The market test will reveal whether the consumer will buy the product and at what price, in what sales units, and in response to what advertising. It can also indicate whether sales are sustained, indicating repurchase, or whether they taper off, indicating one-time buying. It will reveal shortcomings in the package design, if any, and manufacturing problems.

Decision Whether to Proceed. After examining the results of the market test, management must decide whether to kill the program, make modifications and retest, or go into full-scale production.

Full Production. It is not the purpose of this book to discuss how to gear up a pilot program to full-scale production for national sales distribution. It is germane, however, to point out that the packaging development engineer has the most technical experience with this product-package combination and should participate in an active or advisory capacity until the product is well established.

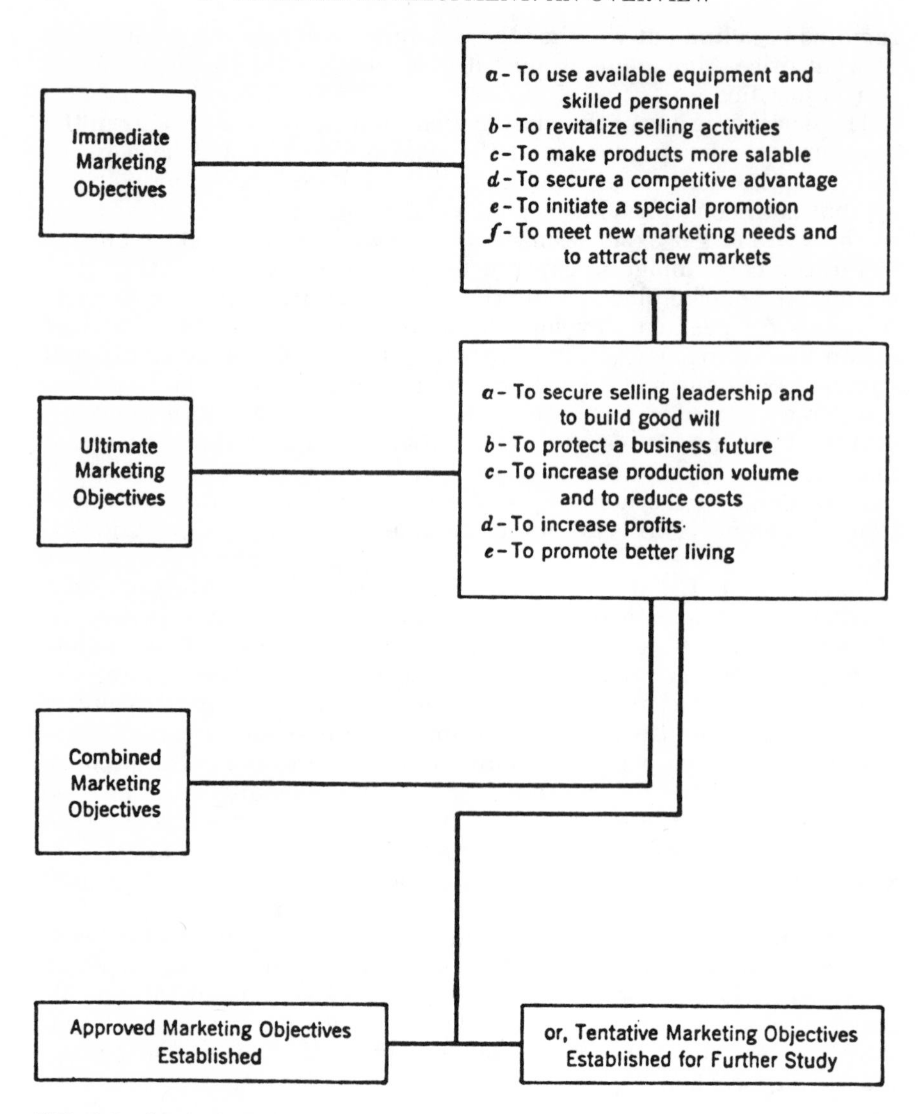

FIG. 2.9 Marketing objectives

BIBLIOGRAPHY

ANON. 1970A. Plastic Bottle Glossary. Society of Plastic Industry, New York.
ANON. 1970B. What's happening in plastic bottles. Mod. Packag. *43* (7), 38–43.
ANON. 1972A. Enter the age of computerized design. Mod. Packag. *45* (4), 28–30.
ANON. 1972B. Aerosols for the future. Plast. Rubber. Wkly. No. 450, 14–15.
ANON. 1973A. A septic packaging: Technology opens new markets. Mod. Convert. *19*, 20.
ANON. 1973B. A–Z of the tinplate can. Tin Int. *46* (1), 7–8.
ANON. 1973C. A–Z of the tinplate can. Tin Int. *46* (3), 75–76.
ANON. 1975A. Evaluating Packaging—In its entirety. Folding Carton Ind. *2* (1), 4, 6–9, 11.
ANON. 1975B. New report out on profile analysis of plastic, non-plastic containers. Mod. Packag. *48* (5), 70.
ANON. 1976A. Adding discipline to creativity pays off for R and D management. Package Eng. *21* (2), 29–32.
ANON. 1976B. Packaging of material preservation, Vol. I. U.S. Defense Supply Agency, (DSAM 4141.2), Washington, D.C.
ANON. 1977A. What users think of flexible packaging. Packag. Dig. *14* (4), 20–21.
ANON. 1977B. Paper packaging. Austral. Packag. *25* (6), 21–22.
ANON. 1977C. Function is the new focus of design. Mod. Plastic Int. *7* (8), 56–57.
ANON. 1978A. A systems approach. Packag. Rev. *98* (4), 43–44.
ANON. 1978B. Pack design in action. Packag. News *98* (8), 29–30, 33, 35.
ANON. 1978C. Evaluating designs for packaging. Can. Packag. *31* (11), 32–33.
ANON. 1979A. The computer as packaging design tool. Package Eng. *24* (1), 39–43.
ANON. 1979B. Flexible packaging: Outlook to 1981. Food Eng. *48* (7), 52–54.
ANON. 1980. Structural design for the 80's faces new challenges. Package Eng. *25* (4), 88–89, 91–97.
ANON. 1981. PVC film and its use in meat packaging. Good Packag. *42* (9), 116.
ANTHONY, S. 1982. Package development. Cereal Foods World *27* (6), 264–266.
ARDITO, G. J. 1974. Designing around the material shortages. Paper, Film Foil Convert *48* (12), 41–42.
BARKER, S. M. 1977. When to change your package and when not to. Can. Packag. *30* (8), 23–25.
BERKA, C. 1970. Use of a computer program to solve dynamic packaging problems. Mater. Handl. Eng. *25* (12), 45–50.
BLOONER, W. J. 1979. The computer as a tool in the design of efficient glass packaging systems. Package Dev. System. *9* (5), 24–28.
BUCHNER, N. 1976. Optimization of packs. Verpack. Rundsch. *27* (1), 1–5. (German).
DARREL, C. G., KALYANPUR, A., and MORROW, D. R. 1977. Computer aids package design. Mod. Packag. *50* (5), 37–39.
DIETZE, V., and GNILKE, R. 1971. The development of packaging methods. Verpack *12* (3), 75–77. (German)
FRIEDMAN, W. F. and KIPNESS, J. J. 1977. Distribution Packaging. Robert E. Krieger Publishing Co., Huntington, New York.
HEINTZMANN, H. 1974. How industrial packaging looks to the consumer in his plant. Can. Packag. *27* (2), 27–30.
HICKINBOTTOM, J. 1981. The role of the package designer. Packing *54* (621), 6–7.
HINES, M. J. 1978. How to establish a cost reduction program for packaging design. Package Dev. Syst. *8* (6), 17–19, 22–23.
HORMGLEN, B. 1980. Structural design in packaging. Packag. Rev. S. Afr. *6* (4), 41, 43.
KELSEY, R. J. 1976. The creation of food packaging facilities in developing countries. Paper presented at First International Congress on Engineering and Food, Boston, Massachusetts.

KLINE, J. E. 1982. Paper and paperboard. Manufacturing and converting fundaments. Miller Freeman Publication Inc, San Francisco, California.
LEE, R. E. 1978. Design vs. Supermarket packaging. Food Drug Packag. *39* (2), 10–12.
LILLQUIST, R. A. 1977. Flexible packaging: The progress and the promise. Paper, Film Foil Convert. *51* (11), 100, 102, 106.
LINDBLAD, K. 1973. How do the designers of packaging think, Packaging *10* (11), 20–21.
MELLETT, L. T. 1977. Flexible package design. Austral. Packag. *25* (9), 24–27.
MELIS, TH. C. J. M. 1979. Innovation in package development. Ned. Verpakkinscent.
OGIIVY, D. 1973. *In* the Silent Salesman: How to Develop Packaging that Sells. J. Pilditch (editor). Business Books, Ltd., London.
POPE, H. 1978. Where do design dollars go? Can. Packag. *31* (5), 34–37.
RAPHAEL, H. J. and OLSSON, D. L. 1976. Package Production Management, 2nd Ed. AVI Publishing Co., Westport, Connecticut.
SACHAROW, S. 1970. Gas or vacuum packing? Food Eng. *42* (3), 83-85.
SACHAROW, S. 1973. Food package—design for disposal. Adhes. Age *16* (7), 10, 12.
STANLEY, C. R. 1976. Paperboard and flexibles. Food Drug Packag. *35* (4), 28.
STEINER, H. 1976. Package design—a professional point of view. Austral. Packag. *24* (4), 42–43.
THALMANN, W. R. 1972. Packaging development and systems. Tara *24* (274), 415–421. (German)

3

Packaging Materials

If you ever do develop the universal monofilm that does everything unsupported and unadorned, take it to the laminator. He will generate your first orders—by laminating it to something else.
D. A. Perino (1966).

Although a package can be used to display a product and encourage its purchase, it is primarily an enclosure used to protect, store, and transport a product. A basic packaging material is that which is used to fabricate the walls of such an enclosure; *auxiliary* packaging materials are those used to combine decorate, adhere, close, cluster, or permit easy opening of the basic package structure. A label would be an auxiliary packaging material attached to a basic packaging material such as a bottle.

The basic packaging materials fall into four major categories: ceramics, metals, vegetable products, and plastics. Ceramics include pottery, chinaware, and glassware. Metals include tinplate (steel), aluminum, and occasionally copper, brass, pewter, and more precious alloys. Vegetable products include wood, wood fiber, other vegetable fibers, cork, rubber, and the like. Plastics encompass a whole family of natural and man-made substances. Most packages are made from combinations of several materials.

In parallel with the development of basic packaging materials and forms, it was necessary to develop methods and materials that could be used to join and fasten them. Thus pegs and nails were developed to fasten wood and rivets were developed to fasten metal. Later, glue was discovered, which is useful in fastening wood, as well as welding and soldering, which are useful in fastening metal. Today we have an entire industry based on production of gums, glues, cements, adhesives, and hot-melt joining materials. Also in parallel with the development of the basic packaging materials was the development of decorative and protective inks, coatings, paints, and lacquers, and the dyes and pigments used to color them. Finally, it was necessary to develop closures for packages. The early plugs, bungs, corks, and lead seals led ultimately to the modern closure industry, which produces a wide variety of caps, plugs, seals, and ties.

CERAMICS

Under the basic term "ceramics" fall those materials which are made from particles of earth substances such as sands and clays. These in turn may be divided between glasses or glazes, which are melted to a complete solution of the solid ingredients and then congealed, and earthenware ceramics, which are fused by heat but retain particulate components bounded within and by a matrix.

Earthenware Ceramics

Pottery is one of the oldest packaging arts, and it takes many forms. The finest forms are called chinaware and porcelain and include the fine bone china used for dinnerware and the exquisite porcelain vases and *objets d'art* of China and Europe. The coarser types are called earthenware or pottery and include terra cotta flower pots and glazed earthenware crocks, jugs, and jars, which are still used today.

The basic raw material of pottery is clay. Kaolin or China clay is chiefly hydrous aluminum silicate, which is a major ingredient of most clays. The name derives from the mountain in China, Kao-ling, where the first clay for porcelain was mined. Since most clays have other ingredients as well as Kaolin, the color of the basic product will vary with the kind and amount of these impurities left after the initial purification steps. Clays used in earthenware that remain porous after firing are called "ball clays." These are purified by washing with water. The fine clay flows out with the water, and the heavier particles of sand, etc., settle. It is particularly important that all iron particles be removed from the clay and other additive ingredients.

In mixing the ultimate blend, various other ingredients are added, depending on the final end product. These include feldspars (sodium, potassium, aluminum silicates), flint or quartz, "China stone" (a mixture of feldspar, alumina, silica, and oxides of alkali and alkaline earth metals), and bone ash (powdered calcined animal bones).

Kaolin and ball clay contribute plasticity to the mix, the latter remaining porous and opaque after firing, the former nonporous and translucent. China stone lends body and translucency to the ware. Quartz and flint, being chiefly silica, resist fusing and lend a skeletal strength to the structure. Feldspar acts as a flux or fusing agent, as does the China stone. In bone china, calcined bone ash is substituted for feldspar to gain more strength and translucency.

While porcelainware, chinaware, and earthenware differ in final properties and basic ingredient mix, methods of manufacture into final formed objects have a great deal in common. Ingredients are blended with water to make a smooth, very thin liquid called the "slip." This is further purified to remove coarse particles, iron particles, and the like, and then may be used in casting or filtered to remove excess water. The filter cake is kneaded into a doughy mass which can then be shaped.

Since the final product is rigid and cannot be reshaped once formed, the converting of clay to a package form is essentially an immediate series of operations following the manufacture of the slip. As will be discussed in Chapter 4, the operations include forming, drying, firing, decorating, and glazing.

Earthenware articles can be porous or nonporous and are rigid and breakable. The greater the fusing of the structure, the less porous it is. Porosity can also be controlled by applying fused glaze coatings. To prevent breakage, walls of commercial packages are usually made rather thick, which means the packages tend to be heavy. Fine china and porcelainware are made with thin walls and are very delicate and fragile. Ceramics are chemically inert and resist very high temperatures. Porcelain is used in sparkplugs, and a modified ceramic is used in rocket nose cones. Alumina, one of the main ingredients of ceramics, is very high on the hardness scale—it is exceeded by few materials other than diamond. This explains the ability of ceramic dishes to withstand the wear of metal cutlery.

Glass

Glass is believed to be an offshoot of pottery. Its earliest beginnings were about 7000 BC. By 1550 BC, glass bottle making was an important industry in Egypt.

Glass is made from limestone (about 10%), soda (about 15%), and silica (about 75%). Lesser percentages of aluminum, potassium, and magnesium oxides may be included. When the components are melted together, they fuse into a clear glass which can be shaped while in a semimolten state. Melting of glass batches is accomplished at a temperature of about 1540°C (2800°F) in a very large furnace, usually gas heated. These furnaces are often described in terms of their surface area, e.g., 800 ft^2 (with a depth of about 3 to 4 ft), and are constructed of heat-resistant refractory materials. Because of the huge amount of heat required to bring such a large mass of material up to temperature, these furnaces are run continuously except for maintenance shutdowns. The life of the furnace walls is perhaps 3 years.

Good glass making depends on maintaining steady-state conditions. Variations in formulation can provide differences in glass properties, e.g., addition of iron oxide to the melt imparts the amber color used in many beer bottles. Broken glass or cullet is often part of the formulation, and in significant quantities it colors or otherwise alters glass properties. Because a batch represents a large tonnage of glass melt, the packager must, unless he is a very large user, accept the basic glass formulation offered by the glass maker. The latter cannot afford to convert an entire batch to produce say, 10,000 bottles of a special glass (unless, of course, the purchaser wishes to pay for the remainder of the batch, a most remote possibility).

Different glasses have different mechanical, thermal, optical, and chemical properties.

Glass is noteworthy for its superb compressive strength. Compressive resistance of glass begins at 500 lb dead weight for even the weaker containers and quickly rises to thousands of pounds with increases in vertical curvature from body to finish.

Impact resistance, on the other hand, is not high, and this is one of the major drawbacks of glass. It is not used when there is a possiblity of impacts. When the glass is marred, impact resistance is significantly decreased below its already low level. Glass has inherently high strength as long as its surface is undamaged. Nonreturnable glass bottles, being lightweight, generally require some surface treatment in order to survive the dual abuses of processing and physical distribution. A number of surface treatments now may be applied in the lehr (annealing oven) to minimize damage from scratching and to provide some resistance to impact through lubrication. Lubricants also assist in moving glass containers on high-speed lines with minimal surface scratching. Among the treatments that have been employed is the application of silicones, but these create problems with label adhesion. Metallic oxides (such as tin or titanium oxide) applied at the hot end of the lehr together with a polyethylene dispersion applied at the cold end are typical glass coatings. Unmarred bottles have good internal pressure resistance, provided that distribution of glass in the walls is uniform and the containers are free from thin spots. Well-designed bottles of appropriate weight (with a minimum wall thickness of about 0.1 in.) can resist steady internal pressures of at least 250 psi, somewhat in excess of the typcial burst resistance of a three-piece metal can. This pressure is well in excess of the maximum encountered in carbonated products today, and so the National Commission on Product Safety is concerned not with internal pressures as much as with potential weaknesses in packages. Glass can be quite dangerous if it bursts into flying missiles; therefore, in actual practice the closure on a glass bottle is designed to give way and release pressure at levels far below the theoretical bursting strength of the glass.

Unless it is specially formulated and treated, glass does not possess a broad range of thermal shock resistance. The maximum sudden differential temperature change (change in the difference between the temperature of the outer skin and that of the core of the glass wall) that glass packages are likely to resist without breaking is about 90°F. Thermal processing of products in glass containers is conducted with care to ensure a uniform temperature rise.

Optical properties are one of the great assets of glass, and, in fact, the word glass is used as a reference for transparency and clarity. Transparency has been promoted as a highly marketable feature, and for some products, visual appearance through glass is doubtless beneficial to sales.

However, by allowing visible and ultraviolet light to pass, transparent glass can promote product deterioration. Light accelerates many color-degradation reactions and some biochemical oxidation processes, such as development of fat rancidity. The use of amber color for beer bottles is predicated on reduction of light-catalyzed reactions that might alter flavor.

Probably the most significant of the several properties possessed by glass is its chemical inertness. More than any other commercial packaging material, glass resists interaction with almost all types of contents.

METALS

The earliest metals used by man included copper, gold, and silver, which are soft and easily workable, in their natural state. Next came metals that were extremely easy to smelt from their ores. These included tin, lead, antimony, and zinc. Later experiments led to the development of alloys, such as bronze, brass, and pewter, and of platings, such as antimony on copper. During these years man also learned how to fashion metal into sheets, rods, and wires by hammering and drawing. He learned how to cast metal into molded shapes. The Iron Age produced a stronger metal still, and men learned to purify iron and to make steel for fine armor and weapons. Tin-plated sheet iron was developed by artisans of Bohemia in 1200 AD; this was the forerunner of the tin-plated steel first used in "tin" cans in the early 1800s. Aluminum metallurgy is a late-nineteenth-century accomplishment.

Precious metals, copper and brass were used by the wealthy for household utensils. Some sheet copper and sheet lead were used for lining wooden boxes and chests for transportation of rare spices and the like. Lead was also used as a seal for glass jars during the Roman period. Widespread use of metals in packaging had its birth perhaps in 1620 when the Duke of Saxony stole the secret of tin-plating. This coincided with the general beginnings of packaging as we now know it, whereby honest merchants sought to produce quality products for sale and wished the package to protect and identify the product.

By the late 1700s, handmade soldered tinplate canisters were in general use for various dry foods. In 1810 Peter Durand invented the cylindrical tinplate sealed container for processed foods, which was then called a tin canister and today is called a tin or a can.

Tin Plate

Steel ingots produced by the steelmakers are first reduced by hot rolling mills to slabs, plates, and coiled sheeting of 0.06 to 0.10 in. thickness. The coiled sheet is pickled, worked, oiled, and then sent to cold-rolling mills, which reduce it under tension to thinner gauges (14 to 32). Tin-plating used to be accomplished by passing the steel strip through a flux and into a bath of molten tin. Modern methods are electrolytic. Thickness of the plating can be varied, and the alloy and

FIG. 3.1. Tinplate can features foil label; many cans are coated inside so that products do not react with the can.
Courtesy of Reynolds Metals Co.

temper of the steel itself are also varied to achieve the desired physical properties. Tin coatings may vary between 20 and 80 millionths of an inch in thickness. Tin-free steel (TFS) has an electrolytic plating of chromium.

Plain or plated steel sheeting may also be coated with organic coatings or enamels to provide further protection against corrosion by the products to be packaged (Table 3.1).

Aluminum

Aluminum was first isolated in 1825, but practical methods for extracting it from its abundant ores were not developed until the late 1880s. First, alumina is extracted from bauxite ores. This alumina is then melted with a cryolite flux, after which electrodes are placed in the molten mass and aluminum is produced electrolytically.

As is the case with steel, ingots are poured after suitable alloying of the original refined pig. The ingots are hot-rolled into slabs, plates, and sheeting. Finally the sheeting is rolled to the gauges needed for the contemplated end use.

Advances have been made in ingot casting and cooling that have resulted in more uniform grain size and fewer particulate or gaseous impurities. The most modern development is the continuous casting process, which delivers molten metal through nozzles into the nip of a rolling mill, which produces sheet stock. This bypasses ingot casting and hot rolling, with considerable savings. Aluminum is used in cans, in semirigid containers, and in flexible packages. In the latter case it may be rolled into foils as thin as 0.00025 in. (0.0064 mm). After rolling, the foil may be annealed or heat treated to alter its physical

TABLE 3.1 Some Typical Can-Lining Resins

Type of Resin	Typical Uses
Acrylic	O/S whites, varnishes
Alkyd	O/S whites, varnishes
Amino-epoxy	Beverage base coats, milk, nonfoods
Butadiene polymer	Beverage base coats, citrus
Ester-epoxy	Varnishes
Oleoresins	Fruits, vegetables
Phenolic	O/S varnish, fish, liver, nonfoods
Phenolic-epoxy	Fish, meats
Phenolic-vinyl	Drawn cans, food and nonfood
Vinyl	Top coats for beverage, nonfood

properties. Soft, fully annealed foils are used for flexible laminations. Intermediate-temper aluminum foils are used for mechanical forming into semirigid containers.

Commercial foil gauges generally range from 0.0003 to 0.0007 in. (0.0076 to 0.0178 mm), but some 0.00025 in. (0.0064 mm) foil is available. In general, foils of thinner gauges, e.g., 0.0003 to 0.00035 in. (0.0076 to 0.0089 mm), which are most commonly used, are not considered pinhole free (the presence of pinholes means moisture and gas porosity). When combined with other materials, however, the effective porosity of the aluminum foil diminishes rapidly. Unsupported foil is pinhole free only in gauges of 0.001 in. (0.025 mm) and above in general commercial practice.

Aluminum foil is impermeable to light, gas, moisture, odors, and solvents and has stiffness and dead-fold characteristics

FIG. 3.2. Foil-laminated packaging for "Mazola" margarine. *Courtesy of Reynolds Metals Co.*

Aluminum foil is impermeable to light, gas, moisture, odors, and solvents and has stiffness and deadfold characteristics not found in any other flexible packaging materials. It is, however, subject to abrasion, scratching, and rupture and so must be protected from stresses and abuses. Further, aluminum cannot be sealed without use of metal-bonding techniques unless a heat-seal coating is introduced. As an unsupported flexible material, aluminum foil, with its greaseproof and deadfold characteristics, finds considerable application in wrapping chocolate and baked goods. But it is as a laminating substrate that aluminum foil has found its major uses in flexible packaging.

PLANT PRODUCTS

Leather (an animal product) was used for many centuries as a nonbreakable container or bottle. Today it is almost impossible to find a leather package except by straining the definition to include shoes, pocketbooks, and book bindings.

Similarly, a wide variety of plant products have been employed for packaging, probably starting with the wrapping of a piece of food in a large leaf. Baskets have been woven from grasses, rushes, leaves, and fibers. Net bags have been fashioned from twisted cords from vegetable

FIG. 3.3. Early tobacco pouches were made of cloth.
Courtesy of Reynolds Metals Co.

fibers. Cloth in the form of sacking and wrappers was used for centuries until the introduction of paper supplanted it. Cloth today is generally obsolete as a packaging material. Some cloth is used to impart burst and tear strength to composite laminates; some is still used to "package" the dirt portion of shrubbery plants; and some is still used as grain sacking. But the last holdouts are gradually being displaced. The tea bag formerly cloth, is now paper. Cloth flour bags have been replaced by bags made of paper or plastic. Even fiber twine is now made from polypropylene fiber, which has greater strength and mold resistance than cloth fiberns. Because of this, in tropical climates, it is making inroads on burlap sacking.

By far the greatest influence on packaging has been by a vegetable product—cellulose. The primary source of the cellulose is wood and its derivatives; some secondary cellulose sources are cotton, linen, and bamboo.

Wood

Wood has been used for many centuries to manufacture containers. It comes in many varieties, from fine hardwoods used to produce beautiful finished articles to the utilitarian softwoods used to fabricate boxes and crates.

Wood is easily shaped and joined and has good strength. Although not as strong as metal of the same thickness, it resists bending, is resilient, and is not excessively fragile, unlike ceramics.

Not all wood is useful for structural purposes. The soft, pithy woods of shrubs, twigs, and leaves are useless. The wood of commerce is rather the wood formed in the straight trunks of trees, which is called timber. When timber is processed from rough logs into simple sawn and planed shapes, it is called lumber. Such shapes are called boards, planks, beams, or timbers, depending on their dimensions. Further processing by sawing, shaping, planing, etc., into more precise shapes is known as millwork production. Millwork shapes include window sashes, doors, door frames, blinds, shingles, poles, posts, flooring, box shooks, doweling, and fancy trimwork. Useful by-products are also produced, including wood pulp, rosin, tar, wood alcohol, charcoal, and creosote.

Some "artificial boards" are made by bonding compressed wood chips or shavings with suitable plastic bonding agents.

Plywood is made by first shaving a thin sheet or veneer from a log, and then gluing several of these sheets together with the grain direction of each sheet alternating 90° to the neighboring ply. Defects formed by knots may be removed and patched. Plywoods are graded according to freeness from such defects, number of plies, thickness, and type of bonding glue used. Some fine cabinet work and interior paneling is made from plywood faced with a fine hardwood veneer.

The "grain" of wood is a result of its fibrous structural formation. Since these fibers are arranged parallel to the long axis of the tree trunk and in concentric rings, the mechanical properties of a piece of wood are related to the orientation of grain direction, and likewise to the pattern of sawing. If a log were to be sawn into boards, the center plank would have grain more or less perpendicular to the face, and in the outer planks the grain would be more or less parallel. These grain differences are minimized by quarter sawing.

Wood has resistance to damage by crushing, stretching, bending, and twisting, whether by slow application or rapid application of force. It is also resilient enough to yield temporarily to a load and then return to its former shape. With regard to compression and tension, wood is strongest along the grain and weakest at right angles to the grain. With regard to shear, wood is strongest at right angles to the grain. A beam will best resist bending if the load is applied perpendicular to the grain.

The fibrous nature and other strength properties of wood give it a unique property when compared with metals or other amorphous materials; that is, it can be split along planes parallel to the grain, leaving a rather smooth surface.

Although there are a wide variety of timber-producing trees, relatively few woods are used in packaging. Usually the cheaper pines or spruces are used for box making and for packing cases. Some oak and pine woods are used in pallets. Some oak and chestnut woods are used in kegs for wine, beer, and whiskey packaging. Cigar boxes have traditionally been made from tropical red cedar. Poplar and basswood have also been used for box making.

Wood Pulp

When wood is chopped into small chips and these chips are digested in chemical solutions, such as calcium or magnesium bisulfite (sulfite process) or sodium hydroxide and sodium sulfide (sulfate process), the solutions dissolve away the noncellulosic materials, such as lignins and resins, leaving essentially pure cellulosic fibers. The latter are worked, bleached (if desired), and then collected into a coarse sheet by screening. The water passes through the screen, leaving the fiber mat behind. Another method of making wood pulp is to grind the fiber from a block of wood by means of an abrasive wheel. Mechanical pulps (or "groundwood") are cheaper and more opaque than chemical pulps.

While some packages are made from molded wood pulp, a far greater number are made from further refinements of the pulp. These refinements include paper, paperboard, and cellulosic plastics.

Wood fibers and groundwood flour are also used as fillers for molded plastics.

Paper and Paperboard

Wood pulp and other cellulosic pulps can be further processed into useful materials called paper and paperboard. Paper was made from

FIG. 3.4. Wood keg used for figs; this package dates back to the early 1900's.
Courtesy of Walter Landor and Assoc.

mulberry bark by the Chinese as far back as the second century BC. The Arabs captured the secret in the eighth century AD. and it then migrated to Spain, France, and the rest of Europe. Early papers were made from linen rags, flax fiber, and straw. Use of wood pulp began in the early to mid 1800s.

First the pulp is "beaten." This consists of subjecting the slurry of pulp and water to a mechanical action, which breaks the fibers into smaller "fibrils" and hydrates the fibers. The finer fibrils make a better mesh or mat, and hydration softens the surface to a gel-like consistency which enhances fiber-to-fiber bonding. Beating also tends to shorten fibers. Shorter fibers give papers with lower tear resistance. Highly hydrated fibers give papers with better burst resistance. Beating may be done in a machine called a "beater" or in a machine called a "jordan," with the latter giving a more refined action.

After beating, the fibers and other additives, such as clay and dyes, are passed to the paper-making machine. On a Fourdrinier machine the pulp enters at a tank called a stuffing box, from which it moves to a mixing box, where more water is added. It passes over a fine screen or riffle, where the last particles of dirt are removed, and then into the head box. From the head box the pulp is poured onto a continuous moving belt of fine wire mesh. The water drains through the mesh, leaving the pulp behind in more concentrated form. The belt also has a side-to-side motion which tends to cross-mat the fibers. As the belt nears the end of its travel, suction boxes placed beneath remove more free water, forming the fibers into a sheet. Next the sheet is transferred from the wire to a felt blanket by means of a suction roll and is carried through two press rolls which squeeze out more water. The pressing action may be repeated several times by means of additional combinations of felt blanket and press rolls. By the time this action is complete, water content has been reduced to about 67%. Another series of rolls and blankets convey the web through a drying section, where moisture is evaporated down to about 10%. A size press may be placed just before the last set of driers, where sizing solution can be sprayed on to seal the surface pores. From the driers the paper passes over chilled calender rolls which "iron" it smooth. It is then slit and wound into reels.

A cylinder-type paper machine differs from a Fourdrinier in that the fibers are deposited on rotating cylinders and then picked up by a wet felt. With a cylinder machine, more than one layer of different fiber mixtures can be laid down to produce duplex papers.

Paperboard is made like paper on a Fourdrinier or a cylinder machine, the difference being only the thickness or weight of the finished product. Papers and paperboards come in many different densities, thicknesses, compositions, and finishes. To make some order out of the many thousands of varieties, they have been grouped into some descriptive categories. One broad classification is according to pulp origin; thus a paper may be a groundwood sheet, a sulfite sheet, or a sulfate sheet or it may have a rag content. A second broad classification consists of the terms "bleached," "semibleached," and "unbleached" (or "natural"). These are self-explanatory. Bleached pulps are whiter but not as strong as unbleached. Other classifications relate either to the appearance or to the end use of the material. Thus the term "glassine" orginally referred to a paper having a high degree of smoothness and transparency, like glass, whereas the term "pouch" in pouch paper refers to an end use.

Paper can be made with a broad range of surface "finishes." Many factors influence the finish; however, the final calendering operation has the greatest influence. Thus there are machine-finished, machine-glazed, and super-calendered finishes, the last of which is the smoothest. Smoothness is also influenced by the fiber mix, the amount of beating, and the type of fillers used.

The following are some typical categories for packaging papers.

Tissues. Lightweight papers from any type of pulp, ranging from 7 to 18 lb per ream (3000 ft^2). These can be hard or soft and may be treated to give wet strength. They may be waxed or laminated to other materials, such as aluminum foil.

Bleached or Natural Laminating Papers. Papers made from sulfate or sulfite pulps ranging from 10 to 90 lb per ream. The finish can be rough or smooth. Generally they are fairly porous. When kraft pulp is used, they are strong; when a high level of groundwood is used, they are smoother but weaker.

Bleached or Natural Printing Papers. Similar to laminating papers, but with a smoother surface on at least one side. They may contain fillers such as clay for this purpose. Porosity and opacity must be controlled for good ink flow and printing contrast.

Pouch Papers. Supercalendered papers made from virgin kraft pulp for maximum strength, usually bleached. They may be pigmented for opacity. These are high-quality papers useful in coating, laminating, and printing.

Greaseproofs. These papers are made from highly hydrated pulp to achieve a high density, good smoothness, and good resistance to grease, oil, and odors. They may have wet strength.

Glassines. Very dense papers made from highly hydrated pulp. They may be very smooth and glossy. "Genuine" glassines usually have a high degree of transparency and are excellent barriers. "Imitation" glassines are not so transparent and have lesser barrier properties due to lower levels of beating and hydrating. Glassines often contain plasticizers to reduce brittleness.

Parchments. Smooth papers that have been treated with acid to gelatinize the surface and close the pores. They are very highly resistant to oils and greases and have very high wet strength, even when boiled in water.

FIG. 3.5. Glassine candy pouch.
Courtesy of Glassine and Greaseproof Manufacturing Assn.

Other papers. There are also many filled, impregnated, and coated papers sold for use in packaging. There are also specially creped papers with high strength and stretchiness, which makes them suitable for industrial sacks.

Some typical categories for packaging paperboards are as follows.

Chipboards. These are low cost boards made from reclaimed fibers. They may be bending (will take a 180° bend), semibending (will take 90°), or nonbending. They are made on cylinder machines and consequently may have one or both surfaces made from a higher-grade "liner" stock. These are described as lined chipboards or lined newsback boxboards. For a finer finish, they may be clay coated on the liner surface. Newsboards are predominantly made of reclaimed newsprint. Chipboards contain higher percentages of other reclaimed fibers, i.e., corrugated and mixed papers.

Solid Manila Boards. These are also cylinder boards. They are made from chemical pulp and reclaim and are also available with a white liner. They are excellent in terms of strength and bending characteristics.

Kraft Cylinder Boards. These are boards made on a cylinder machine from reclaimed kraft fiber and virgin kraft pulp. They have superlative bending qualities and strength.

Kraft Fourdrinier Boards. These are Fourdrinier boards made from 100 percent natural kraft fiber (virgin pulp). They are very strong and have excellent bending and scoring qualities. They may be clay coated or not to give a better printing surface. When they are hard sized and made from bleached pulps, they are frequently called solid bleached sulfate boards. These may be waxed or coated with polyethylene and are used where moisture resistance is necessary.

PLASTICS

The term "plastics" is used rather loosely to describe a host of chemical compounds whose main property in common is very high molecular weight. This high molecular weight is achieved because the plastics are really chains or lattices made from "links" or "building blocks" that comprise distinct molecular units. Chemists call these units monomers and the plastics polymers.

There are many naturally occurring polymers, such as cellulose, rubber, casein, silk, wood, and protein. The truly man-made (or synthetic) polymers originated from man's study of the natural polymers and his desire to find cheaper substitutes. The first man-made plastic coating was pyroxylin—a solution of cellulose nitrate in ether and alcohol. Cellulose acetate was discovered not long after. The first commercial man-made molding plastic was "celluloid" (cellulose nitrate and camphor).

In the search for substitutes for expensive silk, cellulose nitrate and cellulose acetate were put in solutions and then spun into filaments. Later work led to the making of regenerated cellulose (Viscose) rayon. From this it was a logical step to produce the first practical packaging film, "cellophane."

Because of their properties, plastics are useful in different ways. Some form elastic membranes like rubber. Others are permanently sticky and can be used as adhesives. Some form clear, tough films. Others can be drawn into fine filaments. Some are easily molded into complex shapes. Others have low melting points and can be used like waxes.

Basic packaging materials derived from the plastics family are (1) free films, (2) extruded film coatings or laminating adhesives, (3) ex-

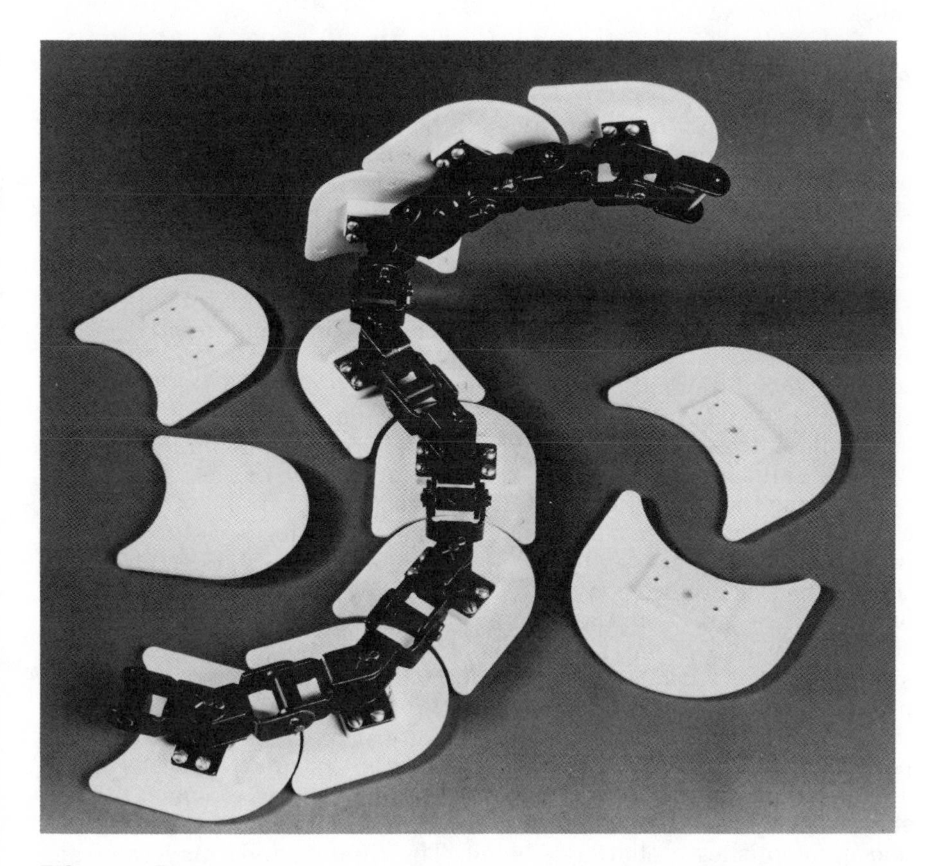

FIG. 3.6. Delrin acetal resin in conveyors; the low friction coefficient of Derlin is essential to smooth operation of the chain.
Courtesy of E. I. DuPont de Nemours and Co.

truded thermoformable resins, (4) injection-molded resins, and (5) fibers and filaments.

Auxiliary packaging materials derived from the plastics family are (1) film formers used in inks, lacquers, and coatings and (2) elastomers and hotmelts used as adhesives.

Although the same basic polymer may be used in any or all of these forms, specific molecular tailoring is needed to get the best results. A nylon designed to be a hotmelt adhesive will not be suitable for use as a film or a fiber. The polyester used to make film is not identical with that used to make fiber. The forming characteristics needed in a blow-molding resin for polyvinyl chloride bottles are not the same as those needed for slot die extrusion of a film. There are literally hundreds of plastic "specifications" within each family of polymers. It is sufficient here to list only the general characteristics of each family (Table 3.2). Detailed properties of plastic films, foams, laminates and molding resins can be found in the *Modern Plastics Encyclopedia* (McGraw-Hill Book Company).

Plastics Processing

The various man-made plastics are produced by chemical reactions which result in precipitated powders or solvent solutions of the polymers. The powder is usually dried, after which, depending on the desired final use of the polymer, it may be retained as a powder, dissolved to form a solution, melted, sheeted, and cut or ground into pellets of varying size and shape. Some methods of fabrication into finished articles require preliminary processing into more useful intermediate forms; others utilize the pellets, powders, or solutions directly.

Polymer Solutions. Solutions of polymers are used in the formulation of coatings, paints, lacquers, and inks. Some films and fibers, such as cellulose acetate and polyvinyl chloride, are made by forcing thick, viscous solutions of polymer through a shaped orifice and then evaporating the remaining solvent or by casting onto a smooth surface, evaporating the solvent, and then peeling off the formed sheet. Coatings, paints, lacquers, and inks may vary in viscosity, depending on the method of application (e.g., spray, knife coating, roller coating, offset, or gravure). Similarly, many adhesives are formulated from solvent solutions of one or more polymers. The advantage of solvent systems is that various additives can be blended intimately and completely to achieve a homogeneous mixture. Often it is advantageous to mix or grind solid additives, such as fillers or pigments, into the solid polymer. This can be done if the polymer is first plasticized by heat and/or the addition of plasticizers or small amounts of solvent. The shearing action of the plasticizing mill or mixer disperses the additive and accomplishes a thorough blend. The final "grind" may then be "dissolved" in a solvent. Actually, the solid particles do not dissolve but are rather suspended, each having been coated by a layer of solvated polymer. A useful modification of the solvent-solution approach is emulsification. Here the polymer is dissolved in a solvent and the solution is then dispersed into a water emulsion through use of suitable emulsifying agents. Solvent or emulsion systems are often used for impregnation of other materials, such as wood, paper, and cloth.

Powder Processing. Powdered polymers can be converted directly into coatings by spraying onto a heated surface, or the heated part or article may be passed over or through a fluidized bed of powdered polymer.

Melted Polymer Processing. A very large proportion of the plastics used in packaging and in other industries are processed and fabricated as follows: heat and pressure are used to soften and melt the plastic, which is then forced through the orifice of a die.

Flat-die Extrusion of Sheet or Film. The polymer is melted, mixed, and deaerated in the barrel of an extruder and advanced to and forced through screens and the die orifice by a rotating screw. The length and diameter of the screw, the diameter and pitch of the lands, and the clearances between the lands and the shaft and the extruder barrel are critical and may differ for different types of polymers. With each type of polymer, rotational speed of the screw and barrel and die temperatures will determine smoothness of flow, output, and sheet thickness.

The molten sheet is cooled by casting it onto a moving, chilled surface. This can be one or more chilled metal drums, resulting in a free film or cast sheet; it can be a substrate web, such as paper, foil, film, or fabric, resulting in the application of an extruded coating; or a second web can be fed in to sandwich the plastic, resulting in an extrusion-lamination.

Ring-die Extrusion of Sheet or Film. When the die is a circular ring and air is blown up through the ring, a tube of plastic film or sheet is formed. After the film has cooled sufficiently, it is passed between a series of pinch rolls which entrap the air bubble and flatten the tube. Later the tube is slit and rewound as a flat sheet.

Parison Extrusion. An extruder can be used to make parisons. These are partially formed containers—usually hollow—made from a gob of molten plastic. The parison may be transferred immediately to a blowing station or it may be allowed to cool, with the blowing operation accomplished later at another location.

Foams. A number of useful and decorative packages have been developed utilizing plastic foams. There are several basic types of foams and methods of manufacture.

TABLE 3.2 Properties of New Plastic Resins

Resin	Density (g/cm^3)	Processability[a]	Oxygen[b] Permeability	Water vapor[c] Permability
Ethylene vinyl acetate (EVA)	0.926–0.937	E, IM, BM	Higher	Higher
Ionomers	0.94–0.96	E, IM, BM, VF	Equal	Higher
Polypropylene copolymers	0.90–0.91	E, IM, BM, VF	Lower	Lower
TPX [poly (4-methyl 1-pentene)]	0.83	E, IM, BM	Higher	Higher
Polybutene-1	0.915	E, IM, BM	Lower	Lower
Halogenated polyethylene (chlorine)	1.11–1.25	E, IM, BM, VF	Equal	Equal
Polyparaxylene	—	E, IM, BM	Lower	Lower
Polyacetals	1.41–1.42	E, IM, BM	Equal	Higher
Vinyl chloride propylene copolymers	1.1–1.7	E, IM, BM, VF	Higher	Higher
XT polymer	1.18–1.19 (approx.)	E, IM, BM, VF	Lower	Higher
Polycarbonates	1.20	E, IM, BM, VF	Higher	Higher
Phenoxies	1.18	E, IM, BM	Lower	Higher
Polyphenylene oxide (PPO)	1.06	E, IM, BM, VF	Higher	Higher
Polyurethanes	1.24–1.26	E, IM, BM, VF	Higher	Higher
Fluorohalocarbons	—	[VF[d]	Lower	Lower

Note: This table is only intended to serve as a general guide to property selection. Resins produced by different firms may vary.

[a] Extrusion, E; injection molding, IM; blow molding, BM; vacuum forming, VF.

[b] Relative to low density polyethylene.

[c] Relative to low density polyethylene.

[d] Other processes possible in properties to some degree.

Resistance to			
Acids	Alkalis	Solvents	Uses
Fair	Good	Alcohol, good; others, poor	Contour wrap, snap-on caps, bottles, coating resins for frozen food, and corrugated cartons (as copolymers)
Poor	Excellent	Good	Skin packaging, vacuum-forming bottles, extrusion coatings
Excellent	Excellent	Good at room temperature	Shipping crates, shrink films, low- temperature applications
Excellent	Excellent	Aromatic and chlorinated hydrocargons, poor; alchols and ketones, good	Hot-filling applications, bottles, steam-sterilizable packages
Excellent	Excellent	Excellent	Fertilizer bags, laminates, and unsupported films
Excellent	Excellent	Aromatic alcohols, poor; others, good	Flexible film for ease in printing, blow-molded bottles
Excellent	Excellent	Good	Film coating, dielectric material
Fair	Excellent	Good	Aerosol containers, bottle closures, aerosol valves
Good	Good	Good	Blow-molded bottles, film wrap, blister packages
Excellent	Excellent	Aromatic and chlorinated hydrocarbons, poor; others, good	Blow-molded bottles, vacuum-formed trays
Fair	Poor	Aromatic chlorinated hydrocarbons and ketones, poor; others, good	Thin walled, bottles, skin packaging, blister packs
Good	Good	Poor	Blow-molded bottles, hot-filling applications, laminates
Good	Good	Alcohols, excellent; others, poor	Blow-molded and injection-molded container, steam-sterilizable packages, flexible film
Fair–Poor	Poor	Alcohols, poor; others, good	Motor oil pouches, formed-melt packages
Excellent	Excellent	Excellent	Vacuum-formed sheets for pharmaceutical packaging

One of these is the two-step process for molding expandable polystyrene. Here the starting material is polystyrene in the form of small beads containing a blowing agent. The first step is preheating the beads to make each one expand as a foam. The second step compacts the expanded beads to a desired shape. Further pressure exerted by the remaining blowing agent accomplishes a bonding of the beads into a firm and continuous structure. Foams can vary in density from as little as 0.5 lb/ft^3 to as much as 20 lb/ft^3. The greater the density, the stronger the molded object.

Another method of making foam is by extrusion. The polystyrene is blended and melted in a primary extruder to uniformity, then a blowing agent is injected. The melt then goes to a secondary extruder for further homogenization and cooling. It is then forced through a die to produce a blown tube, which is subsequently slit into flat sheeting. The sheeting is only partly expanded. After a period of aging, the sheeting is ready for final thermoforming, during which operation it further expands, usually doubling in thickness and halving in density.

A number of processes combine the foaming technique with molding techniques to produce molded structures with foamed interiors and unfoamed solid skins.

Other foam-production methods are used to produce flexible, semiflexible, and rigid foams of polyurethane or polyesters. Here the raw material is a liquid prepolymer, which is mixed with reactants, catalysts, and blowing agents and poured into a moving mold to produce continuous slabs of foam.

Foams have also been made by extruding or pouring of polymers blended with blowing agents, including polyethylene, phenolics, polypropylene, vinyls, ABS, polyamides, and polycarbonates.

Coextrusion. Extruders are now being made with two or more barrels feeding polymer into a complex die in such a manner that two, three, or even more layers of polymer are simultaneously extruded through the die, producing a laminated composite film. By this technique a thin layer of a more expensive polymer with excellent barrier or physical properties can be buried between two layers of less-expensive polymer.

Coextrusion technology is moving ahead rapidly, and more and more composite sheets, films, and coatings will be commercialized in the coming years.

Semimolten Polymer Processing. Some processes involve the forming of the final product, starting with a powder or molten or semimolten mass of polymer which is shaped by applying pressure with or without additional heating.

Compression Molding. This was one of the earliest methods used. It involved placing a charge of preheated polymer powder or a preformed and preheated "pill" into an open, heated, female-cavity mold. The mold was then closed and a matching heated male mold was used to apply sufficient pressure to form the part. With reasonable care, little or no plastic was lost due to "flashing" (exudation beyond the confines of the desired molded shape.)

Transfer Molding. A later development involved preheating a charge of powder or pill in a separate chamber until it was melted, then forcing the melt by means of a plunger to flow into the cavity mold, where it congealed to the desired shape. Products formed by this process carried waste material in the form of attached sprues upon removal from the mold. These were snapped off and discarded.

Injection Molding. This is a modernization of transfer molding. Polymer is continuously fed through a mixing and melting zone to a holding chamber and is then gated under pressure into a clamped mold. When the part is cooled, the mold opens, the part is ejected, and the mold recloses, ready for a second cycle.

Blow Molding. In this process, first a "parison" is formed. This can be done with an extruder, which makes a hollow tube, or with an injection molding machine, which forces a solid mass into a shaped mold and onto some hollow pins. One end of the parison is pinched shut and air is blown into the center, forcing the material into a female mold to create the final desired shape. Some of the latest blow-molding machines transfer parisons to multiple blowing stations, thereby increasing productive capacity. Blow molding is widely used to produce bottles, jars and other types of hollow containers.

Solid-sheet Polymer Processing. Sheet and film materials can be further processed to improve their properties or to form them into useful articles.

Orientation. A film or sheet can be "oriented" to improve its physical strength characteristics or to impart heat shrinkability. Orientation is the mechanical alignment of molecular chains into a more parallel configuration, as opposed to a random entanglement of chains. Flat film is usually oriented by speed of take-off of the film on the extruder. Cross-web or lateral orientation is accomplished in a second operation on a machine called a tenter frame. The tenter frame has one or more series of chains carrying clamps which grab the edge of the film. The chains travel in the direction of the web but also diverge, thereby pulling the web wider. Careful control of temperature is required. If orientation is done at a sufficiently high temperature, the film will not shrink during subsequent processing and is called "heat-stabilized oriented film." However, if subsequent processing exceeds the temperature of the orientation operation, the film will try to shrink back to its original as-cast dimension. This property of "heat-shrinkable" films is used to advantage in overwrapping and bundling of other packages.

Blown film is oriented lengthwise by the speed of take off and laterally by the diameter of the bubble.

During the blowing of bottles from parisons, orientation is achieved in a similar manner.

Thermoforming. Many useful products, including some basic types of packaging, are produced by thermoforming sheet plastic. There are several methods. All involve heating the polymer until it is soft enough to flow upon application of force, forming the material into a desired shape by application of force, and cooling the material. A vacuum can be used to pull the material into a female cavity mold or over a male mold. Pressure can be used to "assist," either by mechanical or pneumatic means.

Thermoforming can be used as a final step after previous forming of a container by compression or injection molding.

The latest adaptation of thermoforming, called solid-phase pressure forming, is utilized with crystalline polymers such as polyester and polypropylene. Here the polymer sheet is carefully preheated to just below the crystalline melting point. It is then placed over a multicavity die, and the lips of the containers are formed as the initial clamping action takes place. Shaping plugs force the sheet down into a cavity, after which high-pressure air is injected (at 80 to 100 psi). This forms the final shape and at the same time achieves biaxial orientation. Containers have strength, transparency, high gloss, good barrier properties, and excellent dimensional control of wall thickness.

OTHER PLASTIC PRODUCTS

ABS

ABS is a term used to describe a variety of polymers which are made from three monomers—acrylonitrile, butadiene, and styrene. The first contributes chemical resistance, strength, and heat resistance; the second contributes impact strength, low-temperature properties, and toughness; and the third contributes component rigidity, gloss, and processability. Special properties can be build in by modifying proportions or through the use of additives. For example, the addition of halogen-containing additives provides flame retardancy. The addition of blowing agents provides a grade used for producing strong, rigid foams. ABS has generally good chemical resistance and is used in the making of dairy cups and tubs, trays for meat or baked goods, thermoformed packages, and closures.

Acetal

Acetal homopolymers, made through the polymerization of formaldehyde, are sold under the trade name Delrin (E. I. DuPont de

Nemours and Co.) They are very strong, hard, stiff, tough, and resistant to creep and fatigue. They have low moisture absorption and good chemical resistance in the range of pH 4 to 9. They exhibit excellent abrasion resistance and low friction.

Acetal copolymers made from the coplymerization of trioxane with small amounts of comonomer are sold under the trade name Celcon (Celanese Plastics & Specialties Co.). The chemical and physical properties and processability of the copolymers are claimed to be better than those of the homopolymers.

Acetals have been used in aerosol containers and valve assemblies. While not generally used in packaging directly, they are used where strength and wear resistance is needed, such as in conveyors or valve stems and in housing for aerosol cans.

Alkyds

These are reaction products of polyalcohols and dibasic acids that are used in coatings and inks.

Aminos

Aminos include urea formaldehyde and melamine formaldehyde resins. They are used in molding compounds for bottle closures, particularly for cosmetics and pharmaceuticals; in adhesives and coatings; and as wet-strength additives for paper products.

Casein

Casein is an albuminous compound derived from milk. It is only swelled by water, but, being weakly acid, it can be dissolved in alkaline solutions. Because it is nonflammable, casein was early used as a molding plastic for objects such as buttons. When reacted with formaldehyde, casein becomes very hard and horny and can be polished into attractive imitation tortoise shell, ivory, or marble. Its principal use in packaging is as a component of water-based adhesives.

Cellulosics

The cellulose and cellulose ester plastics are relatively inexpensive. Cellulose is moisture sensitive unless coated, but the esters are less sensitive, the butyrate being rather resistant and the triacetate very resistant. They are hard, strong, tough, stiff, and transparent and are used as tapes, caps, semirigid tubular packages, films, and blister packages.

Cellulose Nitrate. Cellulose nitrate was an early discovery of scientists attempting to make cotton look like silk. Solutions of cellulose nitrate could be cast onto other substrates to form glossy coatings. The first photographic plates were made using cellulose nitrate cast on glass. Early "artificial leathers" were made by casting cellulose nitrate on heavy paper and then embossing. Cellulose nitrates are used in packaging as heat seal coatings on cellulose films and as film formers in inks.

Cellulose Acetate. Cellulose acetate, produced by hydrolysis of cellulose triacetate, was developed to fill the need for a material less flammable than cellulose nitrate. It soon replaced the latter in photographic film uses, but it was in turn replaced by cellulose triacetate, which has higher heat resistance and is tougher. Dissolved in ketone solvents, cellulose acetate was used extensively as a "glue" or mending "cement" and in thinner form as a "dope" for coating the fabric on World War I aircraft wings. In thicker form it was forced through fine-holed nozzles or "spinnerets" to form filaments, which upon drying could be spun into "acetate rayon" yarns, one of the earliest artificial silks. Similarly, the dissolved resin could be forced through fine-slot orifices to form independent films of cellulose acetate. These films and thicker sheets have been used extensively in packaging as envelopes, carton windows, formed trays or boxes, and blister packaging.

Cellulose. Further advances in cellulose chemistry led to the development of regenerated cellulose plastics. Cellulose from suitable sources, such as wood pulp and cotton linters, is dissolved in alkali. The alkaline soda-cellulose is then put in solution in carbon bisulfide, forming a complex called cellulose xanthate. After aging, this viscous material is dissolved in caustic soda solutions and extruded as filaments or film into an acid bath where the cellulose is regenerated. (The filaments were named viscose rayon and the film cellophane.) The regenerated cellulose is washed, bleached, and treated with softeners and anchoring agents prior to drying. At first cellophane film was useful only as a wrapper. The addition of a cellulose nitrate coating rendered it a better moisture barrier and made it heat sealable.

Cellophane is now manufactured in over 25 coutries, with world production capacity approaching 1 billion pounds.

Nitrocellulose coatings include heat-sealable types (MS); non-heat-sealable types (M); breathable types (LS), which are films with decreased moisture proofness; and one-side-coated heat-seal types (MSBO). There are also PVDC-coated cellophanes which have better moisture- and oxygen- barrier properties. These are supplied with or without a seal jaw release coating on one or both sides. Cellophane films are supplied in a wide range of standard gauges, from about 0.0007 in. (0.178 mm), which yields 25,000 in.^{2}/lb and is designated as 250 gauge, to thicker films with the following yields:

Type	*Yields (thousand in.^{2}/lb)*
M	235, 220
MS	220, 210, 195, 140
PVDC	250, 230, 220, 210, 195, 180, 160, 140, 116

Cellophanes are also supplied in laminated or reinforced versions in which oriented polypropylene or aluminum foil is buried between two sheets of cellophane and the sandwich is coated with PVDC. These are, of course, much more expensive, but they provide excellent durability and moisture barrier properties.

Cellulose Acetate-butyrate and Cellulose Propionate. These materials, useful as films, fibers, or molding compounds, are obtained from hydrolysis of the triesters of cellulose and either butyric or propionic anhydride together with acetic anhydride.

Like cellulose acetate, the cellulose acetate propionate and cellulose acetate-butyrate films and sheets are used in formed containers and in thermoformed blister packaging. They are more easily formed than the acetates because they are tougher and less brittle and flow better. The butyrates have superior low-temperature impact strength and weatherability and can be vacuum metallized.

Cellulose Triacetate. The reaction of cellulose and acetic anhydride produces the triacetate ester, which is suitable as a film or fiber former, and is less sensitive to organic solvents and water than cellulose acetate. Cellulose triacetate products are tough and have good dimensional stability.

Films and sheets have excellent gauge uniformity and optical clarity. They are used extensively in flat form—i.e., in photograph albums, protective folders, greeting cards, and visual aids—but are difficult to thermoform.

Ethyl Cellulose. Ethyl cellulose is the reaction product of ethyl chloride and soda-cellulose. Chemically, it is an ether and is more resistant to bases than to acids. It is less resistant to organic solvents than are the cellulose esters. It is lowest in density of the cellulosics but is not optically clear or colorless. It is not extensively used in packaging except in lacquer or hotmelt coatings.

Epoxies

Epoxy resins are made from two-part systems comprising the basic epoxy compound and a cross-linking curing agent. Since each component can be one of several variations, the family comprises many useful members. The final material may be a thermoplastic with a higher softening point or it may be a strong, tough nonthermoplastic. Curing may be done at room temperature or with some heat—rapidly or slowly. Epoxy resins are widely used in surface coatings as adhesives. In packaging they are principally used as primer coatings for tin-free steel beverage cans. The most noteworthy characteristics are adhesion, mechanical strength, fatigue resistance, and heat and chemical resistance.

Fluoroplastics

Polymers made from chlorotrifluoroethylene are highly flexible and resistant to radiation. They are very inert chemically and maintain properties over a wide temperature range (−400°–390°F). They can withstand liquid oxygen. Moisture absorption is essentially zero. They have excellent transparency, are nonflammable, and are resistant to weathering. They can be injection-or compression-molded and extruded. They are not easily bonded to other materials. Their good electrical properties make them useful as insulating cable wraps. Their excellent barrier and chemical properties and thermoformability have found usage in chemical and pharmaceutical packaging as strip and blister packs; however, their relatively high price has restricted applications in food packaging.

Ionomers

These polymers are unique in that they contain ionized carboxyl groups which create ionic cross-links between molecular chains. Thus the plastic is easily processable as a thermoplastic resin, yet exhibits properties similar to those of cross-linked materials, such as high transparency, toughness, flexibility, and oil resistance. They exhibit a high degree of adhesion. They are useful as films, as extruded hotmelt adhesives, and as injection-molding or blow-molding resins. Their principal limitation is a low softening point. A practical maximum-use temperature is 160°–180°F. Films have good abrasion and puncture resistance and excellent properties at very low temperature (liquid nitrogen). Weatherability is poor. Electrical resistance is good. Moisture barrier properties are good. Ionomers have proven useful in packaging as skin or blister packs and as composite films where the ionomer serves as an adhesive layer or a heat-sealing layer. As a component in coextruded films, they contribute heat sealability, durability, and puncture resistance.

Methyl Pentene Polymer (TPX)

One of the lowest-density (0.83) plastics, TPX has a high, sharp melting point at 464°F and its retention of strength up to nearly this point makes it ideally suitable for injection molding and blow molding. It has chemical properties similar to those of other polyolefins, including low resistance to radiation. Articles made from TPX have a useful

FIG. 3.7. Surlyn A ionomer used in skin packaging.
Courtesy of E. I. DuPont de Nemours and Co.

strength above 300°F (149°C) and readily lend themselves to sterilization by heat. Laboratory and medical glassware is being replaced by nonbreakable, lightweight TPX articles. Injection-molded TPX trays are used for freezing, cooking, and serving foods. TPX is used in packaging in molded closures for glass bottles.

Phenolics

Reactions of phenol-type compounds with formaldehyde result in a family of thermosetting resins. The initial reaction products are thermoplastic and can be injection- or compression- molded to a desired shape. Upon heating, however, with or without added cross-linking agents, such as polyamino compounds, the phenolic resins cross-link to strong, hard (and somewhat brittle) heat-resistant materials. Properties can be modified by incorporating fillers such as wood flour, glass, and asbestos. Phenolics are inexpensive and chemically resistant. They can be decorated via vacuum metallizing. In packaging, their chief usage has been in molded bottle caps.

Polyacrylonitrile

The acrylonitriles include the homopolymers and copolymers, in which acrylonitrile monomer represents 50% or more of total weight, and the comonomers, such as methylacrylate and styrene. The acrylonitriles may be also rubber-modified. These resins are sold under the trade name "Barex" (Vistron Corp.).

The acrylonitriles are intermediate in cost, falling between paper, polyethylene, polypropylene, and polystyrene materials and the higher-cost cellulosics and PVDC-coated polymers. They have excellent properties as barriers to gases and aromas.

The rubber-modified acrylonitrile polymers are suitable for blow-molded bottles and formed containers for low-temperature and room-temperature products, processed meats, cheeses, oils, spices, extracts, coffee, and cosmetic and household products.

The polyacrylonitriles are processed by injection and extrusion molding, by blow molding, and by extrusion and co-extrusion into sheet and film products.

Polyamides

Reaction of diamines with diacids led to polymers of the polyamide type, one of the closest approximations of natural silk. Chemists later found that amino acids could also be reacted with themselves to form "nylons." Nylons are identified by numbers indicating the number of carbon atoms in the monomer. Thus nylon 11 springs from an 11-carbon amino acid, and nylon 66 from a 6-carbon diamine and a 6-carbon diacid. Nylon copolymers are designated 6/11, 6/66, etc. The most widely used commercial nylons are 6, 8, 11, 12, 66, 6/10, and 6/12. These are used as both extruded and injection-molded resins to produce a wide variety of products, including (1) bearings, gears, and cams, in which strength, toughness, abrasion resistance, and grease resistance are important properties, and (2) film, tubing, coatings, and fibers, in which low moisture absorbance, wear resistance, and chemical resistance are required. Stabilizers are required for nylons where prolonged exposure to heat, sunlight, alcohol, or water are to be expected.

Nylon films are principally made from 6 and 66; however, some nylon 11 and nylon 12 films have been marketed. These films have excellent strength, abrasion resistance, and toughness as well as acting as barriers to oxygen and the essential oils found in flavors and aromas. Nylons are readily thermoformed. In packaging, films are often used in composite structures, where they may be laminated, extrusion coated or coextruded. Nylon films are used in food packaging, with the largest use for meats and cheeses.

Polybutylene

Polybutylenes are made from butene-1 monomer and are high-molecular-weight, low-density (0.91), semicrystalline, heat-sealable thermoplastics. On cooling from the molten state, there is a change in type and degree of crystallinity over a five-to seven- day period which results in an increase in strength, rigidity, hardness and density. Films have high tear, puncture, and impact resistance as well as high yield strength, and the polymer has useful properties up to 220°F (104°C), high abrasion resistance, and good barrier properties and resists stress cracking. Both homopolymer and copolymer grades are available. They are suitable for food packaging applications where high performance is demanded.

Polycarbonate

The reaction of phosgene with bisphenol A produces a thermoplastic condensation polymer whose monomer is an organic carbonate. Copolymers are also available The polycarbonates have very high impact strength over the range of −100°–280°F (−73°–138°C). They also have very high transparency to light, making them suitable for lenses and window panes. As molding compounds, they have been used in products in which glass formerly held sway, such as protective globes for candy dispensers and light bulbs. They can also be extruded to provide sheeting and useful packaging films. Both sheeting and film can be thermoformed, and they are used in blister packaging for food and for medical and other nonfood products. Polycarbonate is used in returnable milk bottles in the one-half gallon and gallon sizes. Other molded products include 5-gal. water bottles, small vials for pharmaceuticals and medical products, and bottles for hot-fill products, such as pancake syrups. Polycarbonates do not act as good barriers to oxygen and moisture and are attached by low-molecular-weight amines, esters, ketones and aromatic hydrocarbons. They do have good room temperature resistance to neutral and dilute acid products and to oily substances.

Polycarbonate's extremely high strength, stiffness, and impact strength can be utilized in laminated pouches and peelable membranes used in medical-device packaging, where puncture resistance and sterilizability are important factors.

Polyester

When a dicarboxylic acid is reacted with a polyalcohol, long-chain polyesters are formed. These reactions can be accelerated or inhibited through the choice of accelerators, catalysts, and inhibiting compounds. When one of the monomers contains unsaturated linkages, it is possible to achieve cross-linking of the final polymer. The most commonly used polyester in packaging is PET (polyethylene glycolterephthalate). However, there are a number of copolymers now entering the marketplace. Some of these are crystallizable, as is PET, whereas others are amorphous.

Polyesters have found wide usage as molding resins, fibers, films, foams, and sheetings. When used in molding, they frequently are reinforced with fillers, such as wood, sand, and glass fiber. Structures made from such reinforced polyester plastics have found wide usage in furniture, imitation marble, boat hulls, and automobile bodies. Polyester films have found wide usage in packaging applications and magnetic tapes. The principal advantages of polyesters lie in the ease of fabrication of structural materials and, in the case of films, their excellent clarity, dielectric strength, tear resistance, heat resistance, dimensional stability, chemical inertness, and high barrier properties. PET packaging films are available in both heat-shrinkable and non-heat-shrinkable varieties. They also are available coated with polyvinylidene chloride and/or with polyethylene. Because of their strength and stability, thin-gauge films as low as ¼ to ½ mil in thickness can be utilized. The chief drawback for polyester films in packaging has been their relatively high cost as compared to the less expensive polyolefins and vinyls. Over the past decade this gap has been substantially narrowed. One reason for this is that up to 40% of scrap polyester can be salvaged and reextruded in certain applications, such as thermoformed containers.

Amorphous copolyesters have advantages in processing. They do not have to be chilled rapidly to retain clarity and toughness. PET, on the other hand, is crystallizable and therefore can be oriented and heat set. Orientation can treble the strength of the sheet. When properly crystallized and heat-set, the products can withstand temperatures of 425°F (218°C) without losing dimensional stability.

Orientation also improves the already excellent barrier properties of polyesters, but it does impart a slight haze. Thus PET is one of the best materials for blow-molded, biaxially oriented carbonated beverage bottles. On the other hand, amorphous unoriented copolyesters are used to make blow-molded, crystal-clear bottles for other applications, such as toiletries and medicinals. Bottles made of amorphous unoriented copolyesters can be designed with handles and produced on conventional blow-molding equipment, which cannot be done with biaxially oriented bottles.

Because copolyesters can withstand gamma-radiation sterilization and ethylene oxide gas sterilization—processes used in medical device packaging.

PET-coated paperboard and formed crystallized PET containers are finding use both as packages and as containers in food preparation, food reconstitution, and food serving. Such containers can withstand temperatures of up to 425°F (218°C) for short periods of time.

Polyethylene

One of the first of the polyolefins to be commercialized, polyethylene was developed in England during the 1930s. This was low-density polyethylene produced by polymerization of ethylene at high pressures (15,000 to 45,000 psi). Because of the relatively high degree of side-chain branching, low-density polyethylenes range from 0.910 to 0.935 density. Later researchers working with different catalysts perfected low-pressure (15,000 psi) polymerization reactions, which produced polymers with less branching and higher density 0.940 to 0.965. These high-density polyethylenes are sometimes called linear polyethylenes. The latest important development in polyethylene technology is the low-pressure process for making low-density polyethylene. These polymers are called linear low-density polyethylenes (LLDPE). They characteristically have higher melt strengths and stronger film strengths. As a result, they have an economic advantage in that lower-gauge films can be used in a laminate and yield the same strength levels as thinner gauge low density polyethylene.

Today there are literally hundreds of polyethylenes with a variety of densities, melt indices, and degrees of polymerization. They range from low-melting, wax-like hotmelts to high-melting molding resins. They are noted as a class for their low cost, dielectric property, chemical resistance, and moisture barrier properties. They are extremely versatile, as they can be modified to emphasize a particularly advantageous property through copolymerization or additives or by cross-linking.

Polyethylenes are partly amorphous and partly crystalline. The greater the amount of side-chain branching (typical of LDPE), the less the crystallinity. The lower-density polymers are not as stiff, have lower tensile strength, are softer, have less resistance to heat and chemicals, and have lesser barrier properties, but they tend to be clearer and have better impact strength and resistance to stress-cracking.

The rate at which a molten polymer flows through a die is known as the melt index (MI). This is inversely proportional to molecular weight (MW). High molecular weight will give better physical properties but poorer flow (processing characteristics).

Since each "batch" of polymer is a mixture of molecules of varying chain lengths (and molecular weight), there is in addition to the average molecular weight a distribution around the average. This distribution or variability in chain lengths can be narrow or broad. Narrow molecular weight distribution (MWD) results in better impact strength and low-temperature toughness but poorer processing characteristics.

Polyethylenes are widely used in packaging as coatings, films, and formed sheeting and in molded or blow-molded items. One of the most extensive single applications is the use of high-density polyethylene in blow-molded bottles for milk, antifreeze, detergents, etc. A typical end use for LDPE is an injection-molded snap-on reclosure for bottles or coffee cans.

Polyethylene Copolymers

Strictly speaking, only the LDPE and the very-high-density (0.955 to 0.970) polyethylenes are homopolymers of ethylene. Medium-density and linear low-density polyethylenes are achieved by introducing small amounts of other monomers, such as propylene, butene-1 and hexene-1, as side chains.

FIG. 3.8. Polyethylene milk bottle; such bottles are increasingly common on supermarket shelves.
Courtesy of Sinclair Kopper Co.

Other types of copolymers are made by introducing ethyl acrylate, methyl acrylate, or vinyl acetate as comonomers. These comonomers increase density but retain low crystallinity because of branding. The copolymers of this type exhibit high clarity, flexibility, and impact strength.

Additives can be introduced to the molten polyethylene polymers to improve resistance to oxidation and ultraviolet light, to improve slip and antistatic properties, or to introduce color or reinforcing fillers. Surfaces of films or containers can be treated (for instance by flame) to improve adhesion of inks, etc. When a blowing agent is introduced, soft, flexible foams can be extruded.

Polypropylene

Polymerization of propylene by the Ziegler catalytic system yields a homopolymer which is about 90 percent isotactic (whereby all the propyl groups are arranged on the same side of the polymer chain) and about 10 percent atactic (whereby the propyl groups are randomly arranged on either side of the chain). Because the atactic polymer reduces crystallinity and adversely affects physical properties, it must be stripped out. Newer gas-phase polymerization systems reduce the amount of atactic homopolymer. The melting point of homopolymers is in the 325°–335°F (163°–168°C) range, making them suitable for packages used in steam sterilization processing of food.

Polypropylene copolymers are made by introducing ethylene monomer. When introduced in the primary reactor, the ethylene is distributed randomly in the polymer chain, yielding *random copolymers*. These have better clarity, less stiffness, and better impact resistance than polypropylene. They also have a broader melt temperature range, which makes processing easier.

When the ethylene is introduced in a secondary reactor after some polypropylene chains have formed, the polymer is a mixture of chains of polypropylene and chains of polyethylene called a *block copolymer*. Block copolymers are even better in impact strength than homopolymers or random copolymers.

Polypropylenes can be injection molded, extruded, blow molded, and thermoformed.

Injection-molded articles include food containers, closures, and rigid vials for pills. Thin-walled margarine tubs are examples of injection-molded polypropylene containers.

Thermoforming of polypropylene is now done by forming in the solid state just below the crystalline melting point. This overcomes the problems of low melt strength encountered at higher temperatures and produces better clarity, impact strength, stiffness, and barrier qualities than were achieved at higher temperatures.

Blow-molded polypropylene bottles are used for cosmetics and pharmaceuticals (homopolymers). Random or high-impact copolymers are used for substances with more critical strength requirements, such as detergents, fruit juices, medicines, and shampoos.

Films are extruded by casting or blowing. Cast films can be sold "as cast" or can be further processed by tenter-frame orienting. Blown

FIG. 3.9. Rolls of printed oriented polypropylene film.
Courtesy of Hercules, Inc.

films are principally biaxially oriented. Polypropylene films have captured many packaging market applications for which cellophane was previously used.

Polystyrene

Styrene is basically an ethylene molecule with a benzene ring tacked on. When polymerized, it forms long-chain linear polymers that are lightweight, rigid, transparent, and brittle. Polystyrene was one of the early polymers, commercialized in the late 1930s. Toughness or impact resistance can be achieved through use of rubber modifiers, strength through use of fillers. Other additives may include lubricants, pigments, mold-release agents, heat stabilizers, antistatic agents, light stabilizers, and foaming agents. Polystyrenes are chiefly distinguished by their low cost and ease of fabrication, but they are limited in heat resistance and barrier properties. They can be cast into sheets, thermoformed, and extruded or injection molded.

Strength and flexibility can be improved by orientation, and this property is utilized in biaxially oriented film and sheet, in injection

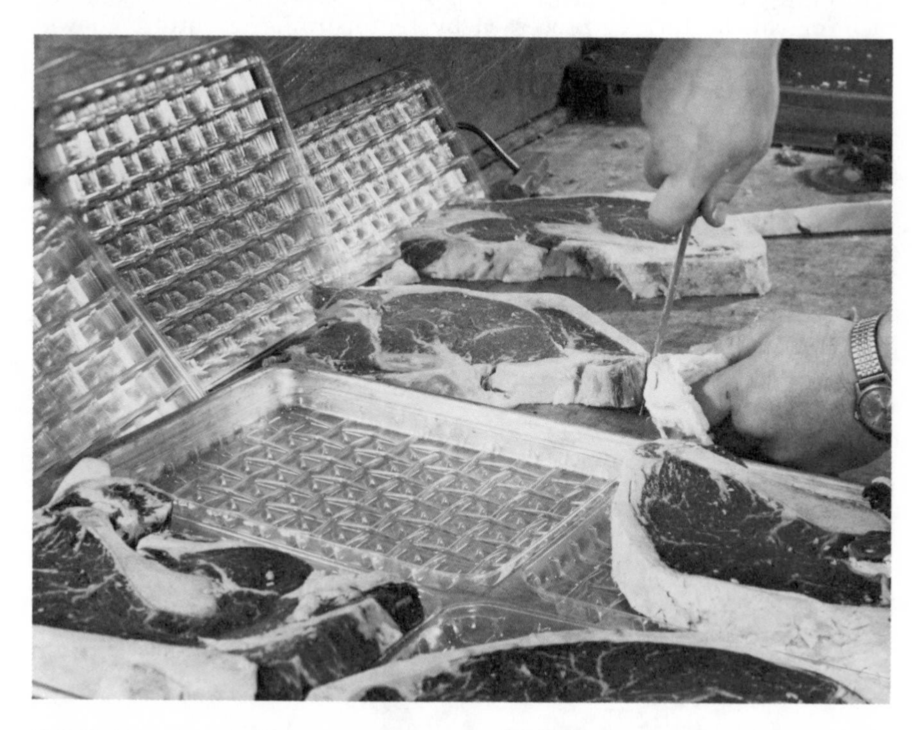

FIG. 3.10. Clear polystyrene thermoformed trays.
Courtesy of Monsanto Co.

blow molding of shaped containers, and in solid-phase forming (thermoforming just below the glass transition temperature).

Polystyrene is being used in conjunction with other barrier resins to produce coextruded composites.

Polystyrene packaging applications include foam containers and trays for meats, poultry, and eggs; insulated drinking cups for hot beverages; and insulated clamshell containers for fast-food take-aways. High-impact (rubber-modified) polystyrene sheet is formed into containers, trays, and tubes for food packaging. The biaxially oriented film and sheet can be formed into high-clarity blisters, trays, and containers. Injection blow molding is used to make laboratory ware and vials and bottles for pharmaceuticals.

Styrene Butadiene

When it was found that copolymer polybutadiene rubber additives toughened polystyrene, it was logical to try copolymerization of styrene with butadiene. Results initially proved to be rubber-like compounds with thermoplastic properties. The polymers remained rubbery and elastic when solid, but could be softened with heat, remelted, and thermoformed as often as desired. They were used as components of adhesive coatings and sealants, as solutions, or as hotmelts and were extruded, injection molded, thermoformed, blow molded, or calendered into useful rubbery packaging components.

A newly family of copolymers, called K-Resins (Phillips Chemical Co.), are crystal clear and have excellent impact strength. They can be readily processed with or without color addition. They appear to be suited to blister packs, fabricated boxes, hinged boxes, injection-molded parts, and closures.

Polyurethane

There are several basic types of polyurethanes, all characterized by the presence of the isocyanate group. Depending on the location of the CNO group in the molecule, curing generally takes place by reaction with water (H-OH) or other hydroxy groups, such as polyalcohols or polyhydroxy esters. Amines may be used to catalyze some reactions. The reaction evolves CO_2 gas and creates a disubstituted amine linkage.

Polyurethanes can be one-part or two-part prepolymers that will cure in air (with moisture, heat, or catalysts) to form cross-linked, thermoset, hard or soft, rigid or elastic, chemically inert, tough, and abrasion-resistant polymers, and they can be used to produce rigid or flexible foams. In packaging they are used as coatings and adhesives. A polyurethane film has been made available, but its cost has prevented extensive use. Injection-molded containers or formed polyurethane structures may yet prove useful.

Polyvinylchloride (PVC)

The polymerization of vinyl chloride and related monomers has led to a host of useful commercial plastics. The homopolymer is hard, tough, and relatively nonflammable. Because of its hardness and its sensitivity to degradation by heat, PVC is usually compounded with stabilizers, lubricants, and plasticizers. In addition, fillers, blowing agents, pigments, impact modifiers, UV absorbers, flame retardants, fungicides, perfumes, antistatic agents, and antiblocking agents may be added to the formulation.

Experimentation with monomer chemistry and copolymerization has led to a family of related vinyl-type polymers, including polyvinyl acetate, polyvinylidene chloride, polyvinyl acetate/chloride copolymer, polyvinyl chloride/polyvinylidene chloride copolymer.

PVC compounding can be effected by dry-blending, melt-mixing, or forming solutions/emulsions.

PVC processing includes calendering, extrusion, injection molding, blow molding, compression molding, slash or rotational molding, powder coating, and spread or roller coating.

Compounding and processing can produce transparent or opaque, soft or hard, and tough or brittle films, sheets, foams, or molded articles. PVC films have intermediate barrier properties and can be classified as relatively breathable (permeable to oxygen, CO_2, and moisture vapor) as compared with polyesters and polyvinylidene chloride.

FIG. 3.11. Polypropylene-modified PVC bottle.
Courtesy of Air Reduction Co.

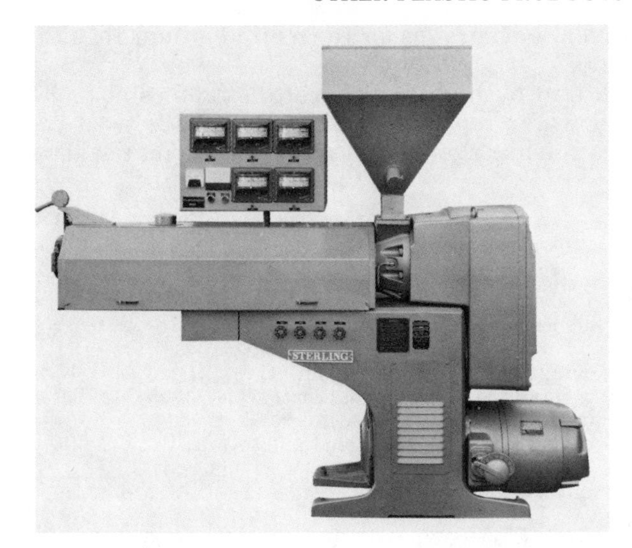

FIG. 3.12. Commercial plastic resin extruder; such extruders are commonly used in plastic processing.
Courtesy of Sterling Extruder Co.

PVC is used extensively in packaging in formed box lids and in blow-molded and injection blow-molded bottles for shampoos, detergents, mouthwashes, syrups, cooking oils, etc. PVC bottles have been used for water and wine in Europe, but regulatory agencies have delayed their use in this area and for whiskies because of possible vinyl chloride monomer retention.

PVC films and coatings have been used extensively in flexible packaging for a variety of foods. Thermoformed blister packaging made from PVC sheeting has also gained extensive markets for use with food, hardware, cosmetics, drugs, toys, novelties, and housewares. Heat-shrinkable PVC film is used in overwraps for a variety of products. PVC plastisol coatings on glass bottles have provided protection against shattering. PVC can coatings and closure seals represent a consumption of more than 14 million pounds.

Polyvinylidene Chloride (PVDC)

The monomer for polyvinylidene chloride contains more chlorine than vinyl chloride, and this produces a higher chemical resistance, a higher melting point, and a substance improvement in barrier properties (grease, oxygen, moisture).

PVDC coatings are used to contribute barrier properties to flexible laminations.

Extruded or coextruded PVDC films are used for high-barrier applications in cheese and meat packaging.

Silicone

Polymers containing silicon are created by producing methyl chlorosilane. Subsequent hydrolysis yields a polymer of dimethyl siloxane.

Substitution of other akyl or aryl groups for the methyl groups results in other, related polymers.

Silicones are characterized by high resistance to extremes of temperature, inertness, and water repellency. They have been used in packaging principally as release agents and antiblocking materials.

BIBLIOGRAPHY

ANON. 1972. Metals. Can. Packag. *25* (4), 32.

ANON. 1973. Coated Cellophane. Mod. Convert. *17* (2), 28.

ANON. 1975. Medical paper. Packag. Rev. *95* (3), 35.

ANON. 1976A. HDPE packaging resin. Mod. Plastics Int. *60* (8), 24.

ANON. 1976B. Outshining them all: Metalized film. Can. Packag. *29* (9), 42–44.

ANON. 1977. ABS plastics. Mater. Eng. *85* (2), 48–54.

ANON. 1978. Plastics in packaging. Plastics Eng. *34* (10), 29–51.

ANON. 1979A. Aluminum Foil. Aluminum Association, Washington, D.C.

ANON. 1979B. Packaging materials: Primed for the 80's. Packag. Digest *16* (3), 50–52.

ANON. 1980. U.S. plastic special report: Materials 1980. Mod. Plastics Int. *10* (1), 37–43.

ANON. 1982. Flexible packagings, fifty years in films. Can. Packag. *35* (4), 22–24.

BAKER, M. 1979. Medical sterilization papers. Packaging *50* (589), 24.

BAWN, C. E. H. 1973. Polymers: Developments for the future, J. Assoc. Off Anal. Chem. *56* (9), 423–429. IEC Prod. Res. Dev. *13* (1), 2–9.

BRIGHTON, C. A. 1982. Styrene polymers and food packaging. Food Chem. *8* (2), 97–107.

BRISTON, J. 1976. New plastics materials for packaging films. Packag. Rev. *96* (3), 71, 73–74.

CROFT, P. W. 1976. Polymers in packaging. Packaging *47* (554), 14–17.

DAY, F. T. 1970. Aluminum foils and foil laminates. Flavor Ind. *1* (9), 607–611.

EVANS, K. A., and FENWICK, S. M. Aluminum in packaging. Sheet Metal Ind. *25* (1), 28–30, 32, 33.

HAAS, L. 1982. Advances in coating. Pulp Pap. Int. *24* (4), 42–51.

LEE, D. M. 1980. Paper and board—What the next ten years will hold. Converter *17* (5), 12, 14, 16–17.

MENGES, G., and BERNDTSEN, N. 1977. Polyvinyl chloride processing and structure. Pure Appl. Chem. *49* (5), 597–613.

MESSER, P. 1977. Low-density polyethylene. Austr. Packag. *25* (6), 25–27.

MILLER, A. 1971. Technology and materials make news in flexible packaging. Plast. Technol. *17* (1), 35–37.

MULDER, J. M. (1976). HDPE Film—A strong contender for paper package markets. Can. Packag. *29* (3), 30–31.

PEARSON, R. B. 1982. PVC as a food packaging material. Food Chem. *8* (2), 85–96.

PERINO, D. A. 1966. Presented at the Fall Meeting of the Commercial Chemical Development Association, Chicago, October 1966.

PRINCE, P. 1982. Metalized film vs. foil. Paper, Film, Foil Convert. *56* (3), 70, 72.

REIMSCHUSSEL, H. K. 1977. Nylon 6. Chemistry and mechanisms. J. Polym. Sci. Macromol. Rev. *12,* 65–139.

SACHAROW, S. 1972. Nylon films: Vital tool for the packaging engineer. Flexography *17* (3), 16, 18, 52–54.

SACHAROW, S. 1973. Edible and water-soluble films. Adhes. Age *16* (1), 10, 11.

SACHAROW, S. 1974. Metals. Packag. India *7* (1), 10–21, 27.

SACHAROW, S. 1975A. Modern practices in plastic film converting. Part II. Package Print. Diecutt. *21* (2), 34, 38, 40.

SACHAROW, S. 1975B. Modern practices in plastic film converting. Part III and IV. Package Print. Diecutt. *21* (6), 10–11, 26–28, 30.
SACHAROW, S. 1975C. Paperboard and corrugated board. Packag. India *7* (4), 24–30.
SACHAROW, S. 1976A. Flexible packaging primer No. 7 Polyester: strong properties account for versatility. Food Eng. *38* (6), 64–65.
SACHAROW, S. 1976B. Polystyrene in drug and cosmetic packaging. Austr. Packag. *24* (9), 29–31.
SACHAROW, S. 1976C. Flexible packaging primer No. 4 Cellulose acetate: a durable pioneer. Food Eng. *48* (3), 80–81.
SACHAROW, S. 1976D. Flexible packaging primer No. 5 Polyvinyliden chloride: the versatile saran. Food Eng. *48* (4), 84, 86–87.
SACHAROW, S. 1976E. Flexible packaging primer No. 6 polystyrene: wraps up longer shelf life. Food Eng. *48,* (5), 74–76.
SACHAROW, S. 1976F. Flexible packaging primer No. 8 The extensible food film. Food Eng. *48* (7), 65–67.
SACHAROW, S. 1976G. Handbook of packaging materials. AVI Publishing Co. Westport, Connecticut.
SACHAROW, S. and GRIFFIN, R. C., Jr. 1973. Basic guide to plastics in packaging. Cahners Books, Boston, Massachusetts.
SALMEN, D. 1978. Corona surface treatment. Paper, Film, Foil Converter *52* (10), 94–96.
SHORTEN, P. W. 1982. Polyolefins for food packaging. Food Chem. *8* (2), 109–119.
TUNBRIDGE, T. 1975. High barrier plastics technology has increased application. Packag. Rev. *95* (10), 47, 49, 76.
WARD, I. M. (Editor) 1975. Structure and properties of oriented polymers. Applied Science Publisher Ltd, London.
WESSLING, R. A. 1977. Polyvinylidene Chloride. Gordon and Breach Science Publishers, New York.
VINCENT, D. 1973. Paper vs. plastics. Paper *179* (1), 3, 26–29.

4

Package Forms

Without anyone having intended it, and when few even noticed, everyone had become increasingly dependent on packaging. Businessmen could well say, "No packaging, no brands—no brands, no business!"

Daniel J. Boorstin (1973)

In prehistoric times, men created the first packages from natural materials of different useful shapes. Leaves and skins became wrappers, and seashells, nut husks, gourds, and animal horns became dishes and bowls and drinking cups. Next man learned to improve on these items by shaping them. Stones, pieces of wood, and animal bones were carved, chipped, or ground into more convenient shapes. As skills improved, other natural materials were fabricated into more complex structures. Grasses were plaited into mattings and baskets, fibers were matted into felt or twisted into thread and yarn from which cloth and rope could be made, and skins were fashioned into bags and other containers. Native metals were wrought into cups or bowls.

As civilization advanced, men learned to convert useless materials into useful ones; clay was baked to make bricks and pottery. Sand was fused into glass. Ores were smelted to produce plentiful supplies of copper, zinc, tin, lead, and iron. Metals were combined to form useful alloys, such as brass, bronze, and pewter. The art of tanning leather was discovered. Paper and paperboard were made from vegetable fibers. With modern technology, man can now tailor molecules and even atoms themselves into more desirable forms.

Fundamentally, a package can be made by taking a solid object and hollowing it, or by taking a thin material and forming it into the desired shape. Sometimes a solid material can be softened, formed into a new shape, and then rehardened, as the casting of molten glass or metals, the forging of hot metal, and the blowing of heat-softened glass and plastics. Still another approach is to reduce a solid material into small particles and shape the particles into the desired form at the same time binding them together, as in the manufacture of clay pottery and paper.

Packaging development is in the final analysis a series of "convertings." A simple example would be converting sand and other chemicals into glass and then converting the glass into a bottle. A more complex example would be the converting of wood into wood fiber, of the fiber into paper, of the paper into a lamination with aluminum foil and plastic, of the laminate into a printed label. The final converting would then be application of the label to a can or bottle. With solid waste recovery, the label might be converted back to fiber and metal again.

Packaging material properties vary when fabricated in different forms, sizes, or thicknesses. The converting of the material to the final package alters some properties. Thus the properties of the original material can only be used as tentative guides in the selection process. Final suitability can only be ascertained through end-use testing of the ultimate combination of materials.

Packages are made from combinations of materials. The materials may be simple, basic items such as glass and aluminum, or they may be more complex, converted items such as corrugated paperboard and printed laminations of films, papers, or foils.

Some packages are completely fabricated by a "converter" and then delivered to the product packager. Other packages may be only partially converted, with final fabrication on the packaging line. Some simpler forms of packaging may be made from the raw material in line with the packaging operation.

Since the procedure is so complex and the end results many times more so, it is difficult to make a simple categorical presentation of package materials and forms. Rather it is necessary to interrelate the two as may be necessary for clarity.

Basic package forms may be grouped into three major categories: rigid, semi-rigid, and flexible. It will be found, however, that there is some overlapping. Rigid packages are formed into a definite shape from sufficiently massive or strong materials, so that they retain their shape when filled with product and are not deformed unless subjected to sufficient force to destroy or severely damage the total structure. Semirigid packages are formed into a definite shape from less massive or weaker materials, so that although they are not intended to be distorted substantially when filled with product, they can be distorted without severely damaging the total structure by the application of a moderate force. They usually recover original shape on removal of the force. Flexible packages may be formed to a definite shape when empty, or they can be made from sufficiently limp or flexible materials that they generally conform their shape to the product contained. They can be distorted or crushed with ease unless supported by the rigidity of the product, and they do not recover their original shape on removal of the force.

Since even very rigid materials become more and more flexible when fabricated to thinner and thinner sheetings, the distinction between rigid and semirigid and between semi rigid and flexible packaging forms according to the above definitions may be hard to determine. In this chapter we have classified package forms as rigid where this property actually exists or where the nature of the package is such that rigidity or near-rigidity in form is a principal desired property, and we have classified as semirigid those packages which either conform to the definition or which, though quite flexible, are intended to replace packages that would be defined as rigid. Thus a plastic bottle is intended to replace a glass bottle. The latter is rigid. The former, because its walls are thinner and more flexible, is defined as semirigid. To make reading simpler, we have included rigid and semirigid packages in one section and flexible packages in the other.

RIGID AND SEMIRIGID PACKAGE FORMS

Glass or Ceramic Containers

Glass containers are one of the major rigid packaging forms. Glass is strong, rigid, and chemically inert and conforms to FDA regulations for food contact, as do many of the standard coatings. Glass does not appreciably deteriorate with age, and because it is an excellent barrier to solids, liquids, and gases, it provides excellent protection against evaporation and odor or flavor contamination. It has a very low cost; however, strength requirements demand a heavier weight ratio than is required with metal or plastic. Its chief disadvantages as a packaging material are weight and fragility. Significant improvements in strength/weight ratios have recently been achieved, and new coatings have reduced damage and made higher handling speeds possible. In 1981 the U.S. glass industry produced and shipped over 50 billion units.

The transparency of glass gives excellent product visibility, and new methods for surface decoration yield highly attractive finishes. Glass is extremely versatile in terms of size and shape of the final container.

Glass containers include the following, any of which may be duplicated in earthenware.

Bottles. Bottles are the most extensively used type of glass container. They may have many different shapes but the neck is always round and much narrower than the body. The neck facilitates pouring and reduces the size of the closure required. Principal uses are for liquids and small-size solids.

Jars. Jars are really very-wide-mouthed bottles, and usually with no appreciable neck. The opening permits them to hold larger solid products and permits the insertion of fingers or a utensil to remove portions of other products. They may be used for liquids, solids, and nonpourable semiliquids, such as thick sauces and pastes.

Tumblers. These are like jars but they are open ended. They have no neck and no "finish". They are shaped like a drinking glass and are used for such products as jams and jellies.

Jugs. These are large-sized bottles with carrying handles. Necks are short and narrow. They are usually used for liquids in ½ gal. and larger sizes.

FIG. 4.1 The Glazier—a woodblock print of a glazier practicing his craft in the Middle Ages. From *The Book of Trades*, Standebuch, 1568 (Dover edition reprint, 1973).

Carboys. These are very heavy shipping containers shaped like a short-necked bottle with a capacity of 3 gal. or more. Historically, they were used with a wooden crate holder. Other outer protective frames are now used.

Vials and ampoules. These are small glass containers. The latter are glass bubbles drawn to a fine tube and fused shut after filling with a product; they are used principally for pharmaceuticals. Vials are flat-bottomed tubes with or without a finish, usually with no neck. They are sometimes used for small quantities of food items, such as spices or food colorants, but are more often used for drugs and medications.

Converting of Glass into Packages

There are four basic processes for making glass articles: drawing to produce sheet, pressing into molds, casting into molds, and blowing. The last-mentioned process is the one used to produce most glass packages.

From the furnace, molten glass flows in a steady stream through a narrow throat and then into a feeder channel. The molten glass is gathered through an opening and chopped off by a pair of shears into "gobs" which fall into blank molds. The amount of glass obtained depends on the temperature, glass composition, size of orifice, shear tim-

FIG. 4.2 Glass bottle for Scotch whiskey.
Courtesy of National Distillers and Chemical Co.

ing, etc. Double-gob blow molding involves extrusion of two gobs at once instead of one to increase the speed. At this writing, triple-gob machines making three containers instead of one are in use for 12-oz beer bottles. The gobs feed single, double, or triple cavities of an individual-section blowing machine (or into molds on a rotating table). The blank mold is the mold in which the first stage of shaping of the container body is performed. In one alternative, the suction process, the finish is at the top and the mold sucks the molten glass from beneath. In another, perhaps more common flow process, the molten glass flows downward into the blank mold with a reciprocating plunger assist.

Air is blown on the molds to cool the glass, and the molded glass is removed from the mold, where further cooling occurs. When the glass leaves the mold, the surface is rigid, but the core is still hot and soft. Were natural cooling to continue, hot inner parts could contract more than the cool outer surface, and stresses would be established with the inner parts stretched and the outer surface compressed. To avoid these dangerous differential stresses, the glass is annealed in a tempering oven called a lehr. Glass is conveyed to the lehr, where it is reheated to a temperature at which it can flow slightly to relieve the internal stresses and then cool in a uniform manner to avoid any further stresses. Annealing requires over one hour of controlled temperature, a process that could, of course, be performed in the mold, but only by tying up the mold for the requisite time period. During annealing, the finish (so-called because in the days of manual glass blowing it was the last operation) is fire polished to smooth any imperfections.

In the blow-and-blow option of the flow process, the finish (the molded portion to which the closure is applied) is at the bottom of the cavity and is formed. The pattern or parison is rotated 180° and transferred into a finishing or blow mold, where compressed air blows the parison into the shape of the mold. The blow-and-blow process is usually used for the narrow-neck or bottle shapes.

In the press-and-blow process, the gob flows into the blank mold and the parison is pressed to an exact shape by a plunger before it is transferred to a blow mold, where the final shape is blown. Press-and-blow is used for wide-mouth containers or jars.

Molds require a costly investment—on the order of $3000 or $8000 for a set, depending on the size and complexity.

A set of molds lasts about 2.5 million bottles. Glass industry practice is for the packager to pay directly for the first set of molds and for the glassmaker to assume the direct cost of subsequent sets (although, of course, the packager pays in variable costs).

Glass containers are inspected to ensure dimensions, roundness, verticality, volume, weight of glass per container, etc.

Costs of glass packages generally increase with weight of glass per package, and so, although mechanical strength increases with glass thickness, an economic balance must be attained. Glass may be shaped to relatively precise dimensions to control quality of contents by height of fill. A total glass package is not simply the glass container itself, but also a closure and an identification. The identification may be fired on the glass after molding or may be affixed as a paper label. The cost of a permanent decoration is individually far higher than that of an adhered label, but this differential vanishes with a multitrip container. A disadvantage of the permanence of a fixed decoration is that it could make the container obsolete before its shape or closure is out of date.

Because of the large investment required in equipment, glass-package making is not an in-line operation; glass packages are completely fabricated elsewhere and then sent to the packager. In anticipation of orders, some glass makers produce both stock and custom packages for inventory and then draw from these inventories. For some stock containers, a single run per year is the only economic quantity. Thus the packager could conceivably find a package that fits his requirements from among the many stock designs. But the capacity and finish (which dictates to a major degree the closure) would be fixed.

Glass has been used for pharmaceuticals and other health aids because of its inertness and its image of cleanliness and purity. Many such products, however, have been switched to other, less expensive packages. Those products that require extensive shelf-life without change of potency are glass-packed to protect against evaporation or moisture. The extensive use of glass for toiletry and cosmetic products is due to the ability of glass to retain highly desirable aromatics and perfumes. Since a major attribute of these products is the fragrance, any loss of fragrance or other change is undesirable. The glass container not only prevents loss by volatilization, it does not allow access of oxygen, interaction of product essential oils with the package, or access of aromas from external sources. Coupled with an effective closure, a glass package can retain the quality of a fragranced cosmetic or toiletry for many years. In consumer use, the glass package can be reclosed to retain the fragrance.

Glass packs are used to a considerable extent in the food industry. Probably the broadest usage is for baby foods, which were converted from metal cans almost completely only a few years ago. Except for baby foods and asparagus spears, very few low-acid food products requiring retorting are packed in glass. In order to ensure that the closures stay on the jars during the cool-down cycles, controlled counterpressure must be applied, and this is a complicating factor. Further, retort loading and unloading requires care to prevent breakage. Unless there is some marked benefit, such as showing intact asparagus spears or whole sausages, few food processors slow their cycles to employ glass for products that are more easily processed in other package forms.

FIG. 4.3 Easy-open glass bottle closure. Courtesy of *American Flange and Manufacturing Co.*

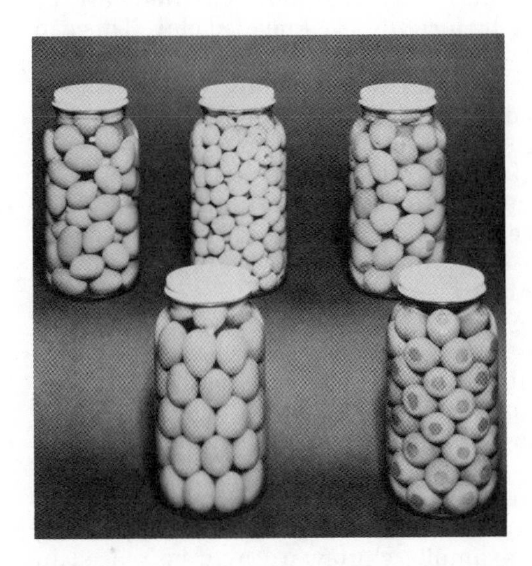

FIG. 4.4 Screw-on glass bottle closures are widely used for large-size institutional packs. Courtesy of *Vita Food Products.*

High-acid foods such as fruits, however, can be conveyed through hot water baths with little fear of breakage during sterilization. Light-colored fruits such as peaches discolor, however, and are rather poor in appearance. Some juices are glass-packed.

Large quantities of pickles are glass-packed, but these require only very mild heat processing. The strong aroma of pickles should be kept away from other products.

Ceramic containers can be made from pottery, earthenware, china, or porcelain. The latter, and more expensive types are used as decorative items, *objets d'art*, and fine dinnerware but rarely, if ever, as packaging. A few earthenware tumbler or jar-type items are still to be found, usually as jam jars or novelty cosmetic packages. Earthenware is porous unless glazed. Because they are not as strong as glass with respect to impact, earthenware packages are thick walled. They do have excellent crush resistance.

Converting of Earthenware into Packages

The converting of earthenware into packages begins with the fluid "slip" formulated from mixtures of clay, sand, and water (see Chapter 3). The slip is formed into shape by one of several different methods.

Casting. Some fine porcelains and chinaware are made by casting the thin slip into a porous plaster mold. The mold absorbs enough water to congeal the mix. Final shaping is done after removal from the mold.

Throwing or Jolleying. Wide-mouthed hollow earthenware or chinaware containers are made by forming the kneaded clay. This can be done by "throwing" on a potter's wheel and forming the object by hand, or it can be done in a semiautomatic or automatic "jolley," where the lump of clay is first pressed into a revolving female mold to shape the exterior, then a revolving "profiler" moves down to cut away and shape the interior. Forming of flatware may be done in reverse. The mold is male and forms the inner face of the object, and the profiler shapes the bottom or outer face. Sometimes "pressing" is used, whereby the soft clay is formed in two halves via plaster molds and the halves are then brought together.

Turning. After the object has been shaped and has lost some water it will become "green-hard"—about the consistency of leather or a hard cheese. In this condition it can be handled, trimmed, embossed, and "trued up." Other features may be added, such as feet and handles.

Firings. After the greenware has dried sufficiently to permit careful handling, it is placed into kilns and fired. Methods vary depending on the ultimate product. For example, chinaware may be placed in fireclay receptacles and surrounded by fine ground flint, which supports the object and evenly distributes the heat. Ovens are closed and the heat raised slowly to drive out moisture. After about 24 hr. the heat is raised further. Porcelains are heated to about 1700°F, cooled carefully, glazed, and refired to about 2500°F, which takes about 20 to 30 hr. This is followed by a gradual cooling, which takes about 3 days. Chinaware pieces are heated slowly for about 24 hr then raised to about 2300°F over a period of 25 to 35 hr. Cooling takes about 50 to 60 hr.

Decoration. Decoration may be added under the glaze or over the glaze on earthenware. Metallic oxide pigments or powdered noble metals are blended with oil and transferred to the object by offset lithography. Low-temperature firing burns off the oil and fixes the color. Overglaze decoration may be done with fluxed colors which fuse to the glaze as a "glass" or enamel. Metals may be burnished or dull.

Glazing. The fired greenware, now called "biscuitware," is sorted and cleaned and then dipped into a glaze, which may be made from alkalies, lead, feldspar, or common salt. These glazes, upon a second firing, fuse into a hard, glassy coating. (The familiar salt-glaze pottery is made by throwing common salt into the kiln. The vapors thus produced form a glaze on the pottery.)

Closures

Ancient closures were whittled wooden plugs, with or without leather seal assists. The Romans also used lead seals on bottles. The whittled cork plug dominated the field for centuries. It was lightweight, easy to shape, and resilient. Until replaced by plastic, cork later served as a liner material for bottle cap closures. The plug cork is still regarded as the traditional closure for wine bottles. Most modern closures are cap- rather than plug-type.

Caps are principally made from metal (steel or aluminum). However, where more expensive products or products with hard-to-mold properties justify the cost, molded thermoset plastic caps are used (usually screw-on). Phenolics and ureas are still used for this purpose.

Nonrigid thermoplastics (polyethylene, polypropylene, and polystyrene, as well as styrene acrylonitriles and ABS polymers) are finding increasing use in special areas where clarity and product compatability are important, e.g., with toiletries, cosmetics, medicinals, distilled spirits, and household products.

A cap closure on a rigid container serves two functions. It seals the opening of the container against loss of contents or against entry of contaminants, and it can be opened with reasonable ease. Beyond these functions are several optional requirements:

1. It may be required to hold contents under pressure.
2. It may be required to hold contents under vacuum.
3. It may require multiple reclosure capability.
4. It may be required to show evidence of tampering.
5. It may be required to open very easily.
6. It may be required to open with difficulty (child-resistant).

In every closure there are two principal components—the seal and the means for holding the seal against the container finish with adequate force.

The seal must be compatible with the product contained and provide the required barrier. It must be sufficiently resilient to compensate for minor imperfections in cap or bottle surfaces. Traditionally made of cork, then of foil laminations, seal liners today are with few exceptions made of plastics.

There are four principal types of cap closures: screw-on, crimp-on, press-on, and roll-on.

Screw-on Cap Closures. The screw cap has a thread molded into its rim which engages with a corresponding thread molded into the finish of the container. This pulls the cover and seal down against the rim of the container with a controllable force, and the container can be opened and reclosed. This cap is used extensively for dry, liquid, viscous liquid, creamy, and paste products, usually in sizes of under 1 qt.

The lug cap is similar to the screw cap but easier to open and close. It is particularly good for vacuum packs and can be applied at high speed. Lugs molded into the cap are forced under projections on the container neck when the cap is twisted.

Crimp-on Closures. The crimp-on closure (the familiar crown cap) has a fluted skirt. During application, the capper presses the cap (and the liner seal) against the rim of the container opening. The fluted skirt is crimped over a projecting ring molded into the finish. Friction holds the lid in place. A tool is used to pry the lid off.

Variations of the crown cap include flip-off, twist-off, and tear-off versions. The flip-off version works because the holding ring does not exert as much force. The twist-off version relies more on a threaded finish than on the crimp ring to hold the cap on. The cap is sufficiently resilient to release from the ring when twisted off. The tear-off version relies on thinner-gauge metal and a scored tear tab to break the crimp force.

Another version is used on containers that require frequent resealing. The cap is fitted with a built-in spring lever that releases or tightens the crimping force as required.

Press-on Closures. Press-on closures can be metal or plastic and rely on a smooth finish, such as that of a tumbler, to achieve a friction fit. The snap fit relies on an extra locking mechanism—a slight ridge or projection on the glass and a ring of indentations on the rim of the cap, for example.

Another type of press-on closure is that used in vacuum-packed processed foods. Here the seal is forced down by atmospheric pressure because a vacuum is formed inside the container during thermal processing of the food. The cap is usually held in place for further security by a locking lug or a screw-on ring. Although the cap may be threadless, a thread molded into the bottle that engages the cap sealant can provide an easier, twist-off release.

Roll-on Closures. The roll-on closure is usually made of aluminum and is supplied as a blank shell with a liner. The bottle is finished with a top retainer ring, a screw thread, and sometimes a bottom locking ring. During capping, the capper presses the cap liner down on the bottle with a predetermined force and tools "roll" the side metal into the threads to lock the seal to the top ring and the cap skirt to the bottom ring, when present.

In the latter instance, the cap shirt is perforated. When the cap is twisted off even partially, the perforations permit a tear above the locking ring. Although the remainder of the cap can be twisted off or resealed, the tearing of the bottom rim of the skirt remains as evidence of tampering.

Child-resistant Closures. Some medicinal and household products are considered sufficiently hazardous to require that they be packaged in containers that small children cannot readily open. Several closures have been developed and marketed which pass test panels promulgated by the Consumer Product Safety Commission. A present stumbling block is the finding that some child-resistant closures present severe problems to aged or handicapped adults.

There are three basic types of child-resistant closures: squeeze and turn, press and turn, and combination lock and tab.

The squeeze-and-turn closure requires sidewall pressure on a freely rotating outer cap to make it engage a threaded inner cap. Or the sidewall pressure may disengage a locking mechanism. The press-and-turn closure requires simultaneous application of downward force and an unscrewing motion. The combination lock requires orientation of components molded into the cap and/or the container before the closure can be pried off.

The type of product, application, and consumer use dictate the closure, which in turn dictates the finish. Typical closures and usages include the following:

Type	*Typical examples*
Plug	Wine, liquor
Crown	Beer, soft drinks
Twist-off	Beer, soft drinks
Roll-on	Beer, soft drinks, wines
Continuous thread	Instant coffee
Lug or interrupted thread	Jams and jellies
Press-on	Baby food

Originally, crowns were intended for single use, but the development of the inexpensive aluminum roll-on for reclosure led to the twist crown (and to an upsurge in the sales of larger bottles that can be opened and reclosed after withdrawal of only part of the contents). Plugs are almost traditional with wine and liquors and so are not being changed too rapidly. Jams and jellies, being reuse items that are packed under vacuum, require a screw-type closure. Instant coffee is not preserved with vacuum, but it still requires reclosure. Baby foods require limited reclosure, vacuum, and resistance to high temperatures as well as very-high-speed application.

Closures may be decorated, with metal lithography being the most widely used method.

Metal Cans

Metal cans have maintained a large share of the packaging market for many decades. This is the result of a continuing technological effort to provide efficient, competitive, high-speed production and to hold costs down. Locating new plants strategically near population centers or even captively next door to can users has reduced haulage and warehousing costs, and technical advances in steel and aluminum alloys and in can design and manufacturing techniques have reduced the amount of metal and energy required. Beverage cans are now made with side walls drawn so thin that the can relies on the liquid content to provide the needed structural resistance to crushing. Over the past 15 years, total steel content in steel cans has been reduced by 25% and the tin content by 50%. During the period from 1972 to 1978, concerted conservation efforts reduced energy consumption 14.6% while production increased by 16%.

In the United States plentiful supplies of aluminum have led to strong penetration of the can markets. The aluminum can is lighter in weight and has a higher recycling value than the steel can. While it takes more energy to make aluminum than steel, 95% of the energy is recovered each time the aluminum can metal is recycled. In 1979, more than $100 million was paid out to recycle 10 billion aluminum cans.

The three principal market categories for metal cans are beverages, foods, and nonfoods. Beverages are the largest category, consuming 61% of U.S. production (54 billion cans) in 1979.

FIG. 4.5 Aluminum beer cans.
Courtesy of *Reynolds Metals Co.*

Classification. Metal cans are classified according to the material used in their manufacture and according to their configuration; however, within each classification there are many variations in size, method of seaming or closing, end use, interior coating, and exterior decoration. There are more than 200 interior protective coatings now in use.

The metals used are aluminum and steel, sometimes in combination. Aluminum comes in several alloys and tempers designed both for the final product and for the method of fabrication. Typical alloys are: 3004–H-19 (Full Hard) and 5082–H-251 (Somewhat softer).

Three types of steel are used in cans: "black plate," which is bare steel; "tin-free" which has a thin coat of electrolytically deposited chromium; and "tin plate" which has a thin coat of electrolytically deposited tin.

Shapes. High-speed can-making equipment is principally suited for cylindrical shapes. Although other shapes would offer the advantage of conservation of space in storage and shipment, cylinders are easier to make and can manufacturers are set up economically to make cylinders.

Many "standard" sizes and shapes are available from can manufacturers with a wide variety of closures.

The familiar round paint can with its tight-fitting multiple-friction lid comes in sizes from ¼ pt. to 1 gal. in size. It has a very large opening, and the lid is tight fitting for reclosure due to its S-shaped flange. On larger sizes there are "ears" and wire bails to permit easy carrying.

For less stringent product requirements, a round, oblong, or square-bodied can may have a round, single-friction lid which readily pries out of the circular opening in the top. Round cans may have a slip cover which slides down over the inset vertical-flanged top of the can body.

Oblong or round cans for liquids frequently are fitted with a pour spout and a screw-cap closure.

Cylindrical cans for aerosols may have dished-in bottoms or tops to better resist internal pressures.

A wide variety of square, round, oval, and odd-contoured cans have double-seamed lids, and opening is accomplished by using a slotted key to tear a strip out of the body side wall along prescored lines. Such cans may have an unflanged body and may be nonreclosable, or they may have a flanged lid and body arrangement that permits reclosure after the strip is torn away.

Many flat metal "cans," sometimes called metal boxes, are fitted with hinged lids and simple snap-open features. These are used for such items as medicine tablets, tobacco, and bandages.

It has been estimated that there are over 600 different shape, size, and closure combinations in standard metal can products.

Cylindrical cans are fabricated in three-piece, two-piece drawn and ironed, and two-piece drawn and redrawn versions. The three-piece can is adaptable to other shapes; the two-piece is much more limited.

The three-piece can is the most familiar and consists of a body and two ends. Body seams can be joined be soldering, cementing, or welding. The two-piece can has a seamless, formed body and bottom to which is attached one end closure.

Converting of Metal into Cans. *Three-piece cans* begin with the making of the body blank. Coils of metal are sheared into large sheets, each representing several body blanks. The sheets are coated, cured, decorated, varnished, and recured. At the can-making line, the sheets are sheared again into individual can body blanks.

For soldering processes, the edges to be soldered are curled, fluxed, gas-flame preheated, and hooked together and solder is applied. Next a second flame is used to smooth the seam, and any excess solder is wiped off. The seam area requires an extra lacquering operation.

For cemented seams, the edges are flame heated and cement is applied. Excess is removed by a chilling and knife-trimming station. The cement is then resoftened and the edges are "bumped" together and fast chilled. As with soldered seams, the seam area requires an extra lacquering operation.

For welded seams, the blank edges are abrasively cleaned to remove chromium and/or oxides, the edges are lapped and tack welded, and the final seam is welded by rolling electrodes.

After the body is made, the bottom edges are curled and double-seamed to the can bottom. A final spray coating is applied to the can interior and cured.

Two-piece drawn-and-ironed cans begin with aluminum or tin plate stock that is formed into a cup. The cups are then forced through progressively smaller die rings, which thin out and lengthen the side wall without appreciably affecting the gauge of the bottom. A final punching operation forms and shapes the bottom. Circular knives trim the excess metal from the top. Cans are then washed, dried, decorated on the exterior, coated on the interior, and baked. The can is then necked in at the top to save material and flanged to receive the top end closure.

Two-piece drawn-and-redrawn cans begin with sheet stock, which can be tin-free steel (not suited to drawn-and-ironed production). This process is slower but simpler than the drawn-and-ironed process, and the bottom and sidewall have the same thickness. The blanks can be coated prior to drawing. Only one or two redraws can be performed, which tends to limit the depth of draw, as does the extensibility of the lacquer coatings. After drawing, the can body is necked, the bottom is formed, and the sidewall may be beaded or indented for added strength.

All cans are tested for leaks and then palletized for shipment.

Aerosols

Aerosols are pressurized containers that were first commercialized in World War II as the DDT "bug bomb" insecticide spray. They are logical improvements on the mechanical pump atomizer, substituting a pressurized propellant for the mechanical piston pump.

Every aerosol is made up of a container and a valve assembly which is fitted to the top of the container. The container may be made of steel, aluminum, or glass. Steel containers may be drawn, soldered, or welded. Aluminum containers are drawn. Glass containers may be uncoated or coated with PVC to reduce chances of breakage.

Glass has dominated the pharmaceutical and perfume market.

The valve assembly is comprised of a number of parts: (1) The dip tube delivers the product from the bottom of the container to the valve. Usually the dip tube is made of polyethylene, and it can range from 0.03 to 0.25 in. (0.76 to 6.35 mm) in inside diameter depending on the spray rate desired and the orientation of the container during spraying. (2) The housing is fastened inside the mounting cup. It can be made of plastic (nylon, Delrin, polyethylene) or of stainless steel. It is like a junction box. The dip tube fits into the housing and an orifice conducts the product to the valve stem. A second orifice may be used to mix propellant gas with the product in the tube. (3) The stem allows the product to flow from the housing to the actuator (nozzle). The stem may also be made of plastic or metal. It further mixes and/or meters flow. It is spring actuated on the return to close off flow. Springs are usually made of stainless steel. (4) The gasket seals the outermost orifice when the valve is in the closed position and also helps to seal the housing against the container cup. Gaskets are made of a product that will maintain strength and resilience, and will not swell over a long period. Buna N and Neoprene rubber are the materials used most often. (5) The actuator is the final nozzle and the means by which the stem is depressed to release the contents. Various shapes and sizes of orifices are used to shape the spray pattern. Various designs are used to make sure that spray cannot be readily aimed in the wrong direction, and child-proof designs are available for hazardous products. (6) The mounting cup is a metal (usually aluminum) cup that holds the valve housing and can be sealed against the rim of the cup container. Sealing is done by crimping, using a flowed-in rubber gasket on a plastic or cut-rubber gasket.

Aerosol Products Systems. There are three basic product systems: emulsion, miscible, and immiscible. Emulsion systems usually include a water-in-oil emulsion. The product may be the oil, it may be dissolved in the oil, or it may be dissolved in the water. Miscible systems include a solution of active ingredients and solvents. The propellant both dilutes and propels the product. Immiscible systems include an aqueous phase containing the product and a propellant. The propellant usually separates into a vapor phase and a liquid phase. Since the liquid phase is insoluble in the aqueous phase, it floats on top and only the product is dispensed.

Aerosol Propellants. The earliest propellants were hydrocarbons, with some use of carbon dioxide and other gaseous propellants. Fluorocarbon propellants (Freon 11 and 12) were developed and quickly penetrated the market, as they were efficient, inert, and non-flammable. Then, in the 1970s, chemists at the University of California theorized that fluorocarbons would react with and deplete the ozone layer in the stratosphere, which in turn would increase the incidence of cancer on Earth as well as adversely affecting climate. Under pressure from consumer groups, the U.S. government agencies involved prohibited the manufacture and use of fluorocarbon propellants for most products. Industry was quick to switch back to the previously tried propellants.

Metal Shipping Containers

Shipping containers have traditionally been made from steel. Because of the higher cost of aluminum and required strength factors, aluminum has not made substantial penetration of this market.

Most end uses can tolerate low-carbon mild steel when suitable lining coatings are applied; however, some end uses require the more expensive stainless steel or nickel alloys.

The principal types of metal shipping containers are drums and pails. Drums are single-wall cylinders made from 26-gauge or heavier steel sheet with capacities ranging from 13 to 110 gal. Pails are single-wall, round containers with straight or sloped sides. They are made from 29-gauge or heavier steel sheet with lock-type or welded seams and in capacities ranging from 1 to 12 gal. By far the largest use sizes are the 55-gallon drum and the 5-gal. pail.

Drums and pails are used to ship quantities of less than a carload of chemicals, petroleum products, paints, lacquers, solvents, foods, adhesives, inks, cleaners, detergents, soaps, edible oils, asphalt products, and the like.

Drums and pails may be made with a tight-fastened head and side or end "bung-type" openings, or they may have a fully removable head. The latter is used for dry, semisolid, or viscous liquid products; the former for free-flowing liquids. Open-head pails are usually tapered to permit nesting when empty. Covers may be fastened by crimping lugs or by locking rings. Pails are usually fitted with lugs for bail-wire carrying handles.

All steel shipping containers fall under control of one or more of three U.S. regulatory groups:

U.S. Department of Transportation (DOT)	CFR 49 Part 1–199
Uniform Freight Classification Committee	(U.S. Railroads) Rule 40
National Classification Board	(U.S. Truckers) Item 260

All dangerous (hazardous) products are regulated by the DOT, which specifies type of container and number of usages. Nonhazardous materials are regulated by one or both of the latter agencies.

Interior linings may be phenolics or epoxy-phenolics, sprayed or roller coated. A newer development is the use of a steel container for strength and an inner, rigid plastic container for product holding and protection.

Exteriors may be decorated both to provide information and for promotion of product or corporate identification.

Since the 1930s when the 55-gal. drum used 18-gauge steel in both body and ends, the industry has continuously experimented and tested the safety of designs utilizing lighter gauges. Today some nonhazardous products are being shipped in a 20/20-gauge steel drum which represents a savings in weight of 12 lb/drum.

In the opposite direction, heavy-duty drums are made from heavier-gauge steel for hazardous products. Side flutes can be reinforced by extra steel plate or even I-bar rolling hoops.

Semirigid Foil Containers

When aluminum foil containers were first introduced in the early 1950s, few would have predicted that the market would grow to over 10 billion containers a year within 3 decades. In retrospect, however, the reasons for this growth can readily be seen.

Foil containers proved to be not only attractive, but extremely convenient and functional as food packages. Because of the barrier quality of the metal, food quality was protected, and because of its good heat conductivity, food could be frozen or reheated in the package. Heating food in the package meant no dirty pans for the housewife. Foil containers could be washed and reused or crushed and thrown away—and today they can be recycled.

The growth of convenience foods and prepared frozen meals followed a similar growth pattern during the same period. It is reasonable to assume that without the foil container, the growth of the convenience foods market would have been substantially slower. The foil container has in more recent years moved into the volume feeding market in the form of steam-table pans for frozen foods, packed by a food processor or by a control commissary.

Types of Containers. Foil containers have evolved through several "generations": folded, die-formed, die-drawn, and air-drawn.

Folded containers are simply imitations of a rectangular tray-type folded paperboard carton. A square or rectangular piece of foil is folded along lines corresponding to the score lines of a carton blank. Corners are given a "simplex" lid to prevent leaking. Due to the bendability of foil and its lack of "springback" when creased, somewhat greater variety in edge (or rim) treatments can be achieved. Disadvantages of the folded container lie in its lack of rigidity even when the foil is tempered. This limits sizes.

Die-formed containers are started from a sheet material blank of greater size than the finished container. Matched male and female dies with sufficient tolerance to allow metal to slip and wrinkle in the sidewalls form the container. It is important to note that very little drawing (thinning and stretching) of the metal occurs in this process. The wrinkles in the sidewalls and flanges contribute to much greater structural rigidity. The method also permits a wide variety of shapes, including round and oval as well as rectangular. Beading of the flanges for trimming also adds strength. Compartmentation of shallow trays proved to be both an advantage to marketing (e.g., the TV dinner tray) and function (the bottom ribs added more strength). Air or vacuum pressure can be used to supplement the mechanical pressure in forming the containers.

Die-drawn containers are the third generation; they represent attempts to further improve the wrinkle-walled container. Wrinkled flanges are not capable of accepting a sealed lid. In the die-drawn method, the flange area is clamped so tightly that slippage is drastically reduced and wrinkling is almost totally eliminated. Hence less metal is made available and the sidewalls are drawn thinner, are work-hardened somewhat, and are less wrinkled in appearance. Tighter tolerances between the sides are required for this method. Air or vacuum pressure can be used to supplement the mechanical pressure in forming the containers.

FIG. 4.6 All-aluminum containers are used to provide air passengers with hot and tasty meals.
Courtesy of *Reynolds Metals Co.*

Foil containers may be used uncoated for some items, but many coatings are often used to provide decoration or protection. Coatings may be clear or colored, or, where extreme heat resistance is required, two-part curing coatings such as epoxies may be employed.

Container closures vary, depending on the product and the type of container. Small frozen pies, for example, are not covered, but items that might spill or become contaminated may be covered with printed foil tuck wraps. Paperboard (with or without laminated foil) die-cut plug lids can be crimped in place with containers having vertical raw-edged flanges. Hermetic seals are possible with smooth-flanged containers, and a variety of peelable heat-seal membranes have been developed for this purpose.

Most foil containers are retailed in printed paperboard cartons to protect them against possible damage and to convey the necessary "copy."

Collapsible Metal Tubes

Under our own definition, collapsible metal tubes ought to be classified as "flexible" because they were invented to replace flexible pack-

FIG. 4.7 Package from Italy's Star.
Courtesy of *Instituto Italiano Imballagio.*

ages made from animal bladders. It would appear, however, that this origin is obscure, and the fact that their shape does not conform to the product would certainly classify them as rigid or semirigid.

The earliest types (1841) were used for dispensing artists' paints, and this use continues today. By far the greatest market, however, was for toothpaste—a material invented in the early 1890s.

The advantages of collapsible metal tubes lies in their deadfold characteristics and product protection. They can be attractively decorated and, through variation in closure design, can provide extreme versatility in dispensing. Plastic and laminated tubes are making strong inroads on markets previously held by metal tubes.

Materials. Selection of the metal to be used depends primarily on the product to be packaged, although liner coatings now available make this rule a little less stringent. Aluminum is now the metal most commonly used for cosmetics, toiletries, foods, and many pharmaceuticals, and it holds 90 percent of the metal tube market. Pure tin is used in cases where exceptionally strict compatibility requirements must be met—as with certain ophthalmic ointments. Lead is cheaper and can be used for products where FDA requirements do not interfere—for example, for glues, greases, and paints. Tin-lead alloys and tin-coated lead are sometimes used to provide a brighter appearance at lower cost than tin.

Manufacture. Tubes are made by punch and die impact extrusion presses, starting with round slugs of metal. The formed tubes are then trimmed and the neck ends are threaded. After annealing, lining materials (usually epoxy-phenolics) are sprayed or flushed in and baked on. The external coating is applied and baked on, followed by decorative printing. The final step is the insertion of a polyethylene neck liner and the application of the cap closure, after which the tubes are placed in divided shipper cartons protected against crushing.

Filling and Closing. Modern filling and closing lines vary in speed, depending on the type of equipment and the production rates desired. Loading can be automated or done by hand. Tubes are then flushed clean, filled, oriented, and crimped closed. Orienting is necessary so that the printed copy will be correctly displayed relative to the flat, crimped closure. Orienting can be done by hand on hand-loaded equipment or automatically by means of an electric eye mark on the printed copy.

Crimping consists of folding the end over at least twice and then applying a crimp by pressure. The crimp creates a mechanical lock. For some products the crimped end is dipped in a lacquer, latex, or cement to create a stronger, tighter seal.

Closures. Nearly all closures are made from injection-molded plastics. These may be polyethylene, phenolics, or urea formaldehyde. Cap liners may be included. These are made from composition cork or from soft paperboard with plastic-film, coated-paper, or metal-foil facings.

Converting Wood into Boxes and Crates

When lumber was plentiful and inexpensive, nailed or wirebound wooden boxes and crates were in common use as shipping containers. With rising costs of lumber, labor, and freight and the availability of cheaper substitutes—fiberboards, composition boards, and the like—use of wood containers has been on the decline. Wood is still used when high stacking strength or wet strength is needed for heavy overseas shipments, and for perishable products requiring ventilation.

The type and cure of wood used will affect the weight, strength, and ease of fabrication of the container. Softwoods are easier to nail but are not as strong as hardwoods. Green lumber is excessively heavy, is weaker than seasoned lumber, and will warp and shrink, causing loosening of nails or other fasteners.

Wooden Boxes and Crates. Wooden boxes are usually solid-walled, rectangular-shaped, nailed containers that vary in construction, with extra cleats and braces used as may be required by the load. The top, bottom, and sides of a box provide the main structural strength.

Crates are similar to boxes but may be of lighter weight and more open construction—that is, spaces may be left between boards or the crate may be fully enclosed or sheathed. A crate differs from a box in that the frame members carry the load. The sheath merely encloses; hence sheathing may be corrugated fiberboard, thin plywood, or lightweight lumber.

Several joining methods may be used for boxes and crates. These include use of metal fasteners, glues, and wires or wire tapes. When wires are used, thinner side, top and bottom sheathing can be utilized, as the wires add strength. Cleated ends and stiffeners provide the structural strength required.

Advantages of wooden boxes and crates depend on the relative cost, strength, and weight ratios involved. Where lumber is cheap or wooden boxes can be reused, they may prove advantageous. However, in most food uses today, wooden containers are being phased out and solid or corrugated fiberboard containers are replacing them. Some wood is still used for reinforcing cleats and bottoms. Certain consumer items use specially built wooden boxes designed for luxury or practical home storage—tool boxes, flatware chests, cigar boxes, and gift boxes for toiletries are examples.

Barrels, Casks, and Kegs. A barrel is a cylindrical container of greater length then breadth, having two flat ends of equal diameter and bulging at the waist. A keg is a small barrel. A cask is a large, tight wooden barrel.

Barrels may be made of wooden staves bound together with hoops and may be tight or slack. Wooden barrels are rarely used now except for aging of beer and whiskey. Plywood drums are made by wrapping thin sheets of plywood around circular reinforced wooden ends. They may be lap seamed or butted with a metal strip reinforcement. Both types are usually supported by metal banding. Plywood drums have been widely used for dry products, but they are being displaced by fiberboard drums.

Wire-bound Boxes. The principal advantages of this old package (early 1900s), which keep it alive in the modern packaging arena, are its stackability (strength), its weatherability (resistance to moisture), its breathability (allowing free circulation of air to the product, where desired), and its relative cost/strength/weight relationship. Wirebound boxes are still widely used in shipping perishable products and in shipping heavy nonperishables, such as bathtubs, automobile windshields, and metal equipment.

Although wirebound containers are difficult to make when smaller than 12 × 6 × 6 in., there is essentially no maximum limit on size. Unitized "mats" can be tied together to form larger sizes. They may be conventionally rectangular in shape, but these containers are sufficiently flexible to conform to octagonal or nearly round shapes.

The wire-bound box is made from wood. The basic section may be a frame of wood which supports a skin of thin wood veneer or thin sawn boards. The skin is fastened to the frame by means of steel bonding wires held in place by galvanized steel staples. Where the "skin" is one continuous sheet, the unit is less flexible and less open. Where the "skin" is a series of narrower "boards" or "cleats," the structure is very flexible and open. Sections are joined by twisting or looping wire ends together. Additional strength may be gained by using stronger end pieces ("picture frame") or solid ends. Various types of woods can be utilized. The latest trends involve use of corrugated paperboard and plastic as substitute materials. Cushioning materials may also be utilized to protect against damage.

Baskets. Baskets are made from overlapped panels of wood, or braided woven splits or staves of thin wood veneers. They come in a variety of shapes—round, oblong, and squarish. Typical examples are the berry tills, mushroom baskets, and tub-shaped bushel baskets. They may be fitted with wire handles or pivoting wooden carry handles. Bottoms may be woven or made of solid wood. Tops are usually finished with a wooden hoop. Some are fitted with wooden lids or covers.

Converting Wood Pulp, Paper, and Paperboard into Rigid or Semirigid Package Forms

Pulpboard Packages. Molded wood pulp can be formed into useful package forms by vacuum forming a water slurry onto one die and then compressing the mat to final shape with a mating die. For convenience in shipping, most such products are nestable. This is primarily because the vertical side walls are angled for ease in stripping from the die.

FIG. 4.8 Wood package wafers.
Courtesy of *Walter Landor and Assoc.*

Pie plates, cake circles, and picnic plates are major applications for molded pulp. Packaging applications include trays for eggs and fruits, flower pots for greenhouses, and protective cushion packaging.

The chief virtues of molded pulp packaging are its low cost and its cushioning ability. Even rough unbleached pulp can be decorated with simple coloring techniques. The smoother bleached versions can be quite attractive.

Paper Containers. Paper can be used to construct rigid containers by winding it into multiple layers. Various sizes can be attained by varying the diameter of the mandrel upon which the basic tube is wound. Smaller sizes are designated composite cans or canister containers; larger sizes are used as shipping containers and are known as fiber drums.

Composite cans were first made in 1905 by winding paper into a tube. Today, can bodies are also made from combinations of paper, metal foil, and/or plastic films. Ends can be made of paper, metal, or plastic. Structures may be simple for dry products, such as salt and sugar, or complex for liquid or gas-tight packages.

Although not suitable for packages that must survive heat processing or high internal pressures, composite paperboard containers or "cans" enjoy a share of the can market. Composite cans have been used for decades in nonliquid applications, such as for cocoa powder and powdered cleansers. Most of these applications have involved simple combinations of a chipboard with a glassine liner and a printed paper label. The introduction of aluminum foil and of polyethylene as packaging materials made it possible to improve the design of the composite can sufficiently to permit it to capture a new market—refrigerated biscuit dough. This product required a rigid container that could be opened from the side and yet would retain moisture and carbon dioxide in a refrigerated and usually moist environment. An end-opening metal can could not be used because of the great difficulty that would be met in trying to remove the contents intact. Scored aluminum body walls, as are now employed for canned luncheon meats, had not yet been developed. Convolute winding, as used on cocoa powder cans, could have been employed, but the entire opening procedure would have involved tearing the body from the double-seamed metal ends. Spiral winding proved to be a method that would enable the consumer to unwind the package along the spirally wound joints of composite fiberboard without having to attempt to break the bond at the double seam. A lamination of kraft paper with a printable exterior sheet and a protective foil inner sheet was wound in a unique manner to effect a moisture proof and gasproof joint. Metal ends were double seamed to the body. Without this package, the product probably could not have succeeded in the marketplace.

The two basic methods of manufacture for composite paper containers are spiral winding and convolute winding. Spiral-wound containers

FIG. 4.9 The Cooper. A woodblock print of a cooper using Scotch pine, fir, and oak to make barrels, tubs, beer and wine casks in the 1500s. From *The Book of Trades*, Standebuch, 1568. (Dover edition reprint, 1973).

are always cylindrical in shape, and are made by feeding continuous strips of materials at an acute angle to the mandrel. The different plies are bonded together by means of adhesives, which accounts for the containers' structural strength. Since the inner ply becomes the interior of the container, it may be specially coated paper or a foil or plastic film. The outer ply often is a printed label which becomes an integral part of the container wall. Convolute-wound containers are wound from a strip of paper whose width is equal to the height (or is an exact multiple of the height) of the container. The strip is fed at a 90° angle to the axis of the mandrel and builds up several layers upon itself—again, bonded by means of an adhesive. A label can be applied to the outside of the can in similar fashion. Convolute winding permits odd shapes (such as squares, rectangles, or octagons) to be wound.

Composite cans used for frozen orange juice concentrate are comprised of a body spiral wound from paper body stock with polyethylene or aluminum foil inner liner and a paper or aluminum foil outer label. Most have one end designed for convenient opening, including full-panel opening and plastic-strip easy-open.

FIG. 4.10 Composite can used for cocoa—early 1900s.
Courtesy of *Walter Landor and Assoc.*

Some other frozen foods, such as sugar-packed strawberries, are packaged in convolute-wound oblong composite cans with metal ends. Shortening, with requirements for greaseproofness and oxygen exclusion, has been commercially packaged in composite cans. Roasted and ground coffee, with gasproof requirements, has been market tested in composite cans.

It is claimed that composite cans are employed for over three-quarters of all motor oil in 1-to 2-qt sizes. The inner liner is generally polyethylene, although some foil and glassine are used on a kraft paper core.

Larger drums can be made in similar fashion. Convolute winding is employed where maximum wall strength or a noncylindrical shape is desired. Composite can ends are varied, depending on end use. They may be a simple formed paper wrap, a friction-plug can lid made of metal or plastic, a screw cap, or some form of dispensing lid. One familiar type is the metal-ended frozen fruit juice concentrate can with a tear-open lid.

Shipping drums vary in terms of types of bottoms and top closures. They may be made from fiberboard, wood, or metal and may be attached by means of adhesive, tape, metal locking bands, metal lugs or clips, or various combinations of these fasteners. Drums made for liquids usually have special coatings or liners and tightly clamped or sealed tops and bottoms; some are equipped with bungs. The latest types have molded plastic liners which are liquid-tight. The drum then serves to protect the liner and add strength.

Fiber shipping drums must also conform to DOT regulations and National Motor Freight, and Uniform Freight classifications for the particular commodities to be shipped.

Paperboard Containers, Rigid "Set-up" Boxes. Rigid paperboard boxes are manufactured and shipped to the point of use in a three-dimensional condition—that is, "set up" rather than "knocked down," as is the case with folding cartons. Standard types are made from paperboard with corner reinforcements, paper covering materials, and the necessary adhesives. More complicated types may include plastics, cloth, metal foils, and fancy trimmings. Shapes can be varied widely from simple squares, rectangles, and circles to complicated ovals, heart shapes, and octagons. Because of these variations, they are commonly described by type and inside dimensions of the base.

Flat sheets of paperboard are first scored and blanked to size, usually in multiples that are then broken into individual blanks. After corners are cut, they are folded to form the tray-shaped bottoms and lids. Corner reinforcements of metal, cloth, or paper are applied to give the box its rigidity and strength. Finally the coverings are glued on. Lids are made in the same way.

FIG. 4.11 "Elvax" coated returnable corrugated shipper manufactured by Oy Tako of Tampere, Finland, is constructed entirely of corrugated board, except for a reinforcing steel rod inside the upper rim.
Courtesy of *E.I. DuPont de Nemours and Co.*

Quality can be varied through selection of cheap, lightweight, rough-finished board stock or more expensive, denser, smoother-calendered boards.

There are many basic styles differing in types of lids and interior arrangements.

The full telescope box has a simple base with the lid fitting completely over it. Such boxes are familiar sights in department stores, used for displaying or carrying home merchandise such as clothing, cosmetics, and candy. Variations include extension edges on the lid top and on the base bottom.

In the neck- or shoulder-style box, the base is fitted with a second insert, which provides greater strength and a shoulder for the lid to rest upon. This permits only semitelescoping of the base into the lid. Inserts may be partitioned or a die-cut platform may be substituted.

Hinged-cover boxes come with a cover attached to the base. The cover may be tray shaped and fitted with a hinge, or it may be like a book cover with the tray glued to a similar "book-cover" base. Platforms may be glued in place or removable.

The tray-style box uses a sleeve of paperboard or plastic as a cover. The tray slides into the sleeve much like the old-fashioned matchbox.

Obviously, styles can be varied in any way desired, with domed lids, cutouts, and fancy linings and trimmings. Equally obvious, the more complicated the box, the more hand labor is required to make it, and thus the more expensive it is. Since only a few of the operations can be done automatically, the machinery required is not expensive and runs can be small without adversely affecting cost. Versatility, quick change-over, and fast delivery are also advantages.

Solid Fiberboard and Corrugated Paperboard Containers. Other extremely versatile types of packages are corrugated paperboard and solid fiberboard containers, which have found use both as primary packages and as secondary shipping containers for other packaged products.

Over 90% of all packaged goods are now shipped in corrugated containers. In 1979, 35% of all paper and paperboard manufactured in the United States—some 23 million tons—went into containerboard used in solid fiberboard and corrugated boxes.

Solid fiberboard is made by gluing plies of board together to the desired thickness. It is usually lined on one or both faces with a higher quality (smoother) paper. Total caliper (thickness) may run from 0.040 to 0.110 in.

Glue is applied to both sides of one liner ply and to one side of each of the other liner plies. Facings are fed in dry. Liners are unwound so that any curl is neutralized by alternating curl direction. The assembled plies are then passed between a series of squeeze rolls, which are set to apply successively more and more pressure and which, due to deliberate variation in diameters, have successively faster surface speeds, thereby pulling out any possible wrinkles. After passing through driers, the board is trimmed and cut off into sheets. Blanks are creased, slotted, and printed on presses. Finally the joined corner is made on a stitcher, although a taper or gluer can be used.

Corrugated containers are made from corrugated paperboard and owe their versatility to the amazing strength afforded by the structure of this material. Corrugated board is made from a corrugating medium which is glued to one or more liners. When glued to only one liner, it is called single-face corrugated. When glued between two liners, it is called single-wall corrugated. Double- and triple-wall corrugated boards are made by adding additional (one or two) corrugating medium plies and the necessary extra liners (one or two).

Other variations in strength are attained by using linerboards and corrugating media of different strength (and weight) and by using different types of flutes in the latter. These are given in Table 4.1.

Since corrugated board is used to provide compressive strength and cushioning to containers, for maximum compressive strength the flutes should be oriented vertically in the case. Because it has fewer columns per linear foot, A-flute normally has greater compressive resistance than the other flutes. With more flutes per unit length, B-flute can support greater weight when force is applied at right angles to the

Table 4.1 Types of Flutes Used in Corrugating Media

	Flutes/ foot	Flute height (in.)	Thickness (in.)
Flute A	35–37	0.185	3/16–7/32
Flute B	50–52	0.105	4/32
Flute C	41–45	0.145	5/32

facings. A-flute is a better cushioning material than the others because of greater depth, but because of its greater thickness it does not crease and bend as easily as the other flutes.

C-flute is the compromise between A-flute and B-flute, and it is often used just because it has fairly good midrange properties. Products that possess inherent compressive strength do not require board with high compressive strength. Canned foods and bottled goods usually employ B-flute. Products that do not themselves have compressive resistance or whose primary packaging has little compressive resistance require cases with compressive strength. For such products, A-flute alone or in double or triple wall should be employed. Products that are lightweight or do not deform with compression, e.g., tissue, can be contained in single-wall cases.

To manufacture the corrugated fiberboard, the corrugating medium is first plasticized by passing it over heated drums and then through steam showers. Flutes are pressed into the board by meshed fluted rolls. While still engaged with the last of the fluted rolls, adhesive is applied to the tops of the flutes. The preheated facing material is then laminated to the flutes. Usually a festooning accumulator is used at this point. To apply the second facing, the single-faced corrugated material is preheated, after which adhesive is applied to the tips of the flutes. Preheated facing material is brought in contact and final lamination is done on a heated table using an endless blanket-belt conveyor to apply gentle pressure. After drying, the material passes through a similar cooling section and is then slit, scored, and cut off. Blanks from the corrugator have the correct outside dimensions for the final container. These blanks are turned 90° and fed into a printer-slotter, which applies desired printing and adds cross-direction scoring and necessary flap cutouts.

Next the joined corner must be made. This is done on a stitcher, a gluer, or a taper, depending on the type of joint.

Special sizes and shapes of "blanks" can be made on plate presses where knives and rules cut and score the desired patterns. This is also where the various inserts, reinforcements, separators, supporters, etc., called "fitments," are produced.

The final product is shipped in knocked-down form, ready for assembly.

A wide variety of container constructions is possible through use of extra folds or flaps and special inserts. Box manufacturers maintain design departments as a service to their customers where containers are custom designed to fit the product. Some liquids, semiliquids, and dry powders are now being shipped in a "bag-in-box," which consists of a plastic bag to hold and protect the product and a corrugated shipper for strength.

Improvements in solid fiberboard or corrugated paperboard containers lie in the realm of making them stronger or more weatherproof through saturation or coating with waxes, plastics, or other materials,

FIG. 4.12 Folding carton for candies, winner of a merit award presented by the Paperboard Packaging Council, was designed for an expensive line of foreign chocolates.
Courtesy of The Paperboard Packaging Council.

such as asphalt. Metal foils have been used to impart strength, barrier properties, and reflective insulation. Plastic films and foams have been used as liners and to provide insulation against heat or mechanical shock. As with rigid drums, there are regulations established by DOT and other agencies that specify requirements for containers used in shipping certain commodities.

Folding Paperboard Cartons. Folding paperboard cartons are three-dimensional rigid containers formed from die-cut sheets of paper-board which may be partially assembled, then knocked down flat. They are not erected until needed and therefore conserve space in the converter's and packager's plant. Folding cartons may be set up by hand or on semiautomatic or automatic equipment.

Materials. The paperboards used for folding cartons must have the ability to score, bend, or fold without cracking; must have adequate stiffness to resist compression distortion or bulging; must have adequate smoothness to accept the desired printing; and must be able to resist environmental changes and weakening by the product contained.

Three types of paperboard are used: virgin pulp single-ply grades, such as solid bleached sulfate and solid unbleached sulfate; multi-ply grades made from reprocessed pulp; and multi-ply grade using both virgin and reprocessed pulps. Improved techniques have improved the quality of recycled pulpboards, and they now represent one half of total U.S. folding carton consumption. Details on the manufacture of paperboards are to be found in Chapter 3.

Solid bleached sulfate boards are very popular for use in food cartons. The unbleached sulfate (kraft) boards are used where greater strength is required, such as for hardware products and beverage carriers. Recycled multi-ply cylinder boards are used to save cost. The outer ply can be varied, depending on requirements. Clay coatings provide brighter color and smoother printing surfaces as well as better suitability for high-speed carton-forming machinery.

FIG. 4.13 The Paper Maker. A woodblock print of a paper maker working in his water-driven mill in the Middle Ages. From *The Book of Trades*, Standebuch, 1568. (Dover edition reprint, 1973).

Special treatments, coatings, or even laminations are now being used to provide improved properties, such as: mold resistance, corrosion resistance, moisture resistance, heat sealability, brighter appearance, greater smoothness, release, abrasion resistance, ability to withstand refrigerated or frozen storage, ability to withstand oven temperatures, grease or oil resistance, wet strength.

Coatings can be applied to the board at the time of board manufacture, later on the converter, or at both times.

Where grease and oil resistance are desired and the product is dry—as with pet food, hardware products, and certain bakery or confectionery products—fluorocarbon materials can be added to the paperboard.

Where a water vapor barrier is desired for a dry product such as soap powder, a paper can be wax laminated to the printing side of the board.

A vinyl-coated paper laminated to the board improves moisture vapor transmision rate (WVTR), grease resistance, and resistance to odor transmission. It is often preferred for perfumed soaps.

Polyethylene coatings on laminations are rapidly replacing wax coatings and laminates because of cost-effectiveness at high speeds. The result is good WVTR and superior oil and grease resistance, but poor odor protection. Polyethylene coatings are difficult to glue, but heat sealing or hot melts can be utilized. End uses include milk cartons, frozen foods, baked goods, soaps, and detergents.

Extrusion of polyester coatings (polyethylene terephthalate) onto solid bleached sulfate board has resulted in boards capable of resisting moderate oven temperatures (up to 400°F); these are used in some

FIG. 4.14 "Pure Pak" cartons for milk, designed for use by children.
Courtesy of *E.I. DuPont de Nemours and Co.*

baking applications and in reheating of frozen, precooked foods and are suitable for microwave ovens.

Aluminum foil laminated to paperboard provides excellent barrier properties as well as visual elegance.

A variety of inks, lacquers, and coatings can be applied in line with the carton-making operation in both all-over and registered patterns. These may be used to contribute barrier properties, abrasion resistance, grease resistance, product release, or attractive appearance, or they may be used to facilitate gluing or handling.

Carton Manufacture

After a prototype sample has been handmade and approved and selection of paperboard has been made, production personnel must:

1. Produce blueprints for die design.
2. Produce drawings for final artwork.
3. Select inks.
4. Make printing cylinders or plates.
5. Make carton-cutting and creasing dies.
6. Make a sample order and test.

Carton stock is printed on web-fed or sheet-fed presses by letterpress, offset, gravure, or flexographic processes.

Die cutting is done on a rotary die cutter or on a platen die cutter. The former is suitable for high speeds but is less versatile and less precise than the latter.

The finishing step for tube types consists of preassembling the body of the carton and making a glue seam. Glue can be applied with straight-line gluers (which apply glue parallel to material flow

FIG. 4.15 Folding cartons coated with hot melts.
Courtesy of *E.I. DuPont de Nemours and Co.*

TABLE 4.2 Market Distribution—Folding Paperboard Cartons

Market	% of Total Tonnage
Dry foods	23.0
Wet foods	12.6
Bakery and confectionery	12.3
Beverages	11.0
Total, foods	58.9
Soap	8.9
Medicinal and cosmetics	7.4
Paper goods	5.7
Hardware	5.0
Tobacco	3.6
Textiles	2.1
Other	8.4
Total, nonfoods	41.1

through the machine) or right-angle gluers (which apply glue in more than one direction).

Finishing may also include application of windows, hot melts, cutting edges, etc.

After the finishing step, the carton blanks are "knocked down" (flattened) for shipment to the packager.

On the packaging line, cartons are set up, filled with product, and closed. Methods differ, depending on the product and the style of the carton.

Carton Styles. There are two basic types and many speciality types of folding cartons. The basic types are tray styles and tube styles. The former present a larger, tray-like opening for filling, and the latter present a smaller end.

Tray styles may be designed with integral hinged covers. They may be delivered to the packager flat or knocked down. When flat, the packager usually must perform the set-up on automatic equipment. When knocked down, set-up can be performed manually. An example of a tray type would be a box for frozen fish fillets or frozen hamburger patties. Examples of a tube type would be a butter carton, a milk carton, and a toothpaste carton. Examples of speciality cartons would include a beverage-bottle carrier, counter display cards, and a windowed bacon carton.

An offshoot of the paperboard carton business is the merchandise card. Cards come in a variety of shapes and carry a printed message. Products (usually small items) are attached to the card by stapling, ties, adhesives, shrink film, or blisters. The cards can be hung on racks or pegs for open displays. Another advantage is that they are pilferproof. Because of their size they are not easy to hide in a pocket or pocket book.

Rigid and Semirigid Plastic Containers

Thermoformed and Fabricated Boxes, Vials, Trays, Etc. Rigid plastic containers are fabricated from plastic sheet 0.003 to 0.030 in. (0.076 to 0.76 mm) thick, or about the thickness of paperboard. It is not surprising that most of the fabricated products are similar in construction and end use to paperboard set-up boxes, trays, folding cartons or platforms, and inserts. Plastic sheet offers variations in transparency and color together with ease in forming and a wide range of useful properties.

Fabricated sheet packages are produced by die cutting, folding, scoring, and beading in much the same way as paperboard packages. Joining can be effected by means of cement, a solvent, heat, or mechanical fastening.

Thermoformed packages made from sheet plastic have proven immensely popular due to ease of manufacture, economics, and versatility. The amount of plastic used for thermoforming has grown 30-fold in twenty years, from 29 million to 900 million pounds.

Sheet plastics used for thermoforming include acrylonitrile butadiene styrene, cellulose propionate, ionomer, polyethylene, polyprolene, polystyrene, and vinyl. Vinyl is used in highest volume for blister packs. Properties of the product to be packaged dictate the selection of the sheet material.

Blister Packs and Skin Packs. These are thermoformed plastic "bubbles" that are fixed to a card. They protect against product loss, give great visibility, provide a handy surface for advertising or instructions, provide a means for rack or hook display, and, because of added bulk, discourage pilfering.

The blister pack is preformed in a contoured mold and can be made "off-premise" by a carton thermoformer, although integrated automatic machinery is available for large-volume in-house operations. The skin pack uses the product as a mold, draping the softened plastic sheet over the product and pulling it down by vacuum. This must be done in line with the packaging operation. This too can be in-house or captive, or it can be done by a custom packager. Skin packaging immobilizes the product and is excellent for delicate or fragile objects as well as larger odd-shaped items.

Thin-wall Containers. These are thermoformed packages made from high-speed roll-fed equipment. Products include cups or tubs, trays, egg cartons, individual-portion cups, and formed halves for luncheon meat packages and large cubic containers.

FIG. 4.16 Thermoformed butter dish made of PVC.
Courtesy of Chemische Werke Huls Aktiengellschaft.

Cups or tubs for drinking or for dairy and delicatessen products may be made from foamed polystyrene or from clear polystyrene, ABS, high-impact polystyrene, or polyolefins. Meat trays and egg cartons are made chiefly from foamed polystyrene, but clear trays are preferred for baked goods. The individual-portion packages can be clear or opaque. Vinyl, polystyrene, and polyolefins have been used. The formed half (bottom) of the luncheon meat package is made from a saran-coated nylon for toughness and oxygen barrier properties. Large cube containers are thermoformed in halves from heavy polyethylene sheet, joined at the flanges, and fitted with a pouring spout.

Multicavity thermoformed inserts are used in boxed candy and cookie packages to reduce product shifting and breakage. Polystyrene, PVC, XT polymer, and polypropylene sheet materials have been used for this purpose.

Expanded Polystyrene Foam. Polystyrene resin is compounded with one or more suitable blowing agents and formed into small beads of definite size distribution. When exposed to heat, these beads are capable of expanding to up to 50 times their original size because of formation of thousands of tiny bubbles. With proper techniques, these expanded beads can be adhered together to form molded articles of any shape and in densities ranging from 0.8 to 10 lb/ft^3.

The process comprises several steps. First, beads are blended to the desired size distribution (color variation may be introduced at this point). A preliminary exposure to heat preexpands the beads by a desired amount. After a holding or aging period, the preexpanded beads are filled into a closed mold. Heat is applied, usually by steam, to complete the expansion of the beads and to bond them together. The mold is then cooled and opened and the part is ejected.

Molded expanded polystyrene foam has been used extensively for cushioning pads, lightweight pallets, drinking cups, and various containers, such as ice buckets and picnic coolers. It has found much usage as a form-fitting protector for delicate or intricate products. As such, it may be taped together and used as a primary shipper for small items, or it may be inserted into an outer shipper, such as a corrugated box or a wooden crate.

Thick sheets or blocks of expanded polystyrene foam are produced by extruding the resin and blowing agent through a slit orifice and then cooling slowly. Blocks or billets can later be cut to desired sizes or shapes.

Thinner sheets and films of expanded polystyrene foam are manufactured by means of ring die extrusion (blown-bubble) technique. These are used in making semirigid and flexible packages.

Blow-molded Plastic Jars and Bottles. When used to make jars or bottles, plastics have some distinct advantages over glass and metal, including the following.

Nonbreakability: Most plastics are less likely to break on impact than glass. With most commercial plastics, unbreakability is a major advantage.

Lighter weight: Plastics weigh as little as one-third the weight of metal and one-eighth the weight of glass in comparable applications, and this is continuing to improve.

Lower noise level: Plastics make less noise, both as packaging and in use, again compared to glass and metal.

Noncorrosive: Plastic containers do not corrode as metal cans do.

Smaller size: Wall thicknesses for equivalent strengths and volumes are less for plastics than for glass, so that equivalent contents can be contained in packages of smaller exterior size.

Flexibility of forming: The capital investment required to produce rigid plastic containers is only a small fraction of that required for glass or metal.

In economic gauges, most plastics do not have sufficient strength to withstand compressive stresses such as would be encountered in warehouse stacking. This requires compartmentalized shipper containers and limitations on stacking height. The light weight of plastic bottles requires special packaging equipment for high-speed handling, filling, and closing.

The major plastic materials used for blow molding are high-density polyethylene, low-density polyethylene, polypropylene, polyvinylchloride, styrene, styrene acrylonitrile, polyacetal, polyester, and copolyester.

More than 16.2 billion plastic bottles were manufactured in 1981. The largest segment of this production was that of the blow-molded 1-gal. milk bottle made from high-density polyethylene (HDPE), which amounted to 3.2 billion units. Poundage consumed for this single use has grown from 315 million pounds in 1975 to 680 million pounds in 1981 and is projected to 800 million pounds by 1988. The second largest segment was that of the 2-liter carbonated beverage bottle made from polyethylene terephthalate polyester (PET), which amounted to 2.3 billion units in 1981. The PET bottle is also manufactured in 1-liter and ½-liter sizes.

Manufacture. There are three basic methods of manufacture of blow-molded containers; extrusion blow molding, injection blow molding, and stretch blow molding.

Extrusion blow molding uses an extruder to form a gob of molten plastic called a parison. A split-cavity mold is closed around the parison, closing off one end (the bottom of the container). Compressed air is injected into the parison through blowing needles, a blowing pin, or multiple blowing heads in the mold. The air forces the plastic to conform to the shape of the mold. The distribution of plastic and final wall thickness are controlled by the shape of the parison, the amount of plastic, the shape of the mold, the temperature, and the air pressure. The scrap rate can be high and scrap must be recycled. Tooling is low in cost, the production rate is high, bottles with handles can be produced, wall thickness tends to vary, and final trimming of the mold flash may be required.

Injection blow molding is a two stage process. First a "preform" shaped parison is made by injection molding. During this step the neck finish is formed. Next the preform is transferred into a blow mold and

FIG. 4.17 Various blow-molded packages. *Courtesy of Chemische Werke Huls Aktiengellschaft.*

compressed air does the final shaping of sides and bottom. This process produces a superior finish and is scrap free, but it is limited to small-size containers with no handles, requires a high tooling investment, and has a lower production speed.

Stretch blow molding is a modification of both processes in that the preform can be injection molded on an extruded tube. The preform is carefully heated to a precise temperature and stretched before blowing into the cavity mold. The result is a thinner and stronger, oriented sidewall and thicker, less oriented bottoms and tops. Orientation improves tensile strengths, impact strength, MVTR, and CO_2 barrier properties.

Polyethlene is used in a wide range of densities, from low (0.918) to high (0.965). The higher the density, the greater the stiffness, hardness, gloss, yield strength, softening temperature, and creep resistance, but the lesser the stress crack resistance, toughness and permeability. The melt index can also vary. This is rate of flow through a standard orifice at a given temperature and pressure. As melt index increases, permeability and gloss are improved, but strength, hardness, grease resistance, and creep resistance are lowered. Polyethylenes resist most acids and salts but are attacked by strong oxidizing agents. Hydrocarbon solvents tend to soften or swell polyethylene, particularly at elevated temperatures, and the lower the density, the greater the attack.

Polypropylene has most of the advantages of polyethylene plus a higher heat distortion temperature. This makes it a better resin for hot-filled products or even steam sterilization. It is not as good in terms of cold-temperature impact resistance. Because of its high melt strength it is more difficult to blow mold. Careful control of temperature is essential in stretch blow molding to avoid too rapid a rate of crystallization. Polypropylene is finding many uses in edible oils as well as toiletries, cosmetics, and household chemicals.

Polyvinylchloride (PVC) has low oxygen and CO_2 transmission and excellent clarity. It also has good resistance to permeability by oily products. While it has good impact strength and chemical resistance, it has a relatively low heat-distortion point. It has been an important factor in packaging edible oils and toiletries but is now being challenged in terms of both price and function by stretch-blown polypropylene (PP) and polyester (PET).

Polyethylene terephthalate(PET) has excellent barrier properties for alcohol and essential oils, good chemical resistance, and fair MVTR. Its permeability to CO_2 is low, making it an excellent bottle for carbonated soft drinks, which is currently its largest use.

Technological improvements are being developed continuously in the blow-molded bottle field. Chief targets are reduction in weight, improved strength, improved barrier properties, and, of course, improved production efficiencies.

FLEXIBLE PACKAGE FORMS

We already have defined flexible packages as those that generally conform their shape to the product contained. Although there is some confusion and overlapping of nomenclature, flexible packaging can be divided into two basic categories: partial enclosures and total enclosures. Partial enclosures are not intended to do the entire job of holding the product. They supplement or assist the holding element. In this category are labels, some types of intimate wraps, carton liners, cap liners, overwraps, and cushioning materials. Total enclosures are intended to hold the product with or without assistance. These include some types of intimate wraps, bags, envelopes, and pouches. In the following descriptions these differences will be noted where appropriate.

Wrappers

Probably the simplest type of flexible package ever devised was the intimate wrap, where a sheet material was used to enclose a quantity of product. Greaseproof paper proved very useful as a wrapper for butter and later margarine, as well as for meats. Waxed paper provided a better barrier to moisture and could be heat sealed.

FIG. 4.18 Ancient Egyptians wrapped poultry in palm leaves for protection against contamination.
Courtesy of Reynolds Metals Co.

Paper and tin foils were the chief wrapping materials until the coming of cellophane and aluminum foil. Cellophane was improved by adding coatings that made it a better barrier and heat sealable. Later, other heat sealable films were introduced (vinyls, polyethylenes, polypropylenes, etc.), and finally the heat-shrinkable films came on the scene.

Composite laminations of paper, films, and foils were devised where higher barrier properties or more advantageous sealing properties were needed.

Twist Wraps. For round or roundish single-unit items such as fruit, candy, or medicinal lozenges, a simple twist wrap may be used. Here the product is placed in the center of a sheet and rolled into a tube shape after which both ends are twisted.

The material must be flexible, strong enough not to tear, and sufficiently flaccid to remain twisted. Where the latter property may be lacking, heat may be used to effect a seal, or twist ties or gummed tapes may be affixed to hold the twist shut.

Gathered or Pleated-bunch Wraps. Similar types of products can also be wrapped as above, but the corners are gathered or pleated and held in place by deadfold property, by heat sealing, or by an affixed label.

This type of wrap can also be used for irregular bar-type products, usually with a back-up board insert. The wrap is first formed into a tube with overlap underneath the product. The overlap may be folded or heat sealed. Then the ends are gathered and folded underneath and again heat-sealed, if desired, to hold them in place, or a secondary band or tube may be used for the latter purpose.

Folded Wraps. When the product is rectangular in shape or contained in a rectangular box or carton, the wrap may be more precisely contoured by folding. When the folds are brought to one of the major facings of a thin box, the wrap is called a "bias wrap" or "envelope wrap." When the folds are made on opposing ends, the wrap is called a "parcel wrap" (or for candies, a "caramel wrap"). The sequence of folding-in each "ear" may vary, depending on desired appearance or on the type of machinery available.

Box or Carton Liners

Simple wrappers may be used as liners for boxes. Their function may be to cushion fragile objects, as with cookies or candies; to protect against soilage or spillage; to help immobilize the product in the box; to lend additional moisture or grease protection; or to divide individual portions.

Cap Liners and Membrane Closures

Screw-cap closures may need a cushion liner to allow for irregularities in the cap or bottle and thus to effect a proper seal. These may be made from flexible composite materials, die cut to shape.

Removable membrane closures provide more effective barriers over wide-mouthed containers. At the point of use, the membrane is torn or peeled off, and subsequent reclosure is accomplished by the screw cap or snap-on lid that initially protected the membrane against mechanical damage. Such a membrane provides visual evidence of tampering.

Cushioning Pads

A partial wrapper may be placed in the bottom of a tray to cushion delicate products, or it may be an absorbent blotter to absorb juices, as with meats. A wide variety of foams, plastic bubbles, creped papers, hair mats, and the like have been made to serve as cushioning.

Labels

When the custom of wrapping products with paper was adopted in the sixteenth century, a logical development was marking the wrapper to identify the contents. Since printing was a known art, it soon became practice to use a printed wrapper. No one knows who first decided to print a smaller label separately and then affix it to the package.

With better and better printing methods coming into existence, labels became more and more attractive and less and less expensive. Special high-grade papers were developed for the best-quality lithographed labels. More recently, labels have been made from plastic films and from laminated paper, film, and foil composites.

Labels can be die cut to any shape and embossed or flocked for special effects. They may be applied with hot melts or aqueous adhesives, or they may be supplied pregummed or with peelable adhesive coatings on the face or the reverse side.

Total wrap-around labels may be used on cylindrical packages such as cans and may provide extra barrier properties and/or become an integral part of the package, as in spiral-wound canisters.

Neck labels are specially contoured to fit the reverse curvature of a bottle shoulder. Crown labels are even more special, as they come down over the bottle neck and cap.

Saddle labels are strips that fold over the mouth of a bag to help effect closure.

Band labels or sleeves are formed into a tube-like shape and may bundle several units together. Individual chewing gum sticks have a printed paper band label or sleeve which helps hold the foil intimate wrap in place.

FIG. 4.19 Labels are used for sales appeal. *Courtesy of National Distillers and Chemical Corp.*

Transparent sleeves may be printed as labels and placed around non-message-bearing printed cartons. In the home, the sleeve is removed and discarded, and the attractive carton serves as a dispenser.

Sacks, Bags, Pouches, and Envelopes

The first three of these package forms are nearly synonymous, although a pouch is usually smaller than a bag and a sack is larger. An envelope is a special type of pouch with a flap closure.

These often are prefabricated packages that, when empty, and laid flat, assume a generally rectangular shape. Three sides are closed and one side is left open for filling. The opening may later be closed by gluing, sewing, tying, or heat sealing.

Early sacks were made from coarsely woven sackcloth and used for carrying bulky articles or coarse grains. Finer-woven cloth was used for finer particles, such as sugar and flour. Side and bottom seams were stitched with coarse twine and corners "eared" for greater strength. After filling, the top could be stitched as well or merely tied shut.

Smaller cloth sacks of similar construction were called "bags."

The substitution of paper was a logical development. Small, single-layer bags could be glued shut along the body and the bottom seams.

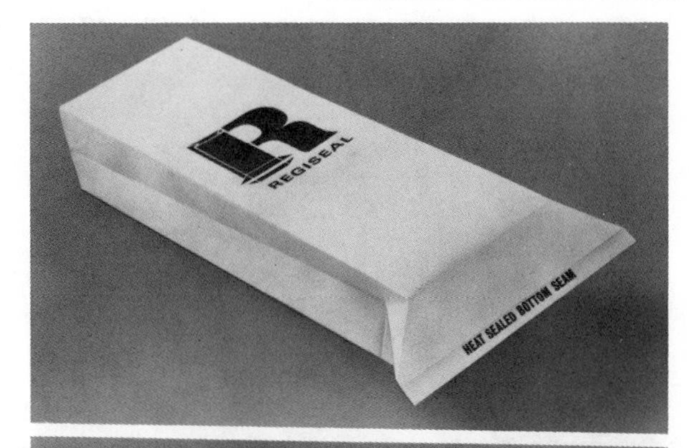

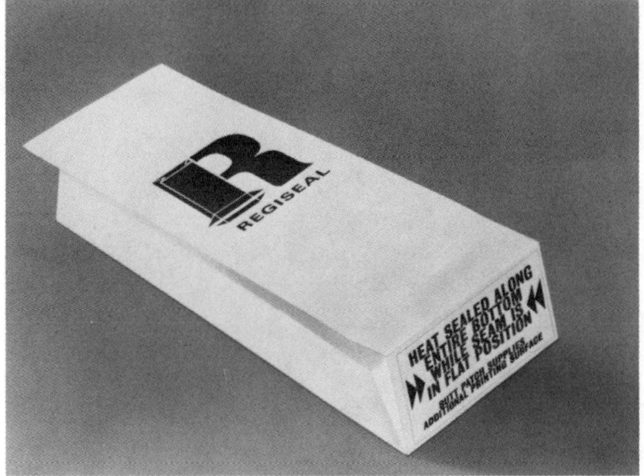

FIG. 4.20 Unique bag for hard-to-hold products; this bag eliminates the channel across the bottom.
Courtesy of St. Regis Paper Co.

Large, multiwalled sacks usually required reinforcing tapes that were glued and/or stitched in place.

The simplest of paper bags is made from one sheet and has one body seam and one bottom seam at right angles to each other. When closure is by means of a flap, the container is called an envelope; otherwise it is called a pouch. Another type of pouch is made from one or two sheets of material and has three seams—two side seams and a bottom seam. One of these is also a fold if a single sheet is used. The three-seamed pouch is called a "fin-sealed" pouch.

One advantage of the two-seam bag or "pillow pouch" is that its sides can be gusseted (pleated). These can be full-length gussets or only partial gussets to form a bottom or top. Gussets allow the bag to open wider and thus hold a greater amount of product without overly straining the bottom or top closure.

Pouches and bags are now fabricated from plastic films and from composite laminates of plastic, foil, and paper. A variety of machines have been developed that take a flexible sheet material, either as a

FIG. 4.21 The Bag Maker. A woodblock print of a bag maker manufacturing various bags for farmers, monks, and priests in the 1500s. From *The Book of Trades*, Standebuch, 1568. (Dover edition, reprint, 1973).

sheeted blank or as a continuous strip, and fabricate pouches or bags at high speeds. Other machines are also capable of filling the bag with product and making the final closure. With all of the fin-sealed pouches and with many of the pillow pouches, when the right material is used, the package can be hermetically sealed and will hold liquids. Such packages can also be vacuum closed or gas flushed, if desired.

Hermetically sealed carton liners or overwraps are merely bags or pouches preformed on a mandrel or formed over the cartoned product.

Another unique pouch is the tetrahedral package. This is nothing other than a pillow pouch or vertical form-and-seal pouch in which the cross seals are alternately perpendicular to one another. This forces the body of the package to assume a tetrahedral shape.

Large pouches are now being used as the inner containers in "bag-in-box" combinations for bulk shipment of both dry and wet products.

Heat-shrinkable Film Packaging

Heat-shrinkable films are plastic films that have been oriented—that is, stretched under such conditions in one or both of the web and cross-web directions that the molecules have assumed a higher degree of crystalline alignment. This imparts improved physical properties. So as long as the film is not heated above the temperature at which orientation took place, it will retain all or most of these properties. However, if it is heated above the orientation temperature, the phenomenon of "plastic memory" exerts itself. The molecules seek to return to their former unoriented state and the film shrinks. If the oriented film is wrapped around a product and then exposed to heat, the film shrinks tightly around the product and will hold it tightly to anything else enclosed, such as a tray, a box, or another unit. Thus heat-shrink films may be used both as intimate wraps and as individual or bundling overwraps. By selecting the right combination of plastic material, molecular weight, molecular-weight distribution, melt index, and degree of orientation, the amount of shrinkage and shrink force (the force exerted on the wrapped product) can be controlled.

CONVERTING OF FLEXIBLE PACKAGING MATERIALS

Because of the almost infinite variety of combinations of materials that can be used in flexible packaging, it is not practical to describe converting operations for each package form. Rather, we shall consider the subject collectively. The package developer must be aware of these operations, as they are vital to the accomplishment of his aim.

Coating

Coatings are applied to flexible packaging materials to make the substrate smoother and more receptive to printing inks, to protect printed surfaces, to introduce color, or to provide improved functional properties, such as scuff resistance, coefficient of friction, moisture resistance, smoothness, wet strength, grease resistance, gas barrier properties, and heat sealability.

There are two basic types of coating equipment systems. In the first, an excessive amount of coating is applied to the substrate web and the excess is then doctored or metered off. In the second, a predetermined amount of coating is applied to the substrate web.

Metered-excess Coating Systems

Coatings are applied by means of roll, pond, fountain, or dipping and the excess is removed by a flexible doctor blade, a rigid knife, an air knife, metering squeeze rolls, or a metering bar or rod.

Flexible Doctor Blade. Flexible blade coaters are used for applying high-solids clay coatings on paper and paperboard at very high speeds (several thousand feet per minute). The method may be used in line with other coating systems or as a separate operation. The coating is applied by means of a roller, fountain, or pond system to the web, and the excess is scraped off by means of a flexible blade held in a rigid holder against a rubber-covered back-up roll. Blade angle is critical and blade wear can alter results. Care must be exerted to avoid particle build-up along the blade edge, which can result in scratches. Coating viscosity and blade flexibility are also important to good results.

FIG. 4.22 Heat-shrinkable PVC film for cocktail multi-packs.
Courtesy of Reynolds Metals Co.

Knife Coating. All knife-coating systems have one thing in common—a rigid steel knife. The knife-edge contour is critical to the coating weight. Sharp edges permit control for light application weights, whereas broad, rounded edges permit heavy coating weights. For porous webs such as woven fabrics, where strike-through is likely, a floating-knife system is used. Here no back-up is provided. Pressure of the knife against the web is controlled by tension in the web itself. A coating weight of up to 3 or 4 mils can be applied in one pass at rather slow speeds. Baggy edges or centers can seriously affect coating uniformity.

For nonporous webs, better control is available through use of back-up blankets or soft or hard back-up rolls. The belt or blanket is used for delicate webs that are weak or that may tend to stretch. Coatings of medium viscosity are used and light-to-medium coating weights can be achieved. Soft back-up rolls are used where coating weight is controlled by blade pressure. Hard back-up rolls are used with an adjustable fixed gap between blade and web. This system is adaptable to medium-high-viscosity coatings run at slow speed and will apply a wide range of coating weights.

Air-knife Coating. Air-knife coating systems entail fountain or roller application of low-viscosity, low-to-medium-solids coatings at speeds as high as 1000 ft/min. A high-velocity stream of air is impinged against the web from a narrow slot opening. The airstream acts like a knife. Pressure of the airstream, angle of approach, web speed, and coating properties all influence coating weight.

Because the solids content is lower, drying can be a problem. If drying is done too rapidly, the coating can be disturbed.

Air-knife coating can be done on all manner of substrates, and aqueous solution, suspension, or emulsion coatings can be employed. Coatings are free of scratches and the "blade" never dulls.

Bar or Rod Coating. The bar or rod coater employs a smooth or wire-wound rod, which is slowly rotated against the web-directional travel to smooth the coating. The coating may be applied by any method; however, usually one or two roll applicators are used. Speeds are usually slow and coatings may be very high in solids—including hot melts. Smooth rods are used for lower viscosities and lower application weights. Wire-wound rods permit heavier coating weights, but coating must flow or streaks will remain. For hot-melt applications, the rod may be heated.

Dip or Saturation Coating. Coatings may also be applied by total or partial web immersion to saturate, impregnate, or coat on one or two sides. Surplus can be removed by blade doctoring or by metering squeeze rollers.

Since this technique is usually employed for saturation, tunnel dryers are employed. Weight pickup is controlled by dwell time, saturant viscosity, speeds, doctoring gap, web tensions, and web wettability.

Predetermined Coating-weight Application Systems

Reverse-roll Coating. In reverse-roll systems, the coating is picked up from a reservoir, either directly by the applicator roll (three-roll system) or indirectly by means of a furnish roll (four-roll system) that transfers the coating to the applicator roll. The furnish roll is usually blade doctored. The applicator roll is doctored by means of a metering roll that removes excess coating. Two-roll systems are also possible but are generally becoming outmoded. A doctor blade is used instead of a metering roll. The measured coating left on the applicator roll is then transferred to the web, which is run between the applicator and a back-up roll. The metering roll is cleaned by means of another doctoring blade.

The back-up roll coincides with web speed. The applicator roll runs in reverse direction and its speed and pressure against the web determine the amount of coating applied—which can be very high. Gaps and pressures of doctoring blades and rolls also affect coating weight and uniformity. Rolls are precisely ground and run in close tolerance bearings to permit gap precision. Speeds are not very high, but coatings are smooth and free of scratches. The system will not level out a rough surface and is not well-suited for fabrics or rough boards. However, use of knife coaters in line to smooth out the rough boards is possible. Use of fountain application reservoirs permit use with more volatile solvents. Most reverse-roll coaters use aqueous coatings or less-volatile solvent systems.

Kiss Coating. A kiss coater is similar to a floating-knife system in that the web makes contact with the applicator roll by means of web tension alone. No back-up roll is used. The applicator roll may pick up the coating from the reservoir with little or no metering, or a second roll may be used to pick up and transfer the coating to the applicator roll. In the latter case, metering is achieved by the nip between the rolls. Here again, the applicator roll travels in a direction that is the reverse of web travel. Kiss coaters are used mostly for application of laminating adhesives or of adhesives to gummed tapes. The kiss coater may be used as the applicator system in air-knife, blade, and bar coaters.

Nip Coating. Nip coaters usually have two rolls—one hard and the other soft or both hard. Low-viscosity coatings are applied by this method, usually to both sides of the web. The coating is flooded on from above, sometimes as a spray. The web passes through the shower and retained puddle, and the nip meters the amount applied. Additional rolls may be employed for better metering or pressure control. For very thin coatings, high speeds are possible. Thicker viscosities are run more slowly.

Another type of nip coater is similar to the two-roll kiss coater except that it uses a rubber back-up roll to hold the web against the applicator roll.

The calender coater is also related in that it combines the two-roll nip coater with a calender system. Coating is flooded into a two-roll nip (with both rolls made of steel). Coating passing through the nip is transferred to the upper roll of a two-roll calender, which applies the coating to the web at the calender nip. This method is used for heavy coating weights of high-viscosity materials, such as vinyl coatings on paper or cloth.

Direct Gravure Coating. In direct-gravure coating, the coating is picked up by impressions engraved in a metal roll. The excess is wiped off the smooth, nonengraved areas by a thin doctor blade. A back-up roll presses the web against the gravure cylinder, and the coating is transferred to the web. The coating then flows together to provide uniform coverage.

Coating weight is controlled by viscosity, solids, and engraved pattern. For the same coating, heavier application requires a different gravure cylinder. Cylinder wear gradually reduces application weight. Clogging of cells can cause drastic reduction. Generally, gravure coating is lightweight (less than 1 mil). This is one of the basic methods for printing. It also is used for applying lacquers, adhesives, waxes, and low-melting hot melts.

Offset Gravure Coating. Offset is very similar to direct gravure, the only difference being that the engraved roll transfers the coating pattern to a rubber-covered offset roll. The latter then transfers the coating to the web. The advantage of the offset system is that the coating has a better chance to flow together and eliminate the line or cell pattern derived from the gravure cylinder.

Slot-Orifice Coating. The slot-orifice coater can be used to apply hot melts or water-based emulsions. The coating material is pumped from holding tanks (heated and insulated for hot melts) and forced upward through a slotted tube against the underside of the web. The web is positioned by a back-up roll. Metering is accomplished by synchronizing the pump to the web speed. Smoothing bars are not required, as there is no turbulence and very little shear.

Cast Clay Coating. Another method for applying clay coatings to paper and paperboard is by casting. This method employs a large, highly polished drum that serves both as a drier and as a means of imparting a smooth surface. Coating is applied to the web by an applicator roll. Excess is pressed off by a small press roll in contact with the drum and is caught by a pan. The wet coating is trapped between the web and the drum and, when dried, is stripped off. Speed is limited by the size of the drum and the drying capacity, but coatings are very smooth and high gloss.

Curtain Coating. The curtain coater is a forerunner of the slot-orifice coater. The coating material is passed through a slotted orifice under pressure but flows down. Very thin material can be applied by overflow of a weir or dam. Materials can be hot melts, lacquers, or emulsions; however, they must be thin and free-flowing, yet capable of maintaining a continuous film. Coating application weight is determined by coating flow rate, web speed, and coating properties. Some curtain coaters operate at speeds of nearly 1500 ft/min.

Extrusion Coating. Extrusion coating differs from curtain coating primarily in temperature, pressure, and viscosity—i.e., temperature of the molten plastic, viscosity of the melt, and the pressure required to force it through the die. An extruder is made of a hopper feeder, a barrel and screw, and a die. Barrel, screw and die are heated. Pellets of resin are gravity fed through the hopper feeder into the barrel.

The barrel is a cylindrical tube made of alloy steel with an especially hard, wear-resistant alloy lining. Its internal diameter is machined to precise tolerances ($\pm .025$ mm), and it must be capable of withstanding pressures of 703 kg/cm^2. The hopper feeder is usually water cooled to prevent premature melting. The barrel is provided with heaters in zones along its length and with means for cooling by air or water so that precise control of temperature can be achieved.

Inside the barrel is a screw that is designed with several zones along its length, each with a specific function. The pitch of the screw flights and the diameter of the screw flights and the channels between them determine the action within each functional zone.

The first or "feed" section picks up the solid particles of polymer and moves them forward into the barrel. In the second or "transition" section the polymer is compressed against the barrel to soften, coalesce, eliminate air pockets in, and begin to melt the mass. The combination of temperature and mechanical shearing action melts and mixes the polymer. In the third ("metering" or "pumping"), section, the final melting and mixing occurs and pressure is developed to force the polymer melt through the screen pack and the die. There are several basic screw designs that have been developed to achieve the best results with specific polymers. The horsepower of the drive mechanism, the speed of rotation of the screw, the screw design, and the length to diameter (L/D) ratio all have a bearing on polymer flow and extruder efficiency.

The shape of the heated die determines the ultimate product. An extruder can produce tubular pipe, parisons for blow molding, a thin-walled tube for blown film, fibers, or, with a slot die, a flat film or an extruded coating.

The setting of the die lips controls the thickness of the melt as it leaves; however, gravity flow plus the take-away speed of the substrate web and chill roll cause a drastic thinning of the melt. Resins must be "designed" to have a high degree of melt cohesiveness to avoid formation of holes. Coating is achieved by extruding the melt between the substrate web and a polished, water-cooled chill roll, using a rubber-coated back-up roll to apply pressure. Gloss of the coating is related to the degree of polish on the chill roll. Matte finishes can be attained by using sand-blasted rolls.

Co-extruded Coatings. Extruders can be made with two barrels and screws and a specially designed die to extrude two polymers simultaneously. Usually one of the polymers is sandwiched between two layers of the other polymer to provide a special property, such as an improved oxygen barrier property.

Tandem Extruding. By placing two extruders in line, one coating can be extruded on top of another to achieve specific results. This can be two layers of the same polymer or two different polymers.

Solventless Coatings. A technology that is beginning to emerge concerns the use of solventless coatings. Here the liquid coating is a monomer or partially polymerized liquid polymer with desired viscosity and blended with whatever additives are required (such as color). After application to the substrate, the coating is passed through a chamber where final curing of the coating to a solid film is achieved through use of ultra-violet radiation.

Laminating

Laminating is the process of adhering one material to another. The laminating medium is the adhesive; the substrate materials are the adherends. In flexible packaging there are two basic types of laminating—wet bonding, in which adherends are brought together before elimination of solvent, and dry bonding, in which solvent is eliminated prior to combining or no solvent is used at all.

Wet Bonding. Wet bonding with a solvent adhesive is possible if one or both adherends is permeable to the solvent vapors. Any of the previously described coating methods capable of handling the adhesive viscosity may be employed. Types of adhesives include aqueous solutions of silicates, starches, gums, and mucilages; aqueous emulsions of latices and other plastisols; and organic solvent solutions.

Heavy paperboard webs may be coated with high-solids aqueous adhesives using a roll, knife, or kiss applicator. The webs are combined

FIG. 4.23 Laminator used for converting flexible packaging materials.
Courtesy of Inta-Roto Machine Co.

FIG. 4.24 Inner view of drying oven in laminator.
Courtesy of Paper Converting Machine Co.

FIG. 4.25 This machine can apply gravure laydowns of various inks and coatings.
Courtesy of Inta-Roto Machine Co.

by passing them through a pressure nip. Drying is accomplished by absorption of solvent by the dry board, with ultimate slow evaporation.

Lightweight paper webs require more precise application of adhesive using a gravure or nip plus smoothing rolls or bars. The laminating nip is a combination of a steel back-up roll and a rubber-coated pressure roll. Drying is accomplished by circulating-air ovens or drum driers.

Dry Laminating. Dry laminating with solvent adhesives is similar to wet laminating of light webs in its application of adhesive (i.e., by gravure or reverse roll), but the solvent is eliminated by drying prior to reaching the laminating nip. The nip may have the steel back-up roll heated to soften thermoplastic adhesives or to convert thermosetting adhesives.

Extrusion lamination is identical to extrusion coating, except that the plastic metal is placed between the two substrates at the chill-roll nip.

Wax or hot-melt laminating is identical to wax or hot-melt coating, except that the second web is added just after the heated smoothing bars.

BIBLIOGRAPHY

ANON. 1970. Simple unit forms, foils, seals bags. Packag. News *17* (6), 3.
ANON. 1971A. Aerosols. Mod. Packag. *44* (4), 14-16.
ANON. 1971B. Cans. Mod. Packag. *44* (4), 24–26, 30.
ANON. 1971C. Tubes. Mod. Packag. *44* (4), 40, 44, 46.
ANON. 1971D. Cartons. Mod. Packag. *44* (4), 50–52, 54.
ANON. 1971E. Closures. Mod. Packag. *44* (4), 84–86, 90.
ANON. 1971F. Drums and pails. Mod. Packag. *44* (4), 104. 108.
ANON. 1971G. New glass/plastic soft drink bottle entry. Mod. Packag. *44* (6), 6, 8.
ANON. 1972A. Boxes: Paperboard, metal edged and components. U.S. Fed. Spec. (PPP-B-665D). General Services administration. Washington, D.C.
ANON. 1972B. Glass Containers, 1972 edition. Glass Container Manufacturers Institute, New York.
ANON. 1972C. Trays and boxes—Plastics vs. wood and board. Packag. Rev. *92* (11), 91-92, 95, 97.
ANON. 1973A. Blister packaging: Putting it all together. Mod. Packag. *46* (5), 34–35.
ANON. 1973B. Egg cartons: New ad medium. Mod. Packag. *46* (7), 13.
ANON. 1974. Skin and blister packaging. Packaging *45* (535), 20, 24, 26.
ANON. 1975A. Pull-tab stays with the can. Packag. *28* (2), 18.
ANON. 1975B. Bags in carton. Food Drug Packag. *32* (4), 24.
ANON. 1976A. New D and R machine for two-piece food cans. Sheet Metal Ind. *53* (11), 385.
ANON. 1976B. Easy-open composite cans. Boxed containers *84* (2), 42.
ANON. 1978. Blister: case studies. Packag. Rev. *98* (9), 41, 43–44, 49, 51.
ANON. 1980. Can technology. Tin Int. *53* (1), 26.
ANON. 1982. Aseptic packaging fever. Food Eng. *54* (1), 59–65.
ABBEY, J. R. 1970. Gas packaging with carbon dioxide. Food Manuf. *45* (9), 37–40.
BOORSTIN, D. J. 1973. The Americans: The Democratic Experience. Random House, New York.
BROOKS, R. 1982. Glass packaging. Austr. Packag. *30* (7), 31, 34.
CROONER, N. H., AND MASON, S. I. 1971. Recent advances in rigid metal containers. Food Trade Rev. *41* (2), 32–36.
GRIFFITH, H. E. 1979. PET beverage bottles—Just the beginning. Food Drug Packag. *41* (5), 18–20.
HARRIS, P. E. 1970. Metal containers. Packag. Technol. *16* (115), 12, 14.
HARTSUCH, P. J. 1970. Flexible packaging. Graphic Arts Mon. *42* (3), 46–48.
HEWITT, J. 1970. Folding cartons. Packag. Technol. *16* (115), 16–17.
HINE, D. J. 1974. Developments in the folding carton industry. Folding Carton Ind. *1* (1), 12–15.
KROGER, M.. AND IGOE, R. S. Edible containers. Food Product Develop. *5* (7), 74, 76, 78–79, 82.
LANE, J. H. 1973. Foil container trends. Food Manuf. *48* (5), 22, 25.
LEMAIRE, W. H. 1978. Paperboard packages for ovens. Food Eng. *50* (1), 62–64.
LILLQUIST, R. A., AND HOLDER, R. G. Flexible packaging—achievements '80. Paper, Film, Foil Converter *54* (3), 84, 86, 88, 90.
MARCUS, S. A. 1982. Multi-layer plastic food containers. Food Drug Packag. *46* (8), 22–25, 36.
NIEBOER, S. F. T. 1976. Retort pouch review. Meat *49* (5), 35, 37.
REED, D. D. 1982. Rigid packaging developments in the 80's. Act. Rep. Res. Dev. Assoc. Mil. Food Packag. Syst. *34* (1), 124–127.
SACHAROW, S. 1970. Trends in collapsible tube converting. Mod. Converter *14* (2), 43–45.
SACHAROW, S. 1971A. Folding cartons protect while they sell. Candy Snack Ind. *136* (3), 46—48.
SACHAROW, S. 1971B. Convenience of metal cans boosts use as a candy, snack package. Candy Snack Ind. *136* (7), 62, 64, 66.

SACHAROW, S. 1972A. The converter and collapsible tubes—A survey. Flexography *17* (5), 24–26.
SACHAROW, S. 1972B. Considering glass containers for confectionery usage. Candy Snack Ind. *137* (8), 26–28.
SACHAROW, S. 1973. How gas or vacuum packs prolong shelf life for confections. Flexography *18* (8), 14–15, 42.
SACHAROW, S. 1974. Rigid and semi-rigid paperboard packaging. Package Print Flexog. *19* (7), 21–24.
SERCHUNK, A. 1978A. Aerosols: On the way again. Mod. Packag. *51* (2), 23–28.
SERCHUNK, A. 1978B. Milk packaging: Still in transition. Mod. Packag. *51* (4), 37–40.
SEYMOUR, R. B. 1979. Progress in polymeric containers. Org. Coatings Plast. Chem. *41*, 313–317.
SLAWSON, D. A. 1976. Lightweight foil containers. Fod Manuf. *51* (3), 39.
TSUTSUMI, Y. 1972. Retort pouch. Its development and application to foodstuffs in Japan. Jpn. Plast. *6* (1), 24–30.
WILKINS, D. R. 1978. Plastic bottles for carbonated beverages. Plastics Rubber Int. *3* (6), 249–252.
WOODWARD, S. 1979. The market potential for ovenable food containers. Packag. Today *1* (3), 42–44.

5

Materials and Package Testing

In the manufacture of printing paper, a large quantity of Plaster of Paris is added to the paper stuff, to increase the weight of the manufactured article.

Frederic Accum (1830)

In order to predict the ultimate performance of a package in the marketplace, technicians involved in package performance analysis must use a wide array of test procedures.

They must be able to identify various materials used in packaging. They must judge the suitability of materials for the intended end use, using knowledge of the materials' properties both when used alone and when used in combination with other packaging materials. They must be able to evaluate the materials' performance in the final package form in contact with the packaged product, and they must determine the durability of materials through normal or even abnormal handling and abuse (including shipping.)

In modern packaging laboratories, many such tests have been adopted from the more advanced sciences of chemistry and physics—for example, chemical identification, surface-reactive properties, strength tests, and permeability tests. Other tests have been adopted from the industries which supply materials to the packaging industry. These tests are important in identifying the materials and in measuring their properties. Still other tests have been developed by the packager to evaluate the particular product in its own particular package. For example, a packaged, sterilized surgical blade might have to be tested to determine whether the package seals were capable of preventing bacteriological contamination of the product. Such tests are usually only applicable to the particular product package involved, and the results are not broadly published to the packaging industry.

Manufacturers of packaging materials have developed quality tests for their products and usually will supply test results, test procedures, and sampling techniques to their customers on request. Test procedures that are applicable to general classes of materials or packages are available and published in standardized forms by such industry associations as:

ASTM	American Society for Testing Materials
TAPPI	Technical Association of the Pulp and Paper Industries
FPA	Flexible Packaging Association
ABA	American Boxboard Association
NSTC	National Safe Transit Committee

QUALITY TESTING OF PACKAGING MATERIALS

To determine whether a package will have the physical properties expected of it, the packaging engineer must be able to measure the properties of the materials used in its fabrication. Testing procedures developed for this purpose have for the most part originated with raw material suppliers and then become accepted on an industry-wide basis.

One of the early packaging materials was paper; thus many of the tests that are now used to measure the properties of packaging materials were first developed to measure the properties of paper.

Quality Testing of Paper

Paper is made from cellulosic fibers which have an affinity for moisture. The moisture content varies in proportion to the relative humid-

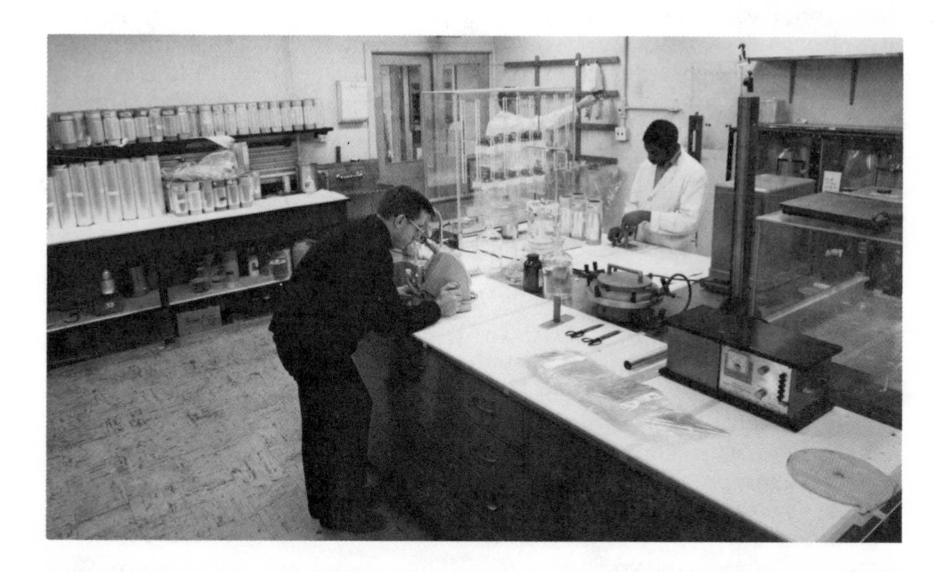

FIG. 5.1. Along with scores of rather routine tests on film products, many special tests are developed and conducted to analyze basic film properties.

ity and temperature of the surrounding atmosphere. Many of the physical properties of paper are affected by its moisture content. It is therefore imperative that all paper testing be conducted in a standardized atmosphere and that the paper be allowed to equilibrate with the atmosphere for at least 24 hr prior to testing. The paper industry has established a standardized (TAPPI 402) testing atmosphere of 73 ± 3.5°F and 50 ± 2% RH.

Before beginning the actual testing of a paper sample, it is important that the sample be established as representative of the whole lot of material, that the test apparatus be properly calibrated, and that the sample be equilibrated with standard temperatures and humidity conditions.

The "sides" of the paper must be determined (TAPPI 455). These have the paired names of wire and felt, MG and wire, coated and uncoated, top and bottom, and others. The actual terms are not important, but it is important to attribute the results of some of the tests to the correct side of the paper. Smoothness, printability, color, brightness, and coefficient of friction (COF) are some of the properties that are affected by the sidedness of paper.

The "direction" of the paper must also be determined (TAPPI 409). Paper has two principal directions: machine direction (MD) and cross-machine direction (CD). Paper comes off the paper machine in a long, continuous web for days on end. The direction of travel of the web as it traverses and comes off the paper machine is the machine direction of the paper. The direction across the web, at right angles to the machine direction, is the cross-direction of the paper. Distinctly different results are obtained in tests performed with specimens cut from these two directions of paper. For many tests, the specimens must be cut with regard to the principal directions of paper. Tensile, elongation, scoring, stiffness, COF, brightness, die cutting, curl, tear, and fold are properties which are affected by direction.

A third direction, Z direction, refers to the direction through the paper—for example, through the page you are reading, front to back. The wax pick, ply bond, and internal bond strength tests are tests for Z-directional strength.

Percentage of Moisture. (ASTM D-644, TAPPI T-412). Not only the strength properties of paper are affected by the paper's moisture content: its running characteristics on machinery are affected as well. Paper is often bought and sold by using weight as a unit of measure. Because moisture content affects the weight, it also affects the cost of paper. It is imperative that moisture content be specified and that this property be measured as a routine quality control procedure. There are instruments which give fair approximations of moisture by measuring electrical conductivity of the paper. The official and accurate method for the determination of moisture is to weigh a sample of paper, dry it in a 215°F oven, reweigh the paper, and calculate the percentage of moisture using the weight data.

Ream Weight or Basis Weight. (ASTM D-646, TAPPI 410). Early paper makers made paper in sheet form and sold it in the following units: the quire (24–25 sheets) and the ream (20 quires). Although most paper in the packaging industry is used in roll form, the historical measure of paper ream weight still persists. Sheet sizes were related to convenience in letter writing or in bookbinding. Thus the weight of a ream is the weight of a designated number of sheets having designated dimensions of length and width.

A sheet size of 24 in. × 36 in. and a ream count of 500 sheets have become generally accepted in the packaging industry in the United States. Thus a 20-lb ream basis weight means that 432,000 in.^{2} (24 × 36 × 500) would weigh 20 lb, and a 40-lb paper would weigh twice as much.

Knowing the basis weight of paper is important, because this is the primary descriptive term for paper, and it is nearly directly related to paper cost and properties.

Thickness. (ASTM D-645, TAPPI 411). Paper is usually described in terms of its basis weight, and paperboard is usually described in terms of its thickness. The thickness of paperboard is expressed in "points" (thousandths of an inch). Thus a 0.025-in.-thick board is called a 25-point board. Paper thickness is called caliper and is expressed as a decimal fraction of an inch. The instrument known as a micrometer is used to measure caliper. ASTM and TAPPI methods give precise specifications for the requirements and operation of this instrument.

Paper thickness affects stiffness and other properties. It is consequential in obtaining maximum yardage in a roll and in fitting the correct number of sheets in a box. Labels are sometimes handled in stacks of 1000 labels. A little variation in thickness per label is thus multiplied a thousand times and is visible as a variation in stack height. Whereas the count and the weight may be correct, the stack height may be cause for concern in buyer-seller relations.

Formation. The formation of a paper is related to its fiber orientation. This can be seen by backlighting a single sheet of the paper sample. A uniform, nonblotchy, even, not-wild (a papermakers' term) formation is desired, but this formation is the hardest and costliest to achieve. Wild formation has weak spots, thin spots, non-uniform ink, and adhesive absorption and is thus less desirable.

Porosity. (ASTM-D 726, TAPPI 460). The instrument known as a densometer measures the porosity of paper, and thus the porosity is often called the density. Care must be taken not to confuse this density with "apparent density." Porosity is measured in terms of the time in seconds required to pass 100 cm^3 of pressurized air through a 1 $in.^2$ area of a sample sheet. A low test result indicates rapid passage of air and that the paper is relatively open and porous. A high test result indicates slow passage of air through the paper, which is characteristic of relatively dense or coated papers.

Greaseproofness. (ASTM-D 722, TAPPI 454). A useful test to measure greaseproofness is begun by laying a paper specimen on an indicator sheet of white paper. A small cone of sand is then placed on the specimen. Strongly colored red turpentine is dropped into the sand, and gravity brings the turpentine into contact with the specimen. At specified time intervals, the indicator sheet is examined for turpentine staining. The time between the start of the test and the first appearance of staining is noted as the "greaseproofness" of the sample.

Greaseproof papers are relatively expensive. Thus it is important to know if a paper is greaseproof, and just how greaseproof it is.

Sizing. Paper is "sized" to increase the property of resisting the absorption of water. Papers naturally absorb water, as evidenced by blotting paper. Many applications require that the paper not readily absorb water, as with building papers, moist and wet foot wraps, labels, shipping cases, and drinking cups. Paper is inexpensively treated to obtain this property with a material called "size." It is important to know if the paper is sized, and to what extent. There are many tests for measuring this property. The selection of the method depends on the type of paper, the amount of size, the application, the problem involved, and the personal preferences of the analyst. A water-drop test involves placement of a drop of water on the surface of the paper, and the time required for the absorption of the water is the reported value. In the ink float test, an inch-square boat of paper is floated on a puddle of ink, and again the absorption time is reported. In an immersion test, the paper sample is immersed in water for a minute, and the results are reported in terms of weight gain. In a sugar-dye test, a boat of paper is floated on water, with the penetration of the water changing the color of the dye. In the climb test, the lower end of a vertical strip of paper is immersed in water for a period of time. The height to which the water climbs is reported. In the Cobb size test (ASTM-D 2045, TAPPI 441), a 100-cm^2 circular area of a test sample of paper is covered with water for exactly 120 sec. At the end of the time period, the excess water is blotted off and the increase in weight of the sample is measured. The more a paper has been sized to hold out water, the smaller the increase in weight will be.

Tensile Strength and Elongation. (ASTM D-828, TAPPI 404, TAPPI 457). Sample strips are cut exactly 1 in. wide and about 10 in. long. These are gripped at each end by a clamp (or "jaw"), and then the jaws are moved apart at a controlled speed until the sample breaks.

The ultimate tensile strength of a paper or paperboard is the amount of force necessary to break the strip. Several types of tensile-testing machines are now available. Some automatically record the results on chart paper. The horizontal displacement of the line represents force applied, and the vertical displacement due to chart speed (which is also calibrated to jaw separation speed) represents change in length of the sample due to elongation.

Tear Strength. (ASTM D-687, TAPPI 414). The Elmendorf tear test measures the energy absorbed by several sheets of a paper in propagating a tear that was initiated by cutting with a blade attachment. A falling pendulum provides the force to tear the sample, and the absorbed energy is read on a calibrated scale and reported as grams required to tear a single ply.

Mullen Burst Strength. (ASTM D-774, TAPPI 403). The Mullen tester is a machine which clamps a specimen between two annular rings. The lower ring's aperture is closed by a rubber diaphragm. When the instrument is activated, a fluid (usually glycerol) is pumped under the rubber diaphragm, causing it to expand to a dome shape. As the diaphragm expands, it presses upward against the paper specimen, causing the latter to expand until it finally ruptures. The rupturing pressure is shown on a gauge and reported as lb ≠ in.^{2}

In general, the Mullen test results are allied with the tensile test results. There aren't many applications for paper in which the paper contains expanding rubber diaphrams, so the use of the Mullen results are limited in that regard, but the Mullen test provides a useful measure of the general strength characteristics of paper.

Stiffness. (TAPPI 451, TAPPI 489). Several instruments are used for measuring stiffness. The Gurley apparatus is used for lightweight and limp materials. The Taber Stiffness Tester is usually preferred for stiffer paperboard materials. This apparatus measures the moment (force) in gram-centimeters (g = cm) that is required to bend the sample through 15° of arc. Stiffness is important for feeding and traversing of printing, converting, and packaging equipment. Stiffness is a factor in the panel bulge of cartons.

Fold Endurance. (ASTM D-643, TAPPI 423; TAPPI 511). A strip of paper is placed under tension and folded repeatedly around a radiused clamp. The folding continues until the sample breaks, and the number of folds are automatically counted and reported.

Brightness. (ASTM D-985, TAPPI 452). Brightness is a measure of the whiteness of a paper and is tested by means of an optical reflectometer against a known white standard. It is expressed in terms of percentage of the reflectance of the white standard.

Opacity. (TAPPI 425). Opacity is a measure of the hiding power (lack of transparency) of a paper and is determined by use of an optical reflecting instrument. Diffuse reflectance of the light which passes through the paper and is reflected back through the paper by a white standard backing is measured, and this is compared to the reflectance with a black standard backing.

Smoothness. (TAPPI 479). There are several testing devices which measure smoothness of the surface of a paper in terms of the rate of flow of air between the surface of the paper and a smooth, flat glass plate. Correlation between these methods is not perfect, nor is there a direct correlation with printability. Nevertheless, papers can be categorized in terms of relative smoothness values; generally, the smoother the surface, the better the printing fidelity.

There are many other tests conducted on paper or paperboard which may be related to specific requirements. These include ash content, Z-directional tensile strength (ply-bond separation force), wet strength, wet elongation, coefficient of friction, "printability" or ink holdout, bending and scoring resistance, mold spore count, etc. It is not practical to list them all. The reader is referred to ASTM and TAPPI standards.

Quality Testing of Plastic Films

The testing of plastic films is usually done in a controlled atmosphere of 23°C and 50% relative humidity. Many of the previously described paper tests are used to measure film properties. The directional properties of a film, as with paper, must be addressed by the analyst. Additional tests that attend to the important properties of plastic films include gloss, haze, transparency or clarity, slip, blocking, static, impact-fatigue, flex, and permeability tests.

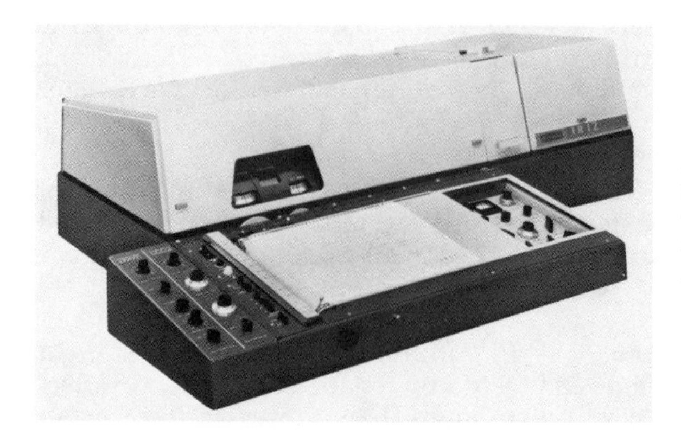

FIG. 5.2. Infrared spectrophotometer.
Courtesy of Scientific Instruments Division, Beckman Instruments.

Gloss. (ASTM D-523). Gloss is measured by determining the amount of light which is reflected from the film's surface at a given angle (usually 45°). The reflected light is detected by a photocell. The quantitative effect on the photo cell is measured with a built-in meter and expressed as a percentage of a standard reflector. A high reading is caused by a large amount of light being reflected, which indicates a high-gloss surface.

High-gloss film usually has more consumer appeal, although it is important to note that excessive gloss may cause annoying light reflection into the eyes of a viewer.

Haze. (ASTM D-1003). Haze measurements are made with specially designed haze meters. The haze meter measures light which is scattered from the incident beam. It consists of a light source and an integrating sphere capable of detecting transmitted light. The sample is placed between the light source and the sphere. By turning the sphere, a measure of scattered light is obtained. The percentage haze at angles greater than 2.5° is obtained. Low readings indicate less scattered light and clear films.

See-through Quality (Clarity). This test is rather subjective and is based on visual appearance. In the test, a standard illuminated wire-mesh grid is photographed through the test film sample. Then the

FIG. 5.3. Flexible film laboratory.

photo is compared to a series of numbered standard photos; the standard photo that most clearly resembles the test photo in clarity is chosen and its numerical designation is assigned to the test sample.

A more objective method utilizes the "Gardner-USI" Clarity Meter. Satisfactory correlation of visual results is obtained. The instrument consists of a light source, a source aperture, a lens system, a specimen holder, and a recording system. A film sample is mounted in the holder and light passes through the sample. Specular transmittance is determined by calculating the ratio of the light intensity *with* the specimen in the beam to the light intensity *without* the specimen in the beam:

$$T_s = \frac{100\ I_s}{I_o}$$

where I_s is the intensity with the specimen, I_o is the intensity with no specimen, and T_s is the specular transmittance.

The clarity of two samples can be compared by placing samples of each film over and in contact with printed matter. To accentuate the results, multiple sheets of each can be used. The total thickness of each sample must be the same, and, if possible, the number of sheets should be the same. The ease of reading the printed material through the films is evaluated and noted.

Slip. (ASTM D-1894). The ability of material to slide or "slip" over a machine mandrel and former plates is of importance in achieving good production rates. Yet in stacking, handling, and shelving packages, too much slip can be an annoyance. Several methods are available for measuring slip or, more accurately, the coefficient of friction (COF).

One method is to cover the top surface of a 6 in. × 20 in. plane with the film, fastened with tape. A smaller and second piece of film is placed in the middle of the first piece of film, and a weight is placed in the middle of the second piece. One end of the plane is slowly raised until the top film starts slipping. The tangent of the angle at which one layer begins to slip is referred to as the "kinetic COF."

Another method utilizes a "slip tester," which consists of a spring and weight attachment. A measurement of the force required to slide one film over itself or another material is obtained. The ratio of the force required to move a film with a fixed weight to the total weight is the COF. This method is often referred to as the "stationary sled" method.

Still another test method utilizes a moving sled. The force required for the sliding effect is calculated in a manner similar to that used in the stationary sled method.

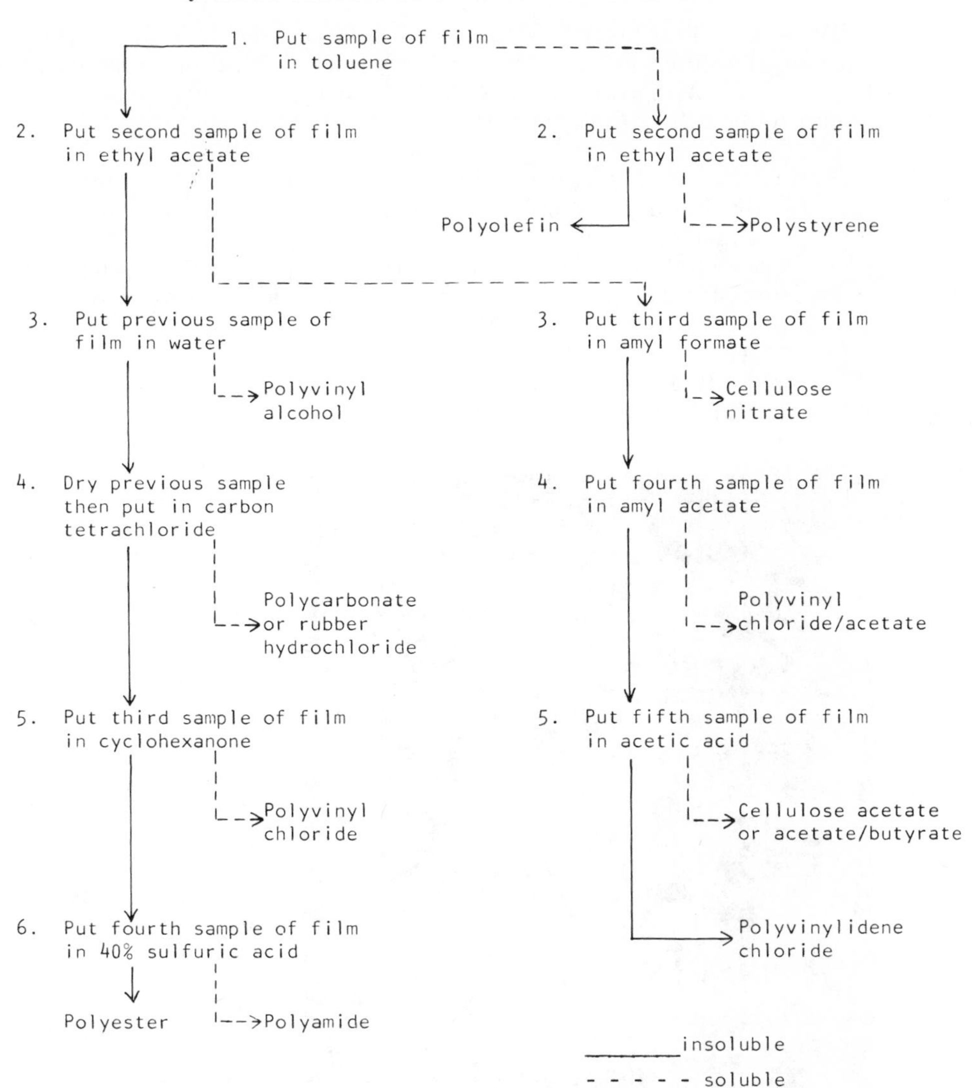

FIG. 5.4. Flow sheet for identification of an unknown plastic film by solvent solubilities.

Blocking. Two layers of material are placed together under pressure for a specific time, and any tendency toward "blocking" or sticking together is noted. Blocking can be caused by plasticizer migration, moisture, or a high level of corona discharge treatment.

Static. Many plastic films have such low moisture content that they readily build up electrostatic charges. These charges cause the film to cling to itself or to feed unevenly on machines. Film manufacturers include antistatic agents in their formulations. A quick way to measure static is to rub two pieces together and determine whether they cling.

Impact Fatigue. A film must not fracture during exposure to shock. When hard and sharp objects are packaged in a brittle film, fractures

FIG. 5.5. A machine test for a snack film lamination in progress. *Courtesy of Hercules, Inc.*

result. Impact fatigue is defined as the force necessary to rupture a film under the stated conditions of the test. It is always preferable to perform the test under actual end-use conditions.

Several different methods are used for determining the impact fatigue of flexible films. One of the most common is the falling-dart method. Darts calibrated in grams by the manufacturer are allowed to fall onto the film's surface from specified distances. Values are obtained in grams because impact fatigue is measured in terms of the weight of the specific dart that breaks the sample 50% of the time. The final value must state the distance involved in the fall.

Another test uses a pendulum impact tester. A puncture pendulum is allowed to swing through a sample. The difference between the energy of the pendulum at maximum height and the energy of the pendulum after a sample rupture is known as the impact strength.

A more recently developed testing mechanism, used in Europe, is the Ball Burst Tester (Fig. 5.10). Again, a pendulum is used; however, the test chamber may be heated or refrigerated in order to simulate actual end-use conditions. The entire tester is automatic, and the results are recorded on a dial.

Flex Resistance. The ability of a film to resist destruction when exposed to severe stresses is important. There are many test methods available, and the selection of the proper method must be based on the end use.

A flex tester folds a film backward and forward at a given rate. A recording of the number of cycles required for the film to fracture is obtained. Flex testing can be conducted in freezing or humid environments.

Permeability Testing. Many types of equipment have been developed to measure the permeation of gases through flexible plastic films. The method based on the measurement of pressure and time involves measuring the increase of pressure with time on the low-pressure side of the film under conditions where the pressure difference between the two sides is essentially constant. This can be done with superatmospheric pressure systems or high-vacuum systems. Water-vapor permeability may also be measured by means of high-vacuum techniques, although there are simple gravimetric methods available with which WVTR can be determined much more easily.

A commonly used test method employs the Cartwright permeability cell. In this method, the pressure changes per unit of time across the substrate are measured. The corresponding volume change is then calculated and expressed as cm^3 (at standard conditions)/100 $in.^2$/24 hr/atm. A picture of the cell is shown in Figure 5.11.

FIG. 5.6.

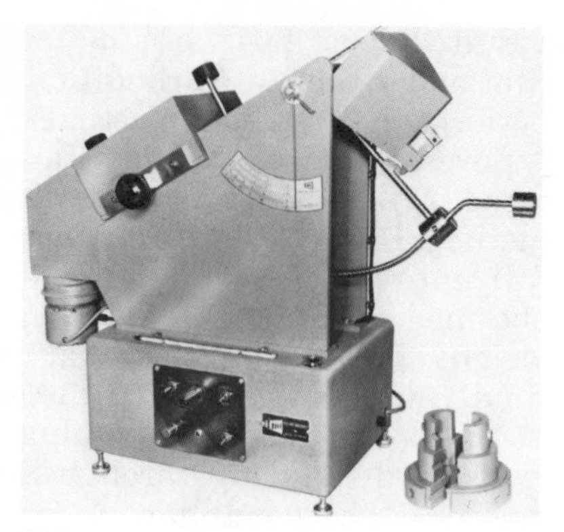

FIG. 5.7.

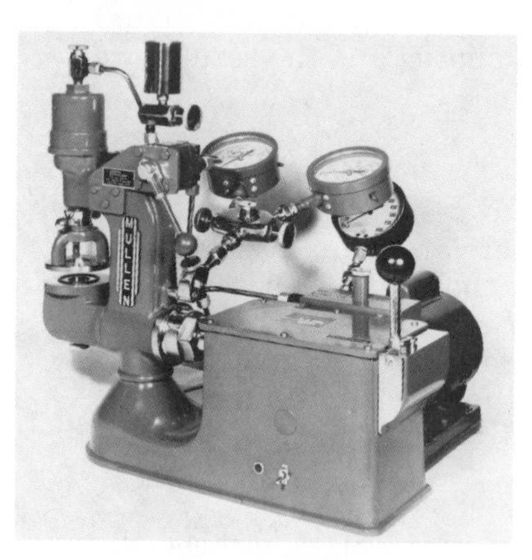

FIG. 5.8.

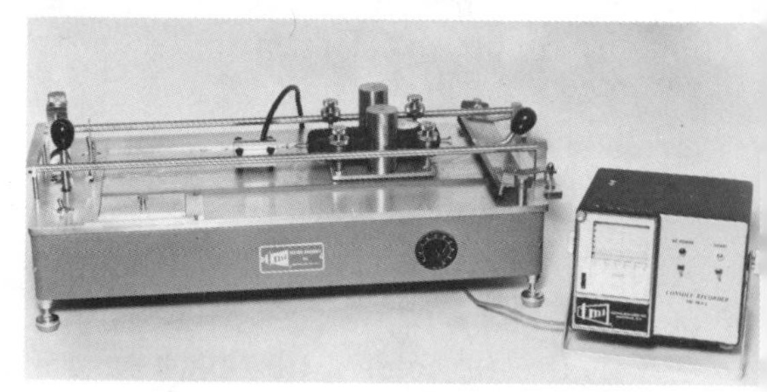

FIG. 5.9.

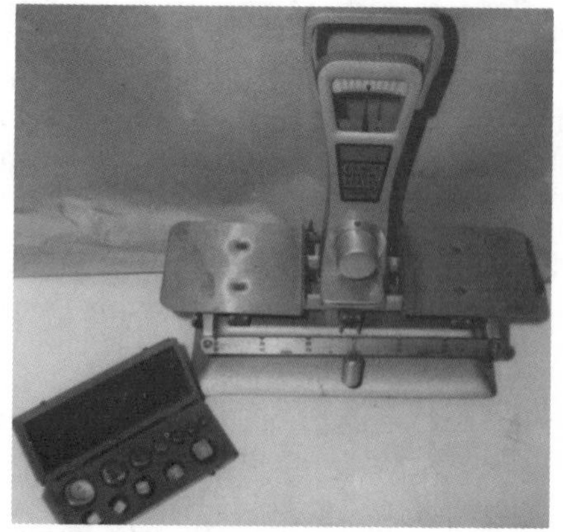

FIG. 5.10.

Water vapor transmission. (ASTM E-96). This may be determined by means of the traditional General Foods method. This method uses low-walled metal circular pans which contain a moisture-absorbing agent such as dry calcium chloride. The open face of the pan is covered by the specimen. The edge of the specimen is sealed to the pan with wax. The preparation is stored in high-humidity cabinets for a recorded period of time. The gain in weight is unitized to g/100 in.^{2}/24 hr/atm.

Other methods in use to measure gas permeability are based on changes in partial pressure. With different gases on the two sides of the sheet, transmission occurs as a result of the differences in total pressure. The permeability may be followed by means of interferometers or absorption techniques. This method is commonly referred to as the "sweep-gas technique."

It must be stated that whatever method is chosen, similar results should be obtained. It is extremely important to construct gastight seals between the film and cell and to minimize film distortion if a high-vacuum system is employed.

Permeability (P) is defined as:

$$P = \frac{qx}{At(P_1 - P_2)}$$

where q = quantity of gas diffusing through a film of surface area A and thickness x in time t with a difference of $(P_1 - P_2)$ between the partial pressures of the diffusing gas on the two sides of the membrane. It has also been shown that $P = DS$, where D is the diffusion constant of the gas in the film and S is the solubility coefficient. Gas permeabil-

FIG. 5.6. Taber stiffness tester.
Courtesy of Testing Machines.

FIG. 5.7. Ball burst tester.
Courtesy of Testing Machines.

FIG. 5.8. Mullen burst tester.
Courtesy of Testing Machines, Inc.

FIG. 5.9. Slip tester.
Courtesy of Testing Machines.

FIG. 5.10. Weight test of finished package; prior to an official commodity measurement, the balance is set to zero.
Courtesy of the New Jersey Division of Consumer Affairs. Trenton, N.J.

FIG. 5.11. Gas cell apparatus provides a way to sandwich the film between two gas tight chambers and measure gas permeability.

ity units may be reported as cubic centimeters of gas at standard temperature and pressure passing through a membrane 1 cm^2 in area and 1 mm thick in 1 sec under a pressure differential of 1 cm mercury ($cc/cm^2/mm/sec/cmHg$). Other units and their conversion factors are shown in Table 5.1.

Water vapor permeability units may be reported as g/m^2 or g/100 $in.^2$ (per mm/24 hr at 100°F with 90% RH versus 0% RH driving force), or the driving force can be expressed in terms of partial pressure gradients.

Some plastic films are naturally inert and unreactive to most ink systems. In contrast to other films, which are extremely amenable to printing, the varied structures of both natural-base and synthetic films serve to complicate their printability.

Treatments are used to improve ink bond to polyolefin films. Most of the techniques applied to polyolefins can be applied to other films. The two treatments commonly used today are surface oxidation by flame and corona discharge. Both of these yield an oxidized surface layer capable of being wetted by an ink. The resultant bond is due to a complex set of physical and chemical reactions.

It is possible for levels of treatment to vary as well as for loss of treatment to occur. For example, the polar groups may be removed from the treated surface by physical abrasion, wiping or brushing of the surface by rollers may induce friction and cause treatment loss, and there may be changes in polarity. In view of this, it is important for plastics processors to have at their disposal testing methods capable of measuring the remaining effective treatment.

TABLE 5.1 Permeability Conversion Factors

Units	Factor to convert to cm^3 (STP) cm^2/mm/sec/cmHg
g/m^2/mm/24 hr/cmHg	$\times \dfrac{2.5927 \times 10^5}{\text{MW of gas or vapor}}$
cm^3/100 in.^{2}/mil/24 hr/atm	$\times$ 5.9957 $\times 10^{12}$
g/in.^{2}/mil/24 hr/vpg[a]	$\times \dfrac{1.0207 \times 10^3}{\text{MW} \times \text{(vpg in cmHg)}}$
g/100 in.^{2}/mil/24 hr/cmHg	$\times \dfrac{1.3430 \times 10^7}{\text{MW}}$
cm^3/100 in.^{2}/mil/24 hr/cmHg	$\times$ 4.5568 $\times 10^{10}$
g/100 in.^{2}/mil/24 hr/vpg	$\times \dfrac{1.0207 \times 10^5}{\text{MW} \times \text{(vpg in cmHg)}}$
cm^3/m^2/mm/24 hr/cmHg	$\times$ 1.157 $\times 10^9$
g/cm^2/cm/hr/mmHg	$\times \dfrac{6.2222 \times 10^2}{\text{MW}}$
cm^3/cm^2/cm/sec/cmHg	$\times$ 10
cm^3/m^2/mil/24 hr/atm	$\times$ 3.8073 $\times 10^{12}$
g/m^2/mil/24 hr/vpg	$\times \dfrac{6.1404 \times 10^7}{\text{MW} \times \text{(vpg in cmHg)}}$
cm^3/100 in.^{2}/mil/hr/atm	$\times$ 1.4390 $\times 10^{10}$

[a]Given vapor pressure gradient.

Several tests may be used, and they are neither complex in nature nor difficult to control. They are (1) the wettability (contact angle) method, (2) the Scotch Tape test, (3) the crinkle test, and (4) the peel test.

The wettability method depends on the fact that oxidation of the film's surface involves the formation of groups capable of creating a water-wettable surface. When a drop of water or ink is placed on a treated surface, it spreads over the surface. A nontreated surface will cause the water to form in beads. Quantitative measurements may be made by determining the contact angle between the liquid and the plastic surface. Instruments are available with which to measure this property.

In the Scotch Tape test, ink is printed on the sample surface and allowed to dry, after which a length of pressure-senstive tape is pressed onto the surface. Subsequent removal of the tape from the surface by peeling then allows any nonadherent ink to be removed. If ink is removed, then treatment level is marginal and a low bond between ink and film will occur in production.

In order for a film/ink surface to be acceptable, no ink should flake or crack. In the crinkle test, the printed surface is folded and pinched manually for a few moments. If no flakes of ink appear, the adhesion is satisfactory.

In the peel test, a sandwich of wet ink is allowed to dry between two films and adhesion is then tested by a peel-block on a tensile tester. Higher values of peel indicate excellent adhesion and satisfactory levels of treatment.

It is possible to have a tape with low affinity for an untreated surface, but with good adhesion for a treated surface. The tape is sealed on a hand-sealer at room temperature. Peel strength is then measured with a tensile tester, values of up to 430 g/cm are obtained with films having high treatment levels.

Aluminum Foil Testing

Since few packaging laboratories are equipped for metallurgical analysis, the principal tests run on plain aluminum foil are caliper and yield (basis weight) tests. Temper can sometimes be determined by measuring tensile strength and elongation, but this is difficult, especially with very thin gauges of metal. The presence or absence of oily impurities can be measured by extraction analysis or, more simply, by wetting the sample with water and determining whether the water "beads."

Aluminum foil manufacturers are usually willing to assist their customers in testing for other properties that may come into question.

Aluminum foil is defined as a solid sheet of aluminum having a thickness of less than 0.006 in. Plain aluminum foil is foil in either roll or sheet form that is not combined with another material; other names

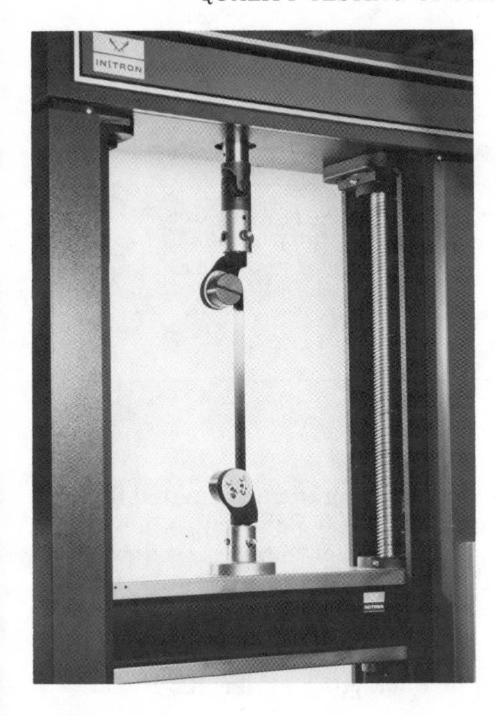

FIG. 5.12. "Instron" tensile tester; the jaws of this tensile tester have been modified to test the tensile evaluation of pressure sensitive tapes.
Courtesy of Instron Corporation.

for this are foil, unsupported foil, free foil, and unlaminated foil. For most converting and many packaging applications, it is used in rolls.

In general, aluminum foil is supplied in the high-purity grades (1000 alloy series). Some additional alloys are also presently being used for special types of products. (Table 5.2).

In most flexible packaging laminations, a 1235 alloy is used. Formed semirigid containers use 3003 or 5005 alloys. The last two digits of the 1000 series of alloy designates the level of purity. Thus 1235 is 99.35% pure aluminum, and 1199 is 99.99% pure aluminum. Alloys of other series contain higher percentages of alloying metals. For example, 3003 contains manganese, and 5005 contains magnesium.

Testing of Inks, Lacquers, and Adhesives

The properties of inks, lacquers, and adhesives are measured by several methods. Weight per gallon is a common measure used to ensure value and uniformity. A small volume of well-mixed and temperature-standardized material is weighed. The results are unitized to pounds per gallon.

Total solids are measured to ensure value and uniformity. A small, weighed, well-mixed sample is evaporated to dryness in a 215°F oven. The residual weight is expressed as a percentage of the original weight.

TABLE 5.2 Alloys of Aluminum Used in Packaging

Aluminum Assoc.		% Other Metals in the Alloy					
Alloy No.	% Aluminum	Copper	Chromium	Iron	Magnesium	Manganese	Silicon
1100	99.0	0.2	—	0.45	—	0.05	0.3
1145	99.45	—	—	0.45	—	—	0.1
2024	93.4	4.5	—	—	1.5	0.6	—
3003	98.5	0.25	—	—	—	1.25	—
5050	98.6	—	—	—	1.4	—	—
5052	97.25	—	0.25	—	2.5	—	—

Measuring viscosity is important in ensuring the anticipated flow of a material on the converter's equipment. The Zahn cup allows the liquid to flow out through a hole in the bottom of the cup; the flow time is the measure of viscosity. The Brookfield viscometer measures the resistance developed by the sample to a rotating spindle.

Testing of Printing Characteristics of Packaging Materials

One of the most important requirements of packaging materials is that they have the ability to be printed and to convey a message. A material devoid of any receptivity to ink has limited value in packaging applications. Over 90% of all converted materials used in packaging are printed.

It is important to test the material for compatibility with the particular printing process that will be used and for its ability to produce the quality of printing required by the purchaser. For example, web-fed presses are more tolerant of curl than sheet-fed presses, gravure printing requires smoother papers than does flexography, offset requires substrates with higher internal band strength than are needed for gravure, and tone or process printing requires smoother substrates than are needed for line work. The customer usually makes the final decision on print quality.

IDENTIFICATION OF UNKNOWN PACKAGING MATERIALS

In the early days of packaging it was relatively simple to identify a packaging material, because fabrication was simple and usually only one or two materials were used or combined. Anyone could differentiate between a glass bottle, a metal box, a wooden barrel, and a paper bag. It remained only for the analyst to ascertain the type of glass, the kind of metal alloy, the variety of wood, or the composition of

FIG. 5.13. Organoleptic testing via jars; the human nose is still one of the best "instruments" for screening aromas in packaging films.

the paper. User demands for product protection, sales appeal, light weight, durability, product compatibility, shipability, and other properties were not as well recognized as they are today. Over the past 50 yr, packaging has become ever more complex and many, many new materials have been developed. The packaging scientist must be able to analyze a sample of packaging material, usually available to him in small quantity, and identify its composition and properties. Fortunately for the analyst, as the new materials are developed and used, new techniques and instrumentation are also developed and used to identify the materials and measure their properties. A variety of useful instruments—such as microscopes, spectrophotometers, chromatographs, and physical property measurement and transmission property measurement instruments—when coupled with considerable ingenuity, can result in excellent analytical work.

A detailed description of all analytical procedures used to identify and measure packaging materials would require several volumes. Let us rather select one more limited area as an example: the identification of the materials in a flexible-packaging composite lamination.

Today's flexible-packaging composite materials frequently contain organic polymeric coatings, inks, adhesives, additives, and films. It is imperative that in analyzing an unknown composite the scientist be aware of the properties of these "plastics." This is usually the key to separating the composite into its component parts. It is also important that the properties of other possible components be known in establishing identification. For example, some metals can be dissolved by acids and others by alkalies. Most plastics can be dissolved by one or more specific solvents.

The first step in analyzing a packaging material is to employ the nondestructive methods and techniques to the sample.

The first of these nondestructive tests is a visual inspection and reporting of the observed properties. This would typically include inspection for means of closure, suspected and obvious materials, type of printing and number of colors, opening devices (tear strip, notches, pull tabs, etc.) special devices (windows, hangers, outer wraps, logos), printed messages, and intended end use of the material.

The next nondestructive tests to run are property-measurement tests. Typical of these tests are the measurements for dimensions, thickness, opacity, color, porosity, haze, gloss, fluorescence of inks, smoothness, slip or coefficient of friction, and brightness. Transmission tests for oxygen, nitrogen, and water vapor could well be selected at this time as nondestructive tests. The actual selection of the tests to perform depends on the analyst's ingenuity, the specific information desired, the requirements of the contemplated or identified end use, and the conclusions made from the visual inspection.

At this point in the analysis it will probably be expeditious to cut the sample into pieces of the appropriate size for the destructive tests. Again, the analyst's experience and objectives come into play to decide which destructive tests are to be made. Tensile, elongation, Mullen, puncture, stiffness, ply-lamination strength, scuff-resistance, curl, electrical-resistance, and other tests require specimens of a specific size. Another specimen of known size will be required for the separation of the composite into its components. The scraps should be saved for the application of spot tests. Spot tests with solvents will show whether there are coatings which can be removed. As the spot tests on the scraps show the solubility and reveal the identity of the coatings, the coatings may be quantitatively measured by weighing the component specimen, removing the coating with the discovered solvent and reweighing the component specimen. The weight difference will be the weight per area of the coating. Finally, the specimen should be delaminated into its separate plies so that they can be separately identified. There are several methods available for separating and isolating laminated components. If the laminate is put together by a thermoplastic adhesive, the material will usually peel back when the edge is heated by a flame. Some coated films will blister when gently heated over a Bunsen burner. Complete separation of other laminates can be accomplished by placing the material in boiling water under reflux. Most laminates can be separated in this manner or by hot solvents. However, for materials that are very hard to separate, suspension in tetrahydrofuran vapor is often effective.

Coating and Ink Identification Tests

If white inks and no overcoat are present, the presence of the coating can be determined by rubbing the edge of a silver coin[1] over the laminate. A black mark from the coin will be made on the white ink. Most overcoats will soften when several drops of toluene are placed on their

[1]This must truly be silver, not a base alloy. Now that silver coinage is becoming obsolete, the analyst would be wise to hoard a few coins for this purpose.

surface. After ascertaining the presence of an overcoat, several tests should be conducted in order to determine the actual formulation. Inks are identified by conducting essentially similar tests.

Spot or Solubility Tests. Many coatings or inks can be identified through the use of simple reagent spot tests, others through flame tests, and others by their solvent solubilities.

For a comprehensive description of many useful spot tests, readers are advised to refer to Feigl (1972).

The analyst should always use the spot-test reagents on known samples so as to be familiar with the results before drawing conclusions from the use of the reagents on samples of unknown composition.

To detect nitrocellulose inks or coatings, the diphenylamine spot test can be used. Diphenylamine in a sulfuric acid solution develops a blue color in the presence of a nitrate-containing coating.

Polyvinylidene chloride (PVDC) coatings are detected by using morpholine. When PVDC is present, a brown-black color results if a drop of morpholine is placed on the surface. An alternate test method utilizes a solution of pyridine coupled with another solution of potassium hydroxide. A dark brown to black color indicates a positive reaction. There are two types of PVDC coatings: those applied by means of a solvent solution and those applied by means of water emulsions. The solvent types are soluble in methyl ethyl ketone (MEK); the aqueous types are not.

Polyvinyl choloride is indicated by a brown color when mixed with pyridine-methanolic solution and by a red color when mixed with pyridine alcoholic KOH solution. Distinguishing PVC from PVDC may be difficult, since both resins tend to yield brownish colors with pyridine-methanolic solution. When in doubt, it is best to perform the morpholine test prior to using the pyridine method.

The Molisch test may be used to indicate the presence of ethyl cellulose. A positive reaction is indicated by a violet ring at the junction of the two reagents used (sulfuric acid and naphthol). It is also possible to dip the sample in alcohol and then quickly wash with water. If "blushing" occurs when the sample is dry, ethyl cellulose is present.

Polyamide coatings and inks may be determined by the resins' solubility in isopropyl alcohol.

Flame (Hot Wire) Tests. All coatings or films that contain halogens such as PVDC or PVC will yield a characteristic green flame color when a copper wire is heated, wiped across the material, and then replaced in the Bunsen gas flame. Wax and polyethylene coatings will melt and burn, giving a typical odor. Waxes melt at a lower temperature than does polyethylene.

Infrared Spectroscopy. Infrared spectroscopy is an important technique for the qualitative analysis of organic compounds. It is especially effective in the identification of solvents, coatings, films, and extruded resins which are used in the packaging industry. If a coating can be dissolved or removed intact, it can be identified through the use of infrared spectroscopy techniques. An infrared spectrophotometer generates and directs a beam of infrared energy through a sample. The instrument detects and makes a record of the amount of monochromatic energy absorbed by the sample. This absorption property is called its spectrum. Each type of material has a unique spectrum. Spectra of unknowns can be compared with the spectra of known materials. Two samples with the same spectrum are the same material. It is not always necessary to remove the coating in order to obtain an infrared spectrum. Special attachments to the instrument are available which will hold a flat sample so as to make a spectrum of the outside coating only. This technique is called ATR (attenuated total reflectance) and MIR (multiple internal reflectance).

Adhesives Identification Tests

Identification of adhesives is similar to identification of coatings and inks, except that first the sample must be delaminated. In selecting the solvent used for delamination it is possible to identify the general class of adhesive. Spot tests can help identify some components and infrared spectroscopy others. However, since most adhesives are complex formulations, general classification is usually as far as the analyst is able to go.

After all the surface coatings and inks have been removed, samples are placed in each of eight bottles of test solutions and stored overnight at room temperature. Results are obtained more rapidly if the solutions are heated gently (under a ventilating hood).

The solutions and adhesive solubilities are shown in Table 5.3.

Substrate Identification Tests

Most flexible packaging substrates are composed of paper (or paperboard), film, foil, or extruded resins. Simple and rapid qualitative analysis is a matter of experience and common sense. Formalized test methods do not effectively replace the skilled hands of a knowledgeable analyst. With coatings and adhesives, certain generalizations are possible from the results of burning, solubility, and heating tests. For more specific analysis of films, infrared spectroscopy is often utilized.

Identification of Paper Substrates. Identification of paper substrates in terms of general classifications can be done by a trained paper technologist through inspection and a few simple tests. A natural kraft paper can be identified by its natural brown color and comparably coarse finish and stiffer feel. A genuine glassine has the unique property of being greaseproof. Some glassines are transparent,

TABLE 5.3 Adhesive Solubilities

Solution	Starch	Wax	Casein Latex	Silicate	Polyethylene	Most "Thermoplastics"
Acetone	I[a]	I	I	I	I	S
Water	S[a]	I	I	I	I	S
Toluene	I	S	PS	I	S (hot)	S
Carbon tetrachloride	I	S	I	I	I	S
Hexane	I	S	I	I	I	I
Skelly Solve (aromatic naphtha)	I	S	I	I	I	S
Ethyl alcohol	I	I	I	I	I	S
A solution of ammonia in water plus a wetting agent such as Dupanol	S	S	S	I	I	S

[a] I = insoluble, S = softened or dissolved. PS = partially soluble

while others contain a pigment which makes them white and opaque. MF (machine-finished) papers are usually white for packaging applications and are of the same smoothness on each side.

Many packaging applications use white papers made on an MG (machine-glazed) paper machine and which are thus called MG papers. These papers have one side that is much smoother than the other. They are used in food applications because they are white and clean and in printing applications because they are smooth and ink receptive.

Clay-coated papers can be identified by their higher ash content, their smooth surface, and the hiding of the fiber by the clay coating. This is observable with about 15× magnification. The clay-coated papers are used where high-quality tone printing is displayed on the packaging.

Groundwood papers, so called because they are made from groundwood pulp, are easily identifiable with the phloroglucinal spot test. The largest use of these papers is for some label applications.

Chemical spot tests can identify starch, sizings, alkalinity, wet strength, polyvinylidene chloride coating, and coating binders of PVA and casein.

Measurement of ream weight is then usually all that is needed to provide the packaging technologist with sufficient information to enable him to choose an equivalent paper. Where necessary, fiber analysis under a microscope can be used to identify the kind of pulp used in the paper's manufacture.

Paperboards are much stiffer and thicker than paper. There is no clear industry delineation between paper and paperboard. However, in general, the measurement of 10 points of caliper (0.010 inch) or a basis weight of 100 lb/3000-ft^2 is used as a dividing point, with paper having lesser values and paperboard having greater values.

Paperboard is described by means of several different and combined classifications. Primary to all descriptions is the paperboard thickness, which is expressed in points (thousandths of an inch). For boards an additional descriptive term is used which identifies the fiber from which the board is made. Major fiber categories are unbleached (natural brown wood color) or bleached (white), kraft or sulphite (pulp preparation process) or recycled (made from recycled scrap paper), and, sometimes, softwood or hardwood (type of tree from which the pulp was made). In addition, the type of paper machine is often included in the paperboard description. This was once more significant than it is now because there were only two types and the properties of the paperboards made on each type of machine were distinctly different. Many advances in papermaking machinery have been made which have provided the industry with machines that make papers and paperboards whose properties cannot be distinctly put into two neat machine categories. However, the categories exist; they are Fourdrinier and cylinder. Simply, if not entirely accurately, "Fourdrinier" describes a single web of fiber compressed into a sheet of paperboard and "cylinder" describes paperboard made of several (5–8) webs of fibers compressed together to form a sheet. Paperboards are frequently clay coated with a paint-like material to improve their appearance, printing properties, and smoothness. These are described as C1S (coated-one-side) and C2S (coated-two-sides).

Identification of Plastic Film Substrates. The identification of an unknown plastic film substrate can most readily be achieved with an infrared spectrophotometer. Characteristic absorption peaks serve to "fingerprint" the material. Where only very small samples are available, a few spot tests or flame tests may be enough to identify the material.

Cellophane. Cellophane burns rapidly and, like paper, will continue to burn after it is withdrawn from a flame. It is also insoluble in the commonly used organic solvents. Acetone will solubilize the nitrocellulose or saran coating commonly found on the cellophane; however, the base film is insoluble in acetone. Surface coatings on cellophane may be determined by means of the rapid tests discussed above. The infrared spectrum of cellophane contains strong absorption bands at 2.9 μm and between 9 and 10 μm.

Cellulose Acetate. This is a strong and clear film which is easy to tear. It does not become opaque upon flexing and burns slowly with an acetic acid odor. A commonly used spot test consists of adding 1 ml water and 4 ml anthrone solution in sulfuric acid to the film. A green to blue color indicates the presence of cellulose acetate. Absorption peaks are located at 5.6 μm [for (C=O)] and 8 μm [for (C=O)] on the infrared spectrum.

Cellulose Acetate Butyrate. This is similar to cellulose acetate. It is soluble in acetone and gives a positive Molisch test result. It has IR absorption peaks at 5.8, 8.1, and 8.6 μm.

Cellulose Propionate. This is essentially similar to its acetate and acetate-butyrate relatives. The one important physical test is infrared spectroscopy. The propionate group has its (C=O) absorption peak at 12.4 μm in addition to the (C=O) peak at 8.5 μm.

Ionomer. "Iolon" films are made from ionomer resins. The latter are similar to polyethylene, but they have metallic ions such as sodium or zinc built into the molecular chain. Flame or infrared spectrophotometry is the best means for identification. A flame spectrophotometer can identify the presence of zinc.

Polyamide. Films may be found in several different varieties. Nylon 6 may be identified by means of a color test employing *o*-nitrobenzaldehyde. The film is heated in a test tube. The vapors that arise contact a strip of filter paper freshly moistened with *o*-nitrobenzaldehyde in 2*N* sodium hydroxide. A deep mauve-black color identifies Nylon 6. The Nylon 11 film (Rilsan) dissolves in concentrated formic acid and melts into pearls upon heating. Absorption occurs at 3.0 and 6.0 μm. Another test uses the Greiss reagent. The film is heated vigorously in a combustion tube with manganese dioxide for 2–3 min. The mouth of the tube is covered with filter paper dampened with the Greiss reagent. A positive response is indicated by a red or pink spot on the filter paper.

Polycarbonate. This film is a smooth, sparkling, nonstretchable material. It burns moderately, producing a phenolic odor. This material may be identified by detecting its phenolic components. This may be accomplished with the iodophenol, ferric chloride, and/or coupling tests. Carbonate determinations may also be carried out on a quantitative basis. The film is soluble in acetone and insoluble in heptane. Bands occur at 5.6 and 8.0 μm on the infrared spectrum.

Polyester. Polyester films are very strong and transparent. They burn slowly with a mildly sweet odor and melt into pearls. The most commonly used polyester film, polyethylene terephthalate, is identified by clamping a strip of the film in a test tube. The lower end is heated to 350°–400°C until heavy vapors form in the tube. A strip of filter paper, moistened with a freshly saturated solution of *o*-nitrobenzaldehyde in 2 *N* NaOH, is put over the tube mouth. A greenish blue color with a yellow margin appears with polyesters and, when washed with dilute H_2SO_4 and moistened with water, becomes a blue or pale indigo.

Polyethylene. This substance has a waxy feel and stretches easily. The odor emitted under pyrolysis is paraffinic. Polyethylene is readily soluble in hot toluene and benzene. Characteristic absorption occurs at 3.4, 6.8, 7.4, and 14 μm.

Polypropylene. This film is tougher than polyethylene. It melts and drys while burning. A distinctly greater clarity and rigidity characterizes polypropylene as compared to polyethylene. Infrared examination shows absorption bands at 3.4, 7.8, and 12 μm.

Polystyrene. This a transparent, sparkling, brittle film which burns with a yellow, sooty flame. It is soluble in toluene, carbon tetrachloride, acetone, and ethyl acetate and insoluble in methyl alcohol. Polystyrene absorbs at 7.0 and 9.5 μm in the infrared range.

Polychloro-fluoroethylene. This film (e.g., Aclar) will give a positive copper wire test but is very inert to solvents. Its high density of 2.0 g/cm^3 helps in identifying it. It is also nonflammable. Spectrophotometer absorption peaks are found at 8.4, 8.9, and 10.3 μm.

Polyvinyl Alcohol (PVA). This film is a weak, opaque material which burns with a sweet, nonacrid odor. It forms a gel with cold water and is insoluble in chloroform and ethanol. To an aqueous solution of film, add a 5 percent aqueous tannic acid solution. The presence of PVA is indicated by a yellow, flocculent precipitate or a milky white turbidity. Another test utilizes a reaction with iodine. Two drops of 0.1 *N* iodine in KCl solution is added to 5 ml of a neutral aqueous solution of the film. After dilution with water, 0.2 g of borax is added until a blue, green, or yellowish green color becomes evident. This is followed by 5 ml of concentrated hydrochloric acid. A positive PVA reaction is indicated by an intense green color. The infrared spectrum reveals the clear presence of hydroxyl groups at 3.0 and 9.1 μm. An additional peak occurs at 6.1 μm.

Polyvinyl Chloride (PVC). PVC may be used in either the rigid or the flexible state. Rigid PVC is a hard material which becomes white at the bend upon flexing. It is soluble in dioxane, tetrahydrofurane, and cyclohexanone. A broad band exists at 14.5 μm on the infrared spectrum.

Polyvinylidene Chloride (PVDC). These films are clear and transparent. The material shrinks violently upon heating and is not soluble in dioxane. Absorption occurs at 9.3 and 9.6 μm.

Coextrusions. Many films are actually multilayers (two to seven) of film materials cohered to each other. These films are becoming very numerous and capable of satisfying many packaging applications. They are popular because layers of resins with specific desirable properties can be used in combination with other layers of resins with other needed properties, which results in a film with better total properties than could be obtained from a resin layer by itself. However, to the analyst, these present a tough problem of separation and identification.

Another method of identification of many of the common plastic films is through a systematic procedure relating to their solubilities. A 1-in.-square sample of the unknown film is placed in a 50-ml flask with 15 ml of solvent. The material is then refluxed and the flow sheet shown in Figure 5.4 is used to identify the material. If the material is soluble, proceed with the arrow marked *S*; if it is insoluble, proceed with the arrow marked *I*. The method is only followed until a result becomes evident.

In analyzing a flexible packaging lamination, the weight per unit area of the foil component is determined before and after removing coatings and adhesive. The final weight identifies foil gauge. The difference is an estimate of coating weights. Table 5.4 shows the typical gauges of aluminum foil and their corresponding yields.

QUALITY TESTING OF FABRICATED PACKAGES

Once packaging materials have been fabricated into packages, it is necessary to measure the properties of these packages to make sure that they are satisfactory for the end use and meet any specifications there may be. These tests involve measurement of critical dimensions and critical properties. Since these tests differ for different types of package, we shall select a typical example.

TABLE 5.4 Foil Gauges and Corresponding Yields

Gauge	Yield (in.^{2}/lb)	lb/Ream[a]	Gauge	Yield (in.^{2}/lb)	lb/Ream[a]
0.00017	60,300	7.16	0.00085	12,100	35.70
0.0002	51,300	8.42	0.0009	11,400	37.89
0.00023	44,600	9.69	0.00095	10,700	40.04
0.00025	41,000	10.54	0.001	10,250	42.15
0.0003	34,200	12.63	0.0015	6,830	63.25
0.00035	29,300	14.74	0.002	5,130	84.21
0.0004	25,600	16.88	0.0025	4,100	105.37
0.00045	22,800	18.95	0.003	3,420	126.32
0.0005	20,500	21.07	0.0035	2,930	147.44
0.00055	18,600	23.23	0.004	2,560	168.75
0.0006	17,100	25.26	0.0045	2,280	189.47
0.00065	15,800	27.34	0.005	2,050	210.73
0.0007	14,600	29.59	0.0055	1,860	232.26
0.00075	13,700	31.53	0.006	1,710	252.64
0.0008	12,800	33.75			

[a]Ream = 432,000 in.^{2} or 500 sheets 24 X 36 in.

Quality Testing of Glass Containers

Glass containers are generally tested for dimensions, resistance to thermal shock, and resistance to pressure.

Dimensional Measurements. Height, body diameter, wall thickness, and "finish" are measured to detect possible variations which may exceed the tolerances which have been established by glass manufacturers. Adherence to these molded tolerances is an important factor in the operation of high-speed filling lines. Since all of these factors influence overall container capacity, this too is measured.

In checking body dimensions, gauges are used which have been specially designed for each specific bottle. Similar methods are used to check the overall diameter of the bottle mouth or "finish." The contour of the finish is checked by means of a shadowgraph. In this method the outline of the finish is projected against a specified contour.

The capacity of a glass container is measured by selecting a sample of 12 bottles at random and checking them for volume.

Resistance to Thermal Shock. This property is needed where bottles are to be filled with hot products. The test involves immersing the bottles in hot water for a specified time and then plunging them into cold water.

Pressure Resistance. Bottles used for carbonated beverages must be able to withstand internal pressure. Devices are available which subject the sample to internal pressure using a gas or a liquid. The pressure required to cause rupture is determined or a predetermined pressure is applied and the number of failures recorded.

PRODUCT-PACKAGE COMPATIBILITY TESTING

Product compatibility is determined by measuring critical product properties immediately after packaging and periodically thereafter until the end of the normal product life and beyond, if possible. The tests conducted vary with the product. For example, a cigarette may become useless after 3 months if its moisture content drops too low, a candy bar after 30 days if it becomes too moist, a luncheon meat after 7 days if it is not protected against oxygen and light, a can of fruit in 1 yr if it dissolves iron from the can, a plastic bottle of rubbing alcohol in 5 weeks if it loses too much by evaporation, a perfumed candle in 2 months if it loses its odor too rapidly, etc. Obvious change in the product caused by the packaging material is undesirable—for example, odor in a food caused by migration of plasticizers from plastic films or of solvents from inks or coatings.

Package compatibility is determined by measuring critical package properties immediately after packaging and periodically thereafter until the end of the normal product life cycle and beyond, if possible. Again, the tests vary. For example, a plastic bottle containing a prod-

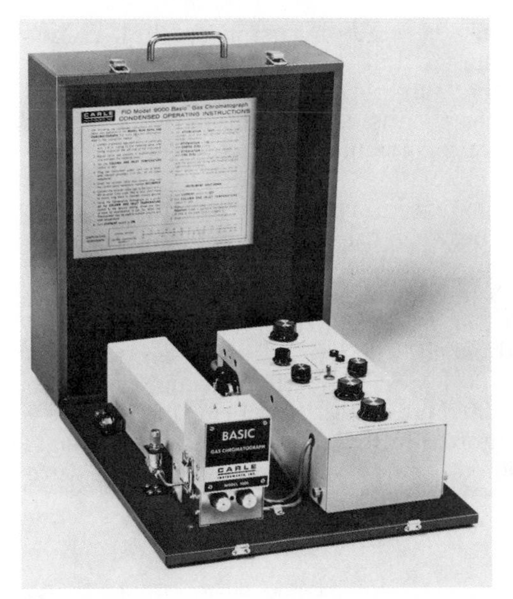

FIG. 5.14. Testing by gas chromatograph; a more objective evaluation of odor is made by a gas chromotograph, such as the model above.
Courtesy of Carle Instruments.

uct which has an essential oil constituent may weaken or stress-crack; a blister pack of small parts may embrittle with exposure to sunlight and allow the parts to spill out; a carton may weaken and collapse on exposure to high humidities.

Accelerated Aging

One of the problems the package designer faces is the length of time required to do product-package compatibility testing under normal conditions. Commercial considerations for the product call for rapid, complete, and accurate evaluations. Much effort has been expended in trying to accelerate shelf-life tests.

Since many product aging reactions are chemical in nature, the old rule of thumb, "The rate of reaction doubles with every 18°F increase in temperature," is sometimes used to justify running accelerated storage tests at high temperatures and then extrapolating the short-term results. However, only when prior experience has demonstrated a correlation between such accelerated tests and normal use conditions can such extrapolations be employed with any degree of reliability.

Where the product is known to be affected by permeability of the package to a gas or to moisture, it is possible to conduct permeability tests and then project probable shelf-life of the product.

Most of the quantitative research relative to shelf-life estimation has been conducted with water vapor permeation. The same considerations that are applicable to water vapor apply to oxygen, nitrogen, and carbon dioxide. The quantitative measurement of gases is fairly cumbersome, and a virgin area exists in estimating shelf-life by using gas permeability data.

An estimation of shelf-life may be made by using the half-value–period method. This method assumes that the rate of moisture gain or loss by the contents of a package is proportional to the difference between the water vapor pressure on the inside and that on the outside of the package. It is also inversely proportional to a factor known as the "moisture resistance" of the package.

The "half-value period" is the time required for the moisture content of the product to move halfway between the initial value and its equilibrium values. As an example, let us assume that five packages of dehydrated soup are packaged in a paper/polyethylene/foil/polyethylene laminate. Initial weights (W_0) are taken and the initial moisture percentages are established. The five packages are then exposed to controlled storage at 60°F and 90% RH. The packages are reweighed at intervals of up to 20 days. After this period of time, they are opened and allowed to come to equilbrium at storage conditions. The final weights (WE) are noted. The mean percentage of moisture (WX) at each time interval is determined.

The log $(WE - WX)$ is plotted against days, and the slope (S) of the line is obtained. Using the equation

$$T = \frac{\log 2}{S}$$

the "half-value period" (T) is obtained.

When the shelf-life of a product is known under specific conditions and a projection is required at different conditions, the same type of calculation may be conducted. An estimate is needed of the average temperature to which the package will be subjected during transit.

SHIPPING AND ABUSE TESTING

In order for a package to do its job successfully, it must arrive at its destination undamaged and still be able to perform its primary function. Today's highly complex materials require protection against the hazards encountered in normal shipment. The less protection is provided by the unit package, the more protection must be provided by the shipping container. The sales appeal of any packaged product is dependent on the condition of the final unit package. A poorly designed and nonprotective unit will not be conducive to consumer purchase.

FIG. 5.15. Instron compression; the compression strength of corrugated board is tested on this machine.

Yet, many converters and package users completely overlook the manner in which a shipping carton arrives at a retail outlet. The testing and development of shipping containers are an important part of a successful retail package. No two unit packages have identical requirements. Should a product be packed 12 to a box or 24 to a box with paperboard dividers? What about shipment to humid climates? An arbitrary decision relative to container construction may spell disaster for even the most highly developed inner package.

Actual Shipping Tests

A simple and inexpensive test to conduct is an actual shipping test. A test quantity of the product is packed into the shipping containers. The shipping containers are then shipped by bus, truck, rail, or air to a destination. After arriving, the packages are examined for damage. One or many units may be shipped at the same time. One disadvantage of this type of testing is lack of knowledge of the actual severity of the transportation conditions. Sensors and recorders can be placed in with the shipment, but the full story can never be told. How destructive were air pockets to the airplane's cargo? What temperature and moisture conditions were encountered? A second disadvantage is that results are usually nonreproducible. And still another disadvantage is the difficulty in choosing an adequate sample size. If no packages failed, does this mean that the package design is good or that the sample is too small? Nevertheless, actual shipping tests do have value when used in conjunction with a planned laboratory investigation. At our present level of package engineering, correlation between shipping tests and laboratory data is often difficult; however, as more information is gleaned from laboratory test data, the risks in making larger actual shipping tests are reduced.

Transportation Hazards and Test Methods

Five factors influence the degree of shipping damages: vibration, jolting, dropping, crushing, and climatic variations.

Vibration Effects. These may be simulated by using a vibration tester. A vibration table consists of a bed driven by two wheels connected in phase with one another. A platform is attached to the top of the vibrating bed and a circular, harmonic type of vibration occurs when the table is operational. The overall frequency may be varied from about 120 cycles/min to 360 cycles/min. In past years, vibration tables were constructed to accommodate a 1000-lb load capacity. A wide range of testers are now available, with up to 20,000-lb capacity.

In order to standardize the test, the frequency of vibration is adjusted until the package slips from the table surface after each cycle. Test conditions vary; however, most vibration tests are run between 15 min and 1 hr. A rough rule to follow is that 1 hr on a vibration table equals a 1000-mile trip by rail.

When successfully performed, the vibration test simulates the vibrational forces of the resonance, bumps, jars, accelerations, and weaving of railroad cars, motor trucks, and airplanes. Since it is an accelerated test, visible points of structural weakness will be detected quickly.

Jolting Shocks. These are simulated by an inclined impact tester. During jolt-shock testing, it is necessary to use a container that has first been vibrated. Most packaged products will not withstand as much shock after vibration as they will prior to their exposure to this hazard. An inclined impact tester consists of a track inclined at 10° to the horizontal. The packaged product is mounted on a movable platform dolly on a presrcibed inclined plane. It is released at varying points to strike a backstop at the bottom. Depending on the distance up the incline from which the dolly is released, impact speeds of up to 8 mph are possible. It is best to test all six sides of a package and record the impacts on an automatic recording device.

Automatic recorders are graduated into five zones. A fifth-zone shock is recorded when the shock is of such magnitude that the recorder stylus moves through all zones into the fifth zone. A fifth zone shock is usually considered to be the minimum shock which a package must pass prior to shipping approval.

FIG. 5.16. Tumble test; an integrated way to test package durability is to tumble an overwrapped box or a premade bag to its destruction.

Drop Strength. This is tested with a drop tester. Two different types of drop testers are commercially available. One form has a release mechanism from which the package is suspended by a sling. Another type is a table-top "trap-door" tester; when the trap door is opened, the package falls onto the floor. A packaged product is simply dropped a specified number of times from a given height. Packages are tested in a series of falls in which all corners and faces are subjected to equal punishment.

Crushing. The ability of a package to withstand crushing during stacking may be tested by using a compression tester. One filled carton is placed between two hydraulically driven plates. Successive compression movements cause the container to be crushed. The failing load of a package is considered to be the weight at which the container exhibits signs of destruction. A safe stacking load is usually about one-quarter of the failing load.

Climatic Variations. These may cause moisture condensation and yield weak packages. Freezing conditions often lead to brittleness and container damage. It is best to test the shipping material in order to evaluate its properties under extreme climatic conditions.

Carton Testing

In a carefully planned laboratory investigation, carton style, flute specifications, and overall carton dimensions are recorded. The entire container should then be conditioned at 73°F and 50% humidity for 24 hr prior to actual testing.

Vibration Test. Vibration tests are conducted first. The package is placed on a vibration tester; a fence may be attached to the tester to prevent dropping. The vibration frequency should be adjusted so that the packaged product leaves the table momentarily at least once during the vibration cycle. Rotation of the package clockwise 90° every 15 min ensures proper vibration.

Following vibration, the same container is subjected to impact. For filled containers weighing less than 50 lb, a drop test should be conducted. Heavier packages must be subjected to an inclined impact shock test.

Drop Test. Using one of the drop testers described previously, the package must be dropped from a flat and firm base. Each face, corner, and edge of the container must be identified in order to know all possible sources of weakness. The overall symmetry of the package should also be known. Facing one end of the container, with the manufacturer's joint on the right, the following six points are marked: (1) top of container, (2) right side, (3) bottom, (4) left side, (5) near end, and (6) far end. Each edge is marked by noting the numbers of the two faces that form the edge. An identification of 2–5 indicates that the edge is formed by the right side and near end. Corners are identified by the numbers of the three faces that meet to form that corner. A corner formed by the top, right side, and near end is marked as 1–2–5. After proper marking of all parts of the test package, the package is dropped following a specified sequence. The first drop height (12 in.) is usually lower than the height anticipated to cause breakage. A recommended sequence is as follows. The package must fall on (1) the base, (2) the longer base edge, (3) the shorter base edge, (4) the base corner, (5) the larger side face, (6) the smaller side face, (7) the vertical edge, (8) the top corner, (9) the longer top edge, (10) the shorter top edge, or (11) the top.

If the container passes the drop test, the drop height is increased by 6 in. If breakage does occur, this fact is noted and the specific point of drop (e.g., the top corner) is omitted in ensuing drop tests. As a result of a controlled drop test, the specific weak points of a container are determined. If improvement is needed, development can be concentrated in the required area.

Inclined-impact Test. For packages weighing in excess of 50 lb, an inclined impact test is necessary. The test container is placed on the dolly with the face or edge to receive the impact projecting 2 in. beyond the forward end of the dolly. A shock recorder is attached to the container in order to record impact. It is placed so that a centerline through the length of the recorder is at right angles to the plane of the back stop.

In order to perform the test, the dolly and container are drawn up the incline to a specified position. The test is repeated so that both sides, front, back, top, and bottom are subjected to impact on the packaged product.

The test results are acceptable if the packaged product in the container is free from damage on examination or if the damage is less than or does not exceed the damage of a control package.

NSTC Testing Procedures. If more detailed testing procedures are desired, the National Safe Transit Committee (NSTC) offers standarized preshipment testing procedures for all packaged products transported by common carrier. The procedures are basically as outlined above and enable a manufacturer to ascertain whether or not his product will reach its final destination safely even before any actual shipment is made. The NSTC was organized in 1948, and its test procedures are widely used. Certification is available from NSTC-certified testing laboratories, and labels are attached to the certified containers.

BIBLIOGRAPHY

ACCUM, F. 1820. A Treatise on the Adulteration of Food. A. B. Small, Philadelphia.

AGARWAL, S. R. 1973. Puncture resistance of flexible materials. Mod. Packag. *46* (6), 47–50.

ANDERSON, G. P., and DEVRIES, K. L. 1982. Evaluation of adhesive test methods. Org. Coating Plast. Chem. *47*, 462–466.

ANON. 1972. Flexible packaging material test method index. National Flexible Packaging Association, Cleveland, Ohio.

ANON. 1973. Corrugated board. Tappi *56* (8), 39.

ANON. 1977. FDA-extraction testing—How and why. Food Eng. *49* (11), 83–84.

ANON. 1978. Checking vacuum levels of cans. Food Process. Ind. *47* (554), 45.

ANON. 1980. Testing protocol for child resistance packaging. Aerosol Age *25* (3), 26–27.

ANON. 1982A. Quality control. Packag. Rev. *102* (7), 29, 31–33, 35.

ANON. 1982B. Annual Books of ASTM Standards, Part 20, Paper; Packaging; Business copy products. ASTM, Philadelphia, Pennsylvania.

ANON. 1982C. Annual Books of ASTM Standards, Part 21, Cellulose; Leather; Flexible Barrier Materials. ASTM, Philadelphia, Pennsylvania.

BAILEY, R. A. 1971. Determining film shrinkage. Mod. Packag. *44* (8), 60–61.

BLAKESLEY, C. N. 1975. Technical note: Organic vapour permiation through packaging films. J. Food Technol. *10* (3), 365–367.

BRICKENKAMP, C. S., HASKO, S., and NATRELLA, M. G. 1981. Checking the net contents of packaged goods. Nat. Bur. Standards Handbk. 133. Washington, D.C.

BRISTON, J. H. 1970. Packaging film properties and their measurement. Converter *7* (6), 16, 18, 20, 21, 46.

BRISTON, J. H. 1975. Package testing over the year. Converter *12* (7), 10, 12–13.

BUCKWALL, C. B. 1976. New developments in impact testing. Plastics Rubber Int. *1* (2), 95–96.

CLARKE, E. K. 1969. Drop testing of plastic containers. J. Soc. Cosmetic Chem. *20* (1), 3–12.

CURTIS, A. J., TINLING, N. G., and ABSTEIN, F. T., JR. 1971. Selection and performance of vibration tests. Shock and Vibration Information Center. Washington, D.C.

CURRAN, T. D. 1976. Seal-quality measurement. Mod. Packag. *49* (1), 30–34.

DANE, E. R. P. 1972. Fragility assessment for packaging. Packag. Rev. *92* (3), 17–18.

DAVIES, E. G. 1975. Evaluation and selection of retail packs. Austr. Packag. *23* (2), 23–24.

DEMOREST, R. L. 1978. Testing for gas and water vapour transmission rates. Food Drug Packag. *38* (9), 10, 18, 22.

ESPOSITO, G. G. and ADAMS, M. L. 1975. Coating. Analyt. Chem. *47* (5), 38R–42R.

FEIGL, F. 1972. Spot Tests in Organic Analysis, 6th Ed. Elsevier, New York.

FLICK, E. W. 1969. Adhesive and coating testing. Padric Publishing Co, Bernardsville, New Jersey.

GILBERT, S. G. 1978. Sorption/desorption theories to migration of indirect food additives. Org. Coating Plast. Chem. *39*, 647.

HALL, C. W. 1973. Permeability of plastics. Mod. Packag. *46* (11), 53–57.

HUNAN, D. B. 1976. Non-destructive testing of plastic bottles. Mod. Packag. *49* (11), 27–29.

LAMPI, R. A., SCHULTZ, G. L., CIANARINI, T., and BURKE, P. T. 1976. Performance of flexible packaging seals. Mod. Packag. *49* (6), 37, 40, 42, 44.

LAYBOURNE, G. T. 1976. Evaluating impacting strength of film. Paper. Film Foil Converter. *50* (8), 46–48.

LOUDEUSLAGEL, K. D., and FLOATE, W. 1970. Whole-package transmission test. Mod. Packag. *43* (9), 78–80.

MALTENFORD, G. C. 1972. Testing a package includes testing the product. Paperboard Packag. *57* (5), 40–42.

MALTENFORD, G. C. 1979. A review of useful (test) methods—Application can help converters, packagers. Paperboard Packag. *64* (3), 56, 58, 60, 62.

MANN, P. J. 1977. High-intensity light detects pinholes in cans. Food. Eng. *46* (10), 57–60.

MILES, G. D. 1980. The reasons for testing and the limitations of test results. Packag. Technol. *10* (5), 4–8.

MONTRESOR, J. M. 1973. Testing of containers and packaging materials. Packag. Technol. *19* (129). Suppl. xxiv–xxvii.

PYE, D. G., HOEHN, H. H., and PANAR, M. 1976. Measurement of gas permiability of polymers. II. Apparatus for determination of permeabilities of mixed gas and vapours. J. Appl. Polym. Sci. *20* (2), 287–301.

RANGER, H. O. 1979. New techniques for characterizing plastics used in packaging. Package Dev. Sys. *9* (1), 21–25.

SACHAROW, S. 1971. Gas chromatography: RX for odor control in flexography. Flexography *16* (7), 16–18.

SACHAROW, S. 1973. How to test flexible packaging films. Flexography *18* (2), 20–23, 30, 32, 34.

SAYLOR, E. R. 1976. The development of ASTM standards on child-resistant packaging. ASTM Symp. STP 609, Child-resistant packaging, pp. 67–76.

SCHAFFNER, R. M. 1981. Lead in canned foods. Food Technol. *35* (12), 60–62, 64.

SCHMIT, M. C. 1978. The Fipago adhesion tester-tack testing of gummed tape. Tappi *61* (3), 53–56.

SERICH, G. A. 1981. Migration to and from plastics. Chemtech. *11* (6), 36.

SPIEHLER, V. 1971. A new developed system for measuring headspace oxygen in gas-filled flexible packages. Food Product Dev. *4* (8), 58, 60, 62.

SPITZ, J. and DANICKOVA, E. 1970. Defects in hollow glassware. Glass *47* (12), 301–304.

SPRING, W. C. 1980. High barrier metallized films. An untapped resource. Paper Film Foil Converter. *54* (2), 79–81.

SULEK, A. M. 1978. Evaluation of lead in raw and canned food. Food Product Dev. *12* (6), 61–62.

TROEDEL, M. L. and KOHLENBERGER, D. 1980. Applying new technology to analyze plastics used in food packaging. Package Dev. Syst. *10* (1), 21–24.

YESINGTON, A. P. 1978. Insects and package seal quality. Mod. Packag. *51* (6), 41–42.

6

The Packaging Process: A Segment of Manufacturing Operations

Look at your spending, not with the purpose of eliminating it, but with the purpose of finding ways to spend the same amount to create more value.

Louis L. Allen (1972).

Packaging is a way to protect the product and facilitate marketing and distribution to manufacturers, distributors, and consumers. The process of bringing product and package together is usually the last step of the product manufacturing operation, although assembling packages within other packages or further units often follows product packaging.

Successful performance of the packaging function depends first on recognizing that the result must serve marketing and distribution purposes. No compelling necessity dictates that product manufacturing and packaging be anything but sequential unit operations of a business enterprise. Product manufacturing operations have been so mechanized and automated that some production plants, such as breweries, sugar refineries, and plastic resin process plants, operate with but a handful of people. Movement of solid, granular, and liquid products by means of pumps, conveyors, and other motive forces has been well studied and engineered. Incorporation of similar or identical products into a broad range of packages, each designed for a specific market segment, has proven more difficult. A single product like beer, for example, might be packaged in returnable and nonreturnable bottles of differing sizes and materials, singly or in multipacks, in metal cans of three or more types in a half dozen or more sizes, in one of several multipacks, in barrels, in plastic bubbles, or in bulk. Although beer-packaging lines have been mechanized, some other products (for example, canned vegetables) have not yet attained the same degree of mechanization and automation.

In many businesses, the unit operation highest in cost is packaging, usually because the costs reflect high labor input due to an absence of directly applicable equipment. Because packaging equipment is often one of a kind, development by machinery engineers for a small industry is usually not warranted. Not many machines would be sold. Since packaging operations often have the highest cost manufacturing, it is logical that operations managers and industrial engineers should focus considerable effort on reducing these costs through improved efficiencies. Similarly, those in packaging development should always bear in mind that package selection must be related to available manufacturing operations, so that the final product-package combination will best fit the market need, maximize efficiency throughout, and minimize cost.

Just as cost-effective processing is sometimes absent, so also has efficient integration of packaging and production functions eluded some businesses. The separation of packaging development functions into materials, graphic design, structural design, maintenance, and engineering groups has led to widely divergent approaches and loss of efficiency. The singular benefits derived from employing materials and machines suitable to each other and to the product have only been achieved through the crossing of organizational lines to achieve market-established objectives.

Those manufacturing firms that place marketing, production, packaging, distribution, development, and measurement in their appropriate perspective are generally both cost-effective and expansive. An organization that is largely materials-production-oriented and deemphasizes other key business components often finds itself with excessive inventories, even larger than those experienced as a result of poor planning. An organization that stresses packaging development at the expense of product processing can find itself with a low-quality product—and, contrary to the popular myth, product quality and quality uniformity are indispensable to a firm's viability. Similarly, a good product that does not attract the customer because of poor packaging or excellent sales with spotty distribution are cost wasteful. More product and manufacturing problems are probably caused by disproportionate emphasis on the many and varied elements that contribute to business success.

Like all other elements, packaging must be put in perspective; because this volume deals with packaging, this perspective will now be unfolded.

THE MANUFACTURING PROCESS

Production is the operation of creating goods that will be purchased by further manufacturers, distributors, or ultimate consumers at a profit to all in the system—including some perceived benefit to the consumer for which he or she is willing to pay. Management of production involves decision-making processes that result in final products that meet specified quality criteria and provide the quantities demanded at the appropriate time and within acceptable economic limits.

The introduction of the computer, with its ability to perform multiple calculations at high speed, has led to the ability to reduce production processes to mathematical terms and the subsequent ability of management personnel to approach optimization, should they be so inclined. What was in the past performed instinctively by managerial people or entrepreneurs or after much laborious clerical calculation has now been converted to an automated, scientifically based series of sequential steps, each with calculable inputs and outputs.

Inputs to the manufacturing process include materials, equipment, people, parts, ingredients, paperwork, orders, cash, credits, packaging materials, and, above all, control. They are modified by a series of production operations whose sequence and number are specified for each input. These may be only one or a number of different operations. They might be characterized as mechanical, electrical, chemical, assembly, reduction, inspection, transit, receiving, shipping, warehousing, recording, etc. Outputs of the total system are completed packaged products, invoices, payroll checks, records, waste packaging, satisfied employees, taxes, a cooperative community, etc. Each system has provision for storage after receipt of the input and probably between each operation in the system, with storage time varying according to the subsystem need.

Control underlies the entire operation, which means that at all times, someone or some mechanism has planned an action which is occurring under a specified condition and that the next step is known or anticipated. Control means knowing where everything is and where it should be at all times and being able to change, if conditions warrant.

Thus, as an example, an oversimplified manufacturing process for a candy might contain the following inputs:

1. Ingredients
   a. Sugar
      (1) Granular sucrose
      (2) Liquid sugar
   b. Corn syrup
      (1) Conventional
      (2) High-fructose
   c. Color
      (1) Natural
      (2) Artificial
   d. Flavor
      (1) Natural
      (2) Artificial
      (3) WONF
   e. Water

2. Electricity
    a. Voltage
    b. Cycles
3. Steam
    a. Culinary
    b. Line
4. Packaging
    a. Primary packaging film
    b. Cast polypropylene
    c. Waxed paper
    d. Cellophane
    e. Lamination
    f. Converted polyethylene bags
    g. Printed paperboard cartons
    h. Printed corrugated fiberboard cases
    i. Adhesive
    j. Primary packaging equipment
    k. Secondary packaging equipment
    l. Tertiary packaging equipment
    m. Pallets
    n. Palletizing equipment
5. Orders
    a. From brokers
    b. From syndicate buyers
    c. From retailers
6. Labor
7. Government regulations
8. Community regulations
9. Traffic

Each element could be expressed in mathematical language, as, for example, cast polypropylene film in unit cost per unit of finished goods.

Outputs of the total system, in this simplified example, might be:
Hard sugar candy, individually wrapped, dump filled, in printed polyethylene bags, cartoned, cased, and palletized
Steam to the atmosphere
Invoices to the customers
Salaries to the workers
Payments to the utility suppliers
Waste to the scrap recoverers
Payment of manufacturer debts
Taxes to the government
Reports to government agencies
Profits to the manufacturer

To attain these outputs, numerous interrelated operations would have to be performed on the inputs. Again, staying with the example:

1. Ingredients
   a. Sugar
      (1) Receive
      (2) Measure
      (3) Report
      (4) Hold
      (5) Transport (pump)
      (6) Inventory
      (7) Measure
      (8) Hold
      (9) Mix with water and corn syrup (use electromotive power)
      (10) Cook (use steam)
   b. Corn syrup
      (1) Receive
      (2) Measure
      (3) Report
      (4) Hold
      (5) Transport
      (6) Inventory
      (7) Measure
      (8) Mix with sugar (use electromotive power)
      (9) Cook (use steam)
   c. Sugar/corn syrup mixture
      (1) Hold
      (2) Transport
   d. Flavors
      (1) Receive
      (2) Measure
      (3) Inventory
      (4) Transport
      (5) Hold
      (6) Measure
      (7) Hold
      (8) Mix with cooked sugar/corn syrup (use electromotive power)
   e. Colors
      (1) Receive
      (2) Measure
      (3) Inventory
      (4) Transport
      (5) Hold
      (6) Measure
      (7) Hold
      (8) Mix with cooked, flavored sugar/corn syrup (use electromotive power)
      (9) Heat

(10) Indirect steam
(11) Evaporate water
(12) Mix (use electromotive power)

2. Cooked candy
   a. Cool
   b. Transport
   c. Hold
   d. Cool
   e. Form
   f. Hold
   g. Transport to first wrap machine
3. Polypropylene film
   a. Receive
   b. Measure
   c. Inventory
   d. Transport
   e. Hold
   f. Place on cut and wrap machine
   g. Test
   h. Wrap cooked candy
   i. Inventory
   j. Transport to pouch wrap machine
   k. Discard scrap material
4. Converted (i.e., printed, cut, formed) polythylene.
   a. Hold
   b. Transport
   c. Place on pouch-making machine
   d. Thread
   e. Test
   f. Pouch polypropylene—wrapped candy (use electromotive and pneumatic power)
   g. Discard scrap material
   h. Transport to cartoning machine
5. Paperboard cartons
   a. Receive
   b. Inventory
   c. Remove from pallets
   d. Transport to cartoning stations
   e. Stack
   f. Set up (electromotive, pneumatic, and vacuum power)
   g. Fill with pouched candy (electromotive, pneumatic, and vacuum power)
   h. Close (electromotive, pneumatic, and vacuum power)
   i. Hold
   j. Transport to casing operation
6. Adhesive
   a. Receive

   b. Inventory
   c. Transport to cartoning stations
   d. Inventory
   e. Introduce into equipment
7. Corrugated fiberboard cases
   a. Receive
   b. Inspect
   c. Inventory
   d. Transport to casing operation
   e. Set up case (electromotive, pneumatic, and vacuum power)
   f. Fill with cartoned product (electromotive, pneumatic, and vacuum power)
   g. Close (electromotive, pneumatic, and vacuum power)
   h. Seal
   i. Transport to palletizing
   j. Record
8. Adhesive
   a. Receive
   b. Inventory
   c. Transport to casing station
   d. Inventory
   e. Introduce to casing equipment
9. Palletize; record
10. Warehouse as finished goods
11. Remove from warehouse
12. Check against orders
13. Select mode of transportation
14. Select from finished-goods inventory
15. Place on transportation vehicle

FIG. 6.1. Candy-packaging machine can package foil-wrapped confections at rapid speeds.
Courtesy of Swiss Industrial Machine Co.

FIG. 6.2. A relatively simple glassine confection package.
Courtesy of Glassine & Greaseproof Manufacturing Association.

This very crude listing of just a few of the relevant steps is illustrative of the manufacturing process for just a single product in a multiproduct plant. As incomplete as the list is, it should still serve to demonstrate that the total manufacturing system consists of smaller subsystems, each of which influences the whole. For example, if the cast polypropylene film for the candy first-wrapping operation were slit to a wrong width, it would not run properly on the equipment provided. At best, this would lead to reduced efficiency, with consequent high labor costs resulting from the increased attention given to the equipment. At worst (and not at all improbably), the error would lead to a shutdown of the equipment. If no alternative machine were available to wrap the candy pieces, then the candy would not be wrapped. The mixer output would then be diverted to inventory or to another customer's order, or the mixer would be shut down, with resulting holds placed on sugar, corn syrup, color, flavoring, cartons, cases, labor, etc. The warehouse would not receive, hold, and ship the finished goods, and there would be no input of cash from repeat orders because this order would not have been filled.

This unfortunately typical example of failure of a whole system when one element fails illustrates the need for yet another subsystem: the information subsystem. Without information in the previous example, the first-wrap station would not have operated, leaving workers and casing and cartoning packaging materials waiting at later operations. The cookers and mixers would have continued to produce candy that could not be wrapped, and, ultimately, the accounting department might have wondered why no bill had been received from the transportation company or no check from the customer. If the machine operators struggled and made the equipment run with the inferior material with no input to the information system, orders for the wrong-cast polypropylene film could be placed again, and the inefficient machine operation could become a standard operating procedure.

The information system thus serves to smooth the flow of goods and operations in the physical system and to highlight exceptions which require corrective action. It is the nerve system of the production operation, generating the interconnecting data needed to start the next step in sequence or needed for a management decision. The information subsystem operates in parallel with the physical system and determines the sequence, the timing, quantities, personnel requirements, transportation vehicles, ordering periods for packaging materials, etc.

On the other hand, the packaging subsystem, with its own series of packaging minisubsystems, operates within the total manufacturing system. Packaging could not exist without the prior product's processing or without the subsequent physical distribution system, which includes storage, transportation, and customer operations. The information subsystem provides data that circumscribe the parameters for the packaging subsystem by providing schedules for quantities, time, size, etc. For example, the marketing subsystem would determine that the candy in the illustration is preferred by consumers in 3 oz. portions, and that consumers would purchase these portions in a specified group of retail outlets for a given price. This information would be employed to calculate the optimum quantities of packaged candy to be produced per unit time. Marketing and production inputs would indicate higher quantities for different reasons—marketing because of eternal optimism, production to achieve 100% efficiency. The physical distribution subsystem would determine the retailer and wholesaler requirements for unit loads and the probable optimum ordering quantities for each. The small retailer, for example, might purchase in carton quantities, and the retail chain syndicate buyer in case lots. This information, in turn, would establish the physical distribution network, which would dictate the transportation modes purchased and the scheduled times for transportation vehicles to appear at the manufacturer's plant.

The system of distributing products from the processor to the point of sale has an important effect on the package's function. The two basic distribution systems in the United States are (1) direct delivery by truck from point of manufacture to retail store, with a driver-salesman policing shelf stocks and delivering fresh supplies, and (2) central warehousing and common carriers, with policing and ordering left to the store manager. (A few products are processed or manufactured at the point of sale, but these labor-intensive products constitute only a small proportion of products sold today.) Producers of snacks, cookies, breads, dairy products, and other products with a short shelf-life consider direct delivery a necessity, even though it is expensive.

Direct distribution system time from processing plant to shelf can be relatively long—1 day to 1 week. The package must therefore be made more protective, the product more stable, or both.

Physical distribution encompasses *all* the steps involved in moving packaged products from the packaging line to the consumer. Sophisticated mechanisms are required to make sure that products reach consumers when and where they wish to have them—and that a small convenience store does not receive a truckload while a large supermarket receives but a single six-pack. Effectiveness must be built into the transfer of goods, and so we have the new discipline of physical distribution, the totality of moving goods: packaging, pallets, warehouses, inventory, and control. Some definitions of terms used in distribution follow.

Unitization is assembly of a number of different packages into one package so that they may be moved as a single entity. *Warehousing* is storage, where the product stops and waits. *Transport* is movement only. *Distribution* encompasses movement, storage, and the control. *Break-up* is the disassembly of large units into smaller groupings better suited for the destination.

Because direct delivery from the packager's plant to the store is expensive, storage points exist from which product can be retrieved.

Control underlies all of the other operations and means knowing where the packaged product is at *all* times and being able to direct it. A distribution system cannot be used until it can be controlled; this control is not merely over the dispatching of trucks, but over production schedules, output volume, speed, inventory, and delivery.

Moving products from the plant to the store requires stops and starts, and each has a purpose.

Packaging and physical distribution cannot function without each other. Packaging may be subdivided into four categories: primary, secondary, tertiary, and quartiary. A *primary package* is in immediate contact with the product—e.g., a bag or pouch. A primary package functions to contain the product, to protect it against the environment, and to exclude dirt and microorganisms.

Secondary packaging is the next layer out from the primary package. It may carry just one primary package, such as a tube, or act as a multiple package. Secondary packaging usually is used to assemble a group of primary packages into a unit. Secondary packaging reduces distribution costs, because rather than having to move 2, 4, 6, 12, or 24 primary packages 1 at a time, 2, 4, 6, 12, or 24 primary packages may be moved as a single unit.

By virtue of its own appearance, the secondary package may act as a retail marketing package, helping the consumer to find the product on the shelf, as in cracker, cupcake or unit portion breakfast cereal packaging. In such applications, the secondary package tells the consumer in larger graphics than are on the individual packages themselves about the product contained.

The next level, *tertiary packaging*, is generally the physical distribution package—the package not often seen by the consumer, but handled by the packager, distributor, and retailer. This unit—usually a corrugated fiberboard case, a corrugated fiberboard tray plus shrink film, or a Kraft paper bundle—carries 12 or 24 primary packages or 12 two-packs and moves through the distribution channels. Tertiary packaging unitizes primary and secondary packaging and therefore is a distribution package. Until recently, the shipping case was the principal source of protection against stacking in the warehouse and against transportation stresses.

Distribution appears complex because food processors and packagers collectively are trying to supply 44,000 supermarkets, 300,000 retail stores, and 500,000 hotel/restaurant/institutional outlets with 10,000 different food products, and each outlet handles only some of the products; in addition, manufacturers and packagers of *nonfood* products are distributing to another half million or more retail establishments through varying channels.

Distribution costs money. Transportation energy and highway taxes each carry a price tag. Warehousing involves not just the capital cost of a building, but the cost of inventory and insurance. The labor of removing the case from a pallet has costs.

With multiple handlings to move product from source to user, physical distribution represents one of the largest single costs involved in the food processing and packaging industries. Most persons do not view distribution as carrying a high cost, because distribution is divided among packagers, wholesalers, warehousers, and retailers. The National Commission of Productivity has stated that physical distribu-

FIG. 6.3. Thermoform-fill-seal operation. *Courtesy of Chemische, Werka; Huls Aktiengell Schaft.*

tion costs represent 25–45% of the retail cost of food—up to 45% for direct-delivery items such as potato chips and bread. Improvement in distribution productivity is one of this country's greatest industrial profit opportunities. The effective use of secondary and tertiary packaging reduces distribution costs.

It is important that we view distribution as a system, a unified whole, beginning at the product and ending with the consumer's disposal of the package. A system is made up of components that fit together and complement each other, forming a single unit.

Regulations dictate certain requirements for packaging. These regulations include Rule 41 of the rail industry and Rule 222 of the trucking industry. Insurance for loss due to damage cannot be granted by a common carrier to any shipper that does not adhere to the regulations. These rules govern the shipping packages for products.

Obviously, it would be absurd to pick up and move bags, cartons, or pouches one at a time. We assemble or unitize primary packages into more easily moved units—multiple packages. Distribution costs today would be much higher were products not moving in units of 6, 24, or even 100 or more.

Primary packages in their secondary and tertiary packaging help to protect each other. In packages of 6 or 24, bags, pouches, or cartons help to protect each other to a greater extent than they would if they were loose. The primary packages are tightly bound to one another and do not vibrate against each other. As a unit, they have greater vertical compressive resistance. Further, secondary and tertiary packaging protects the primary packaging against damage.

Information from the marketing subsystem describes the economic and physical parameters of the package—i.e., the maximum costs permitted and the protection requirements as defined by the environment of the market. This is then used to develop the package, which is the first step in developing the packaging subsystem, including the materials, the various pieces of equipment, power and labor requirements, space needs, etc. Requirements of the various packaging elements that cannot be met economically by the organization's manufacturing capabilities result in a demand for a package modification. Such changes are allowed only within the parameters of the nonvariable requirements—e.g., the highest price customers appear to be willing to pay, the need for packages to bear graphics meeting legal requirements, trucker regulations on corrugated fiberboard weight, the maximum case weight that can be handled by a worker, and inventory requirements. Other parameters would be considered variable—e.g., the number of colors printed on the pouch to achieve marketing desires, the speed of packaging machines, alternative packaging materials that could protect the product and still be run on the available equipment.

Describing packaging up to this point as a subsystem has been a purposeful means of highlighting the need to view packaging as but a single element in the total scheme to produce product for the consumer. But being a subsystem within the manufacturing process does not mean that packaging is just part of manufacturing. Rather, packaging is part of the total business system, with inputs from consumers, physical distributors, marketing, suppliers, etc., and as such is a subsystem equal in importance to the manufacturing subsystem.

Regardless of the nature of the business concern performing the packaging operation, the major errors that arise daily in packaging and as a consequence of packaging, (directly or indirectly) as stated above, result either from excessive or insufficient emphasis on the packaging needs.

THE PACKAGING PROCESS

Having established the concept of packaging as a subsystem within the coherent whole of producing and marketing products, it is now possible to view packaging as an entity—i.e., as a total system with *its own* inputs and outputs. As in the overall system, which has an objective to be met as a profitable operation or measured as a budgeted cost center, packaging must serve a useful function. The objective of the packaging operation is to bring together the product and its package and deliver finished packaged products in the quantities demanded in specified places at the appropriate time within predetermined costs. In most instances, the product is formed first and then brought to the package for enclosure. A few products, such as aerosol whip toppings, become products only after they are integrated with the package. On occasion, as with specialty chocolate candies, the package serves the added function of acting as a mold for the product. The package is often integral to the product—e.g., beer and carbonated beverages would retain their integrity for only minutes were it not for their containers. It is also a function of the packaging process to bring product and package together in such a way that the product retains its initial quality, within the limits of the environment. In today's technology, the package cannot be expected to insulate against temperature abuses that, for example, might ruin ice cream or chocolate bars. The package, however, can be expected to make sure that other packages would not be damaged by defects arising from temperature abuse. With many products whose quality might not seriously deteriorate as a result of excessive temperatures, such as baked goods and chocolate bars, the package is expected not to reveal the resulting product damage, such as fat seepage.

Simple enclosure of the product in the package is no longer acceptable, since the package usually serves functions other than protection. In concert with other packages in a unit, the strength of the totality permits lower cost and lesser strength in other components—e.g., cracker cartons bundled together in Kraft paper are stronger than individual cracker cartons. Although acting as a barrier against the environment is still a principal function of packaging, the need to act as communication tool in the marketing function must be included as another primary function. Thus retention of form and color must be a major consideration. Both product and package must be protected. Both must be appealing—the former to the consumer in use, the latter to the consumer when making the purchasing decision. Only a relatively few consumer goods, such as gasoline, tires, major appliances, and automobiles, are marketed unpackaged, so that the consumer makes the decision to purchase on the basis of the product alone. With the overwhelming majority of products, the consumer decision is based on the information and image signaled by the package. Thus products such as perfume and coffee are marketed to communicate the ultimate consumer benefit, with the consumer being unable to sniff the aroma or taste the flavor. Implicit in this concept is the need for each package to be uniform, so that consumer familiarity with the product's attributes will lead him or her back to the package that provided the product generating the previous satisfaction. Two basic failures in the marketing system arise from distributing unacceptable product in outstanding packaging and distributing excellent product in nonuniform or inadequate packaging. In the former instance, consumers attracted by the packaging, quickly learn of the product's defects and decline to repeat purchase. In the latter, consumers have difficulty finding the product or mediocre packaging causes package failure and the consumers turn in frustration, to one of the many alternatives always available in the marketplace.

It is therefore incumbent upon the packaging development personnel to ensure an output of packages that perform totally and are uniform in time and place. Implicit is a responsibility to influence other elements of the system to create products that fit the quality level of the package construction and graphics designs—as well as meeting legal, marketing, distribution, and cost requirements.

Three fundamental types of packaging functions are in common use in manufacturing operations:

1. Integral: as part of product manufacture, product and package must coexist. The package and product cannot be separated without serious detriment to the product, as with aerosol whip toppings, beer, and carbonated beverages.
2. Formed on the product: the packaging material does not exist in form until it envelopes the product, as with Form, fill, and seal pouches for beverage mixes and thermoform, fill, and seal cups for portion packs of liquid coffee whitener.
3. Preformed packages: the package is formed at one site and the product is inserted later, as with some blister packaging for hardware items, ready-to-eat cereal cartons, and most corrugated fiberboard shipping cases.

FIG. 6.4. Gin-bottling line at the Gilbey's Gin plant in Cincinnati, Ohio.
Courtesy of National Distillers & Chemical Corp.

Between these categories are types that can fall into one or another category, depending on the viewpoint of the reporter or technologist. Canned vegetables, for example, in present thermal-processing technology, require a vacuum-sealed glass or metal container so that the next step of the process, thermal sterilization and cooling, can be performed. Thus a vegetable or fruit can may be viewed as integral because the package plays an essential role in the process. On the other hand, the can or glass jar might be viewed merely as a container for the product, with requirements for temperature and pressure resistance seen as part of the package specifications. The latter view fails to take into account the total system requirements and thus may have deterred major advances in the past several decades. Is the pouch of a retort package an inseparable packaging component, or is it just a surrounding envelope acting as a boundary between product and environment?

Integral Packages

In the integral sector of this classification scheme, the product and package cannot be separated in some or all of its process, distribution, or consumer use. Few, if any, options are open to the manufacturer for his product in terms of number of components, although materials can be varied. The package must occupy a key process role if the manufacturer is to market the product in the specified form. Often, as in food canning and thermal sterilization, the product and package are mated early in the process and the two proceed through manufacturing together. The package thus has some direct influence other than mere protection and marketing—it cannot be separated from the product without adverse consequences. Among the product categories that fit this description are:

Aseptically packaged juices and milk
Beer
Carbonated beverages
Canned fruits, vegetables, and meats
Fresh red meat, the packaging of which allows controlled access of oxygen for color retention
Vacuum-packaged cured meats, which retard microbiological deterioration by limiting access of oxygen
Vacuum-packaged fresh meats, which prevent access of oxygen
Pressurized personal products, such as deodorants
Pressurized whip toppings
Frozen boil- or bake-in-package food products in which the package is a cooling or cooking and serving container
Multipack carriers for cans and bottles, which unitize multiples and thus enable the processor to transport in units in his plant and through distribution channels
Multipack bottle carriers designed to assist the return of the empty bottles after use—e.g., solid fiberboard cases for long-necked returnable beer bottles (for use in on-premises distribution)
Wines which improve on aging in the bottle
Multicomponent toothpaste
Roll-on deodorants
Photographic film, immediate package only (the cartridge)
Recording tape cassettes
Pharmaceutical dispensers
Transparent tape dispensers
Adhesive and other controlled liquid–dispensing packages
Shipper displays for point-of-purchase merchandising

Packages Formed on the Product

For many years, the expanding population has led to increasing demand, and this has led to development of mechanized equipment for combining product and package. In this category, the product virtually functions as a form for the package, and the latter is fabricated simultaneously with the product. The most dramatic changeovers of past years were the paperboard milk carton and the vertical form, fill, and seal lamination pouch for potato chips and similar dump-filled products. Formerly, all milk was packaged in glass bottles that were manufactured in one factory, transported to the dairy, inventoried, sorted, cleaned, filled, sealed, distributed, recycled, and reused. The paperboard milk carton is manufactured in the flat, side-seam sealed, and transported to the dairy, where erection occurs in line with filling and sealing. The carton is not reused. Until the 1960s, potato chips and similar snacks were filled into bags that were completely fabricated at a flexible material converter's plant. Now flat-roll stock laminations are received and placed on packaging equipment which sequentially forms the pouch and fills and seals it. In both instances the savings in storage space are obvious. In addition, the costs of package fabrication by an intermediary profit-making manufacturer are elminated. Equipment for handling flat stock is generally more efficient than that for making preformed containers, and so packaging costs are reduced. Now, of course, a significant proportion of fluid milk packaging, especially in larger sizes, is in blow-molded high-density polyethylene bottles, often fabricated just upstream of the filling and closing operation. More recent examples of fabricating the package around the product include blow mold, fill, and seal, as developed in Western Europe and used for some sterile liquid pharmaceuticals on both sides of the Atlantic Ocean; thermoform, fill, and seal plastic packaging of portion packages of dairy products; aseptic packaging of ultrahigh-temperature sterilized milk and juices; and the use of multiple cans or cartons as a mandrel around which flat, die-cut corrugated fiberboard blanks are wrapped to form a tight-fitting tertiary package.

The several dramatic and profitable successes with these types of operations have led to many attempts to develop analogous systems for all high-volume products. In some instances, such as portion packaging of bread spreads, condiments, and coffee whiteners, great successes have been achieved. In other areas, such as in-plant blow molding of plastic bottles and in-plant adjacent plant manufacture of metal cans, the results are sufficiently good to have generated a market share for in-plant manufacture of 35% of the total cans used in the United States in food and beverage plants. The advantages of preforming containers at a fabricator's or convertor's plant are associated with two factors: (1) lower cost of volume fabrication and (2) the inability of many food, toiletry, detergent, etc., makers to install a totally different manufacturing operation—i.e., a package making operation. Disadvantages include the frequent lack of development, which locks the food and/or beverage processor into mechanisms or structural designs for prolonged periods, thus limiting the processor's ability to take advantage of advances or cost reductions resulting from new technologies.

Some processors have effected a compromise by establishing independent package-manufacturing operations within their corporate structures. Some packaging suppliers have established dedicated lines. Thus several food companies manufacture more metal cans than all but two of the major can makers.

Preformed Packages

The vast majority of packagers rely on merchant package fabricators and suppliers to manufacture preformed package components for filling and closing at the packager's plant. But the majority of packagers are small businesses. As in most industries, a few larger producers manufacture the greater proportion of product. For example, fewer than 100 food processors (of more than 25,000 in the United States) produce about half of the food volume. This does not mean in any way that the larger processors produce only one product in one package. In fact, initially all produce a broad range of packaged products. The concentration of volume among processors, such as brewers, means that the larger packagers can exert pricing leverage that enables them to force development of mechanized equipment or to reduce prices of preformed packages. Spillover from the developments, of course, assists smaller suppliers or packagers, who are able to form packages on their own lines through licensing or imitation. These arrangements still leave enormous volumes of cans, bottles, preformed pouches, set-up cartons, and knocked-down corrugated fiberboard cases to be fabricated by merchant packaging convertors. The vast majority of packages are still fabricated by convertors and will continue to be in the projectable future.

The volume and diversity of prefabricated containers available from convertors and merchant suppliers provide the packager's package developer with a wide selection. Most of the major packagers in the United States have expanded their professional packaging staffs significantly in the past decade to the extent that they often have larger staffs than any of their suppliers. Thus, significant packaging developments have stemmed from packagers rather than from suppliers; for instance, two-piece aluminum cans were developed by a major brewer. Nevertheless, package suppliers are still the major source of packaging innovation as package users broaden their viewpoints and search for packages outside of the normal framework of their operations. Thus, companies that were formerly dedicated to glass moved with hesitancy into plastic. Such moves forced the glassmakers into improving their wares and even into plastic fabrication. Moves into plastic fabrication did not stop with the obvious (blow molding). Glassmakers also started ventures in thermoforming and spin welding of plastic containers and in composite and paperboard containers; there was a recognition that their business was packaging, not glass manufacture. Most American packaging suppliers have oriented themselves away from materials or converting processes and towards their customers by dedicating themselves to packaging specialties.

The existence of the many thousands of smaller packagers, with their needs for preformed packages, has thus been a driving force for packaging innovation. Larger packagers, recognizing this dynamic force, have employed preformed packages and contract packagers for new product and new package initiation. This, in turn, has driven larger packaging suppliers and manufacturers to expand their development activities to produce new packaging forms, although in the late 1970s and early 1980s, the developmental activities of American packaging suppliers were severely curtailed by economic issues and by managerial orientation towards short-term plans.

Whether the system is to be an integral package, a preformed package, or a package formed on or around the product, basic elements are common to all:

1. Obtain product.
2. Measure quantity, weight, volume or count.
3. Obtain primary package.
4. Mate product to primary package.
5. Close primary package.
6. Collect primary packages for secondary packaging.
7. Pack unit packages in tertiary packaging such as shippers.
8. Combine shippers into shipping units, such as pallet loads.

In the case of an integral package, some elements of the product manufacturing subsystem interlock with the above steps. Thus, canned foods may be heat sterilized and cooked after the package has been closed. In the case of the package formed on the product, steps involving measuring, filling, and closing of the primary package may be performed in one operation by one piece of equipment. In the case of the preformed package, there might be several alternatives for acquisition of the primary package:

1. The package could be made in the plant of the product packager and delivered directly to the point of product addition—e.g., a captive can making or bottle blowing program.
2. The package could be made by an outside package supplier or manufacturer—e.g., a bottle manufacturer.
3. The package could be partly made elsewhere and finished on the packager's line—e.g., a milk carton blank made elsewhere and completed on the filling line.

Further up the line from the package-making operation are the package materials manufacturers and convertors. These businesses supply the basic raw materials (such as paper, paperboard, aluminum foil, tinplate, steel, and plastic resin) to convertors, which in turn con-

FIG. 6.5 Production of TV dinner tray.
Courtesy of Reynolds Metals Co.

vert these materials into packages or package components—e.g., plastic film; adhesives; plastic extrusion lamination foil; inks, which may be combined to make a label or a can body; or metal, plastic, and paperboard, which may be combined to make a bottle cap closure. Since the raw materials ultimately become packages or package components, the materials manufacturers and convertors invest most of the considerable effort expended on package development. By developing new or improved packages, they expand markets for their own products and can extend their profitability through propriety developments. By adapting or otherwise acquiring packager developments, convertors can expand their propietary packaging lines.

Historically, package making and package filling and closing operations were handcrafts, or manual operations. Boxes, barrels, bags, bottles, and even tin cans were hand fabricated at excruciatingly slow output speeds. Products were measured out and filled manually. As time went on, various inventive minds created tools to make the job easier. Tools led to simple machines. Some of the earliest packaging machines were labelers and product-dispensing devices such as weighers or counters. Later came machines that could fabricate a bag, sachet, or paperboard container. In 170 years, we have come a long way from a few hand-soldered tin cans per man-hour to machines that fabricate hundreds or even more than a thousand cans a minute.

Mechanization has saved money and increased productivity but has required capital investment in machinery—and an obligation to improve the equipment.

Whether a packaging operation is a manual, a semimechanized, or a totally mechanized line, one thing is always necessary for efficient operation: a proper layout.

Layout

The objectives of layout are to arrange the elements of the packaging operation to form a system which meets capacity, speed, output, and quality requirements in the most economical manner.

As with other elements of packaging, layout begins with definition of objectives and variables—e.g., costs, the space available, the maximum cost of labor allowable, the size of the equipment that has been selected, energy requirements, inputs, and outputs. Provision must be made for storage of incoming product, packaging materials, scrap, etc., all of which can accumulate rapidly if the system or a component breaks down or slows. Provision must be made and space allowed for transportation of all materials into, through, and out of the system. Free access must be available to all equipment for both preventive maintenance and repair. And, of course, the layout must be safe and quiet and meet the requirements of the law, workers, environmental groups, etc. Further, there should be sufficient operating procedures for and instruction on the equipment to permit workers to understand and use it.

Since packaging is dynamic, materials suppliers, fabricators, and equipment developers and manufacturers constantly seek to improve the existing situation. Layouts should be flexible enough and industrial engineering open-minded enough to take advantage of improvements. Some beverage-bottling plants still have enormous bottle washers in their lines despite conversion to nonreturnable glass bottles some years ago. These vestiges of the era of returnable glass bottles have added variable costs to the bottlers' packaging operations because they are so difficult to remove once installed. However, with the pressures exerted by consumerists, environmentalists, and laws and regulations to revert back to returnable or at least recyclable packaging, perhaps the inability to remove these washers will prove a blessing in disguise.

Layouts must further take into account forecasted future capacity and anticipated cost-reduction needs. When space is efficiently employed in the original layouts and equipment is selected only for present needs, it is extremely difficult to expand capacity or reduce costs with better or newer equipment. On the other hand, forecasts are not always accurate, and space, equipment, and utility bays cost money. One means of increasing capacity in incremental modules is to allocate space, but use it for inventory rather than for production. Equipment can be purchased when the additional capacity is actually required. Meanwhile, the extra overhead carried in the accounting records is for the space not used for production rather than for unused production equipment.

Layouts should be carefully planned, because they tend to dictate the packaging operations for some time. Flow patterns are dictated by the layout and the built-in product-transfer methods.

As previously mentioned, packaging lines are generally engineered by unit operations. These might be, for example, paperboard carton set-up, fill, close, overwrap, and case or palletize operations. A major problem is that a breakdown in any single element closes down or slows the entire line. Speed and output are limited by the slowest component. Such bottlenecks can be eliminated by adding duplicate components in parallel incorporating bleed or side lines, running multiple shifts, building in-process inventory, etc. For example, some bottlers, limited by the slow speed of filling liquid into narrow-neck glass bottles (as opposed to closing), add a second parallel filler to the line, thus effectively doubling line speed without adding any more labelers, multipackers, or casers. On the other hand, high-speed canning lines might require downstream multiple packers and/or casers to be doubled, with two six-packers, two casers, etc., per filling and double-seaming set-up.

Mechanized or automated line layouts should be undertaken only when the following requirements are met:

1. Volume or output is sufficient to ensure that mechanized equipment will be used.
2. There is uniformity of product and packaging, so that changeovers are not frequent.
3. A continuous supply of both product and packaging materials is reasonably well assured and the output can be removed rapidly and effectively.

Although the boundaries are blurred, differences exist between mechanization and automation in a utopian sense. Mechanization is assistance from nonmanual power or motive sources, such as electricity, air pressure, vacuum, and water. Because so many packaging lines are manual, mechanization can range from adding a jig to facilitate forming or filling to adding a motorized conveyor to establishing full automation. Automation implies a concerted effort to actuate and control unit operations without human involvement—whether the human involvement is mental or physical. Separation of low-fill packages by a combination of electric eye sensing, comparison to a fixed reference, and actuation of a go/no-go gate that either permits passage or shifts to another channel is an example of automation. True automation is the absence of human involvement—humans design and machines do all the work and routine thinking. Humans monitor the activity and act in the case of exceptions only.

In mechanized or automatic line layouts, each operation is performed at a specific place and at a specific time in the sequence. Materials may flow continuously rather than in batches, with flow either steady or intermittment (incorporating stops and starts as the operations are performed). Control is simplified because defects can be rapidly traced to an exact source.

The major problem in mechanized line layouts is balancing or coupling of all of the elements. Not all machines are designed to function at the same speed or output, unless all are engineered together at the outset of design. Balance means equality of output of each of the successive operations of a line. Since few functioning layouts are in perfect balance, provision must be made in coupling the components to provide areas for surge storage; this takes up the output of a given component that cannot be accepted by a suceeding component. Mechanized conveyors of various types are often used to transport units from one unit operation to another.

Conveyors alone do not make a mechanized line layout or guarantee line efficiency. Fully mechanized line layouts require direct mechanical, pneumatic, or vacuum transfer between components. In addition, conveyors cost money, and so their use must be justified economically. Time can also be important. Layouts should be designed to reduce time, since every unit in transit, in a surge area, or unprocessed represents an inventory cost. Further, in the case of perishable products, waiting time or storage can result in quality deterioration.

At one time, no questions existed about where packaging lines should be located. The entrepreneur made the decision on the basis of his own judgment and the precedent was set. Thus carbonated beverage bottlers are most often franchises of large parent firms that produce syrups and/or concentrates, and they are generally located near markets to reduce the costs of transporting water. Brewers originally distributed from their bottle shops over their marketing areas, which paralleled the distribution system of the carbonated beverage bottlers, to reduce transportation costs. However, brewers have largely changed their strategy to use of highly efficient, cost-effective plants that are centrally located and that often (not always) distribute regionally or even nationally. Canners and freezers were formerly all located near the source of raw materials because of food perishability. The introduction of bulk aseptic processing and individual quick freezing for bulk packaging, however, has allowed some canners and freezers to move bulk commodity food to market-located plants for packaging into consumer-size units, such as cans and pouches.

Analogous situations exist in many industries, for the traditional means of doing business are giving way to the plan that makes the most economic sense from a business standpoint. In a similar vein, as

FIG. 6.6. Plastic milk bottles are an example of a preformed package. *Courtesy of Uniloy Corp.*

manufacturing plants are built or reengineered, the location of the packaging within the building is being reassessed. Some manufacturers have opted for all packaging operations in a multiproduct plant being located in one area, regardless of the product mix. This system places all operations of a similar nature (with similar labor and maintenance needs) together, and presumably close to the packaging materials and finished-goods inventory area. Where packaging lines are liable to be changed often, locating all packaging operations at a single site has advantages. On the other hand, placing packaging directly in line with product-manufacturing operations reduces product flow and minimizes the probability of distorting layouts in order to make sure that the packaging unit operations are all in a single location. Separation of the many packaging lines in large, multiproduct plants can lead to difficult or expensive maintenance problems and to long transit for packaging materials, goods in process, and finished goods. Only when the plant is packaging a large variety of totally different products are different packaging areas completely justified.

Something is to be said for both systems. If the manufacturing operation is efficiently planned, then the several packaging operations will probably be in line with the product production and located adjacent to each other. Situating multiple lines close to each other minimizes supervision and labor costs. Product losses and inventory buildups by short runs are reduced by efficient layout.

Selection of Equipment Types for the Packaging Operation

A wide variety of packaging materials and fabrication procedures are available for forming packages. These materials can be either combined at convertors' plants or brought together in the packager's factory, but filling, closure, and assembly of primary and secondary packages are almost always performed by the packager or the packager's contractor.

Thousands of man-years have gone into developing successful combinations of materials and machines to perform these functions. Assuming that, simply because no one has ever made a package *exactly* like the one being developed, no suitable commercial equipment is available can be and usually is erroneous. Except for high-technology systems such as that for aseptic packaging, equipment for almost every unit operation in packaging has been designed and made available commercially. It is far better to assume that commercial equipment is available and undertake a search for it than to enter immediately into an expensive machine development program. Even if no stock equipment is found, some modification of existing or analogous equipment might be far less expensive than starting from nothing. Bread-bagging equipment apparently evolved from hardware and soft-goods packaging machinery. "Controlled-atmosphere" red-meat packaging was an indirect outgrowth of cheese packaging, which was derived from processed-meat packaging.

For the most part, the package should not be altered to fit an existing packaging machine. Such compromises should be effected only when the stipulated marketing objectives can still be satisfied. No worse marketing failures are found than when the packaging equipment or a packaging material has dictated the package.

In the search for appropriate packaging equipment, a single piece of equipment should not be expected to perform the entire packaging function. A packaging unit operation can nearly always be broken down into a cluster of machine elements. For example, in the packaging of soluble coffee, conveyors transfer glass jars from one operation to the next. The jars are removed from corrugated fiberboard reshippers in one unit operation, unscrambled and aligned in a second, cleaned in a third, filled on high-speed rotary fillers in a fourth, capped and affixed with an inner glassine lamination seal on a fifth machine, labeled on a sixth piece of equipment, and recased on a seventh. The cases are sealed on an eighth unit, and the filled cases are palletized in a ninth operation. The equipment used for unscrambling may be found on carbonated-beverage, baby-food and cold-cream packaging lines; the machinery for casing may be found in carbonated-beverage and household-chemical factories. Few equipment developers prosper in business by serving only a single-user industry, even though all specialize in unit operations. Thus the concept of transferring application from one industry to another is valid, although not always appropriate. Some packagers have special requirements that can only be met by specially designed equipment.

Descriptions of stock equipment that might be used by many different user industries can be obtained in the packaging literature. Descriptions of machinery that is designed for special purposes can also be found in special packaging issues of user-industry trade journals.

The major packaging sources include the *Packaging Encyclopedia*, an annual that includes articles on recent advances in various aspects of packaging systems equipment; the *Packaging Engineering and Packaging Machinery Catalog Issue*, another annual devoted to packaging equipment, with a classified directory and advertisers' literature; the *Packaging Digest Machinery Materials Directory*, an annual devoted to packaging equipment, with a classified directory and advertisers' literature; and the *Packaging Machinery Manufacturers Institute (PMMI) Packaging Machinery Directory*, a biannual advertising book with a classified directory in which listings are limited to institute members. In addition to these directories, there are a host of manufacturers' brochures, specification sheets, and publicity releases that are available on request. The major packaging journals—*Packaging Engineering*, *Food and Drug Packaging*, *Packaging Digest*, *Packaging* (English), *Packaging Review* (English), *l'Emballages* (French), and *Verpackungs* (German)—also carry articles on packaging machinery and applications in almost every issue. The shows, meetings, and exhibitions sponsored by a large number of groups—including Pack Expo, sponsored by the American Management Association, the Society of Packaging and Handling Engineers, The Packaging Institute USA, PMMI, Western Packaging Association, and a host of trade associations and foreign groups—also constitute major sources of information. In addition, the off-shore packaging shows, such as Interpack in Dusseldorf, West Germany, and Salon de l'Emballages in Paris, France, often dwarf American shows because they are truly international. One problem with meetings and shows is that attendance at all is a major undertaking, and attendance at only one could impart only a superficial view of the totality available. Further, a few manufacturers do not exhibit at any show.

The principal underlying problem with many of these sources is that they are all based on packaging equipment manufacturers' information. No critical appraisal exists on packaging equipment and its real applications, costs, merits, and drawbacks. Rare indeed is the packaging equipment manufacturer's literature that focuses on the specific needs for which the equipment design is best suited. This is not to fault the manufacturers, who want to display their best side, or packaging journals, which derive a large portion of their income from manufacturers' advertising. Experienced user packaging developers employ these sources as a starting point for an in-depth search and to obtain an understanding of design, action, and interactions. Critical evaluations evolve from interchanges with contemporaries; with competitors, when performed judiciously; and with manufacturers' marketing and development engineers.

Sometimes, manufacturers' literature is on equipment still in the engineering or prototype stage.

The dynamic nature of the packaging industry means that any single specific equipment listing can become outdated within a few years,

FIG. 6.7. Webtron 8000 rotary letterpress, flexographic printing, and converting capabilities are combined in this completely modular tag and label product system. *Courtesy of Webtron, Inc.*

and a catalog of specific equipment can be employed as a basis for searching for equipment for only a relatively short period of time. Basically, however, packaging equipment does not change in a major way.

The packaging developer is also faced with another dilemma. New equipment introduced by a supplier may appear to offer some benefits, but only by a complete reappraisal of the existing methods can any accurate economic computation be made. Unfortunately, suppliers sometimes develop equipment or improvements in a vacuum, without full and intimate knowledge of packagers' problems. In some instances, this activity results in packaging equipment for which no true need exists, or, as is more frequent, developments for which no one really knows if a need exists. The dedicated packaging developer has a responsibility to investigate all potential improvements, but so many major development claims are made each year that time might not be available for meaningful evaluations of them all. Further, since many developments are not necessarily useful to all packagers, there is considerable waste of effort in marketing and in evaluating the equipment. The developer may not have done his homework; the packager may have invested so much time in evaluating marginal developments that meaningful developments receive only brief attention, and the orientation of the evaluator may weigh heavily on the evaluation.

FIG. 6.8. View of baking line.
Courtesy of AMF Corp.

The other extreme, waiting for the packager to state requirements, can be equally frustrating. Packagers with one-of-a-kind needs often custom engineer their own packaging equipment. User industries that have common needs do not band together, do not always have talented and articulate spokesmen, and often differ among themselves quite widely on specific details. The potential equipment developer has an obligation to become intimately involved with the packager's problems and needs so that equipment development will lead to innovation.

Integration of Equipment into the System

The development of a package involves a multitude of decisions, each of which interacts with the others. To produce a package by selecting materials, graphics, machinery, forms, speeds, etc., from catalogs would be analogous to attempting to construct an automobile from parts in a junkyard. It could be done, but it is not within reasonable mathematical probabilities.

Despite the irrationality of trying to develop packaging without sound structural design, graphics, or engineering, or with little knowledge of the marketing requirements, the process is not uncommon. Many packages are developed on the basis of a material or a competitor (who, as it inevitably turns out, has quite a different distribution method, marketing objective, means of financial measurement, etc.) or a machine that the accounting department does not want to write off. In fact, a careful examination of packaging on the marketplace today would indicate that much of it has been developed with little regard to a total system. Undue emphasis has been given to design product protection, a certain machine, or some other factor, and little or no consideration evidently has been given to one or more equally important parameters. Many packages have been developed not through effective interchanges among persons trained in varying disciplines, but by one or another department with its own specific emphasis, with some secondary inputs from other persons. Review of contemporary packaging shows that much of it is too costly, does not fit the marketplace, does not effectively protect the product, or is assembled on equipment not suited for the package. Although the package can enter the distribution system, it is not an optimum design, and in fact, such packages are not infrequently a negative force on a business and its profitability.

There is no need to sell the concept of systems to those who employ the concept. Neither is there much hope of selling it to that large group of people who falsely believe they are already using it. Eventually, the systems concept will be taught as an initial elementary subject at every packaging school. We believe the reluctance to do so has stemmed from a belief that it is impossible to be knowledgeable in a sufficient number of areas in packaging to involve all parts adequately. But someone has to bring all the pieces together, or else no package could be developed. Why not bring the parts together in sequence? Why not make sure that all the parts are present? Why not place appropriate emphasis where it is needed? Why not make certain that everything is directed towards a profitable objective? This is a system—a unification of all parts in proper order with proper weighting to achieve a given objective.

The package (i.e., the container) and its materials, seal, form, graphics, other parts (closures, labels, etc.), and its secondary and tertiary packaging, as defined by the product and its market and physical distribution system, are basic material ingredients of the system.

Manufacturing-process continuity, speed, and capacity are basic action elements of the system. Other factors, such as the nature of the product, its perishability, and its condition before packaging, all help to specify the packaging equipment needed by the system. The money available, labor skills, product costs, output, speed requirements, capital investment, space, etc., help to circumscribe the economics of the system.

All of these bits of data provide the first approximation of requirements for packaging equipment. It is then necessary to specify the requirements in detail and to determine whether any stock commercial equipment is available to fit the requirements. It may be that

FIG. 6.9. Atlas-Vac model thermoforming machine forms trays from plastic sheet.
Courtesy of Atlas-Vac Machine Co.

no single piece of machinery fits all the requirements. Several elements may have to be clustered together. The specifications for each of the equipment units should be studied to determine how well they fit together in terms of size, speed, and capacity. The equipment cluster is blocked out, with inputs and outputs on each block, to determine the match. Two pieces of equipment will probably not match if they must be run at the extremes of their capability.

If the simplified elementary block diagram appears to fit together, the complete details are incorporated to make sure that there are no inconsistencies especially relevant to product and package. In an idealized situation, no blocking occurs without information inputs from each of the potential packaging equipment suppliers. As the details are developed, some potential alternative packaging machines will be rejected and others will emerge as more appropriate.

Detailing will be the combined effort of packager and supplier engineers, along with the advice of the packaging developer who coordinates the program. If the packager has no engineers (as is frequently the case), the packaging machinery supplier's engineers must work closely with the packager to fit the equipment to the packager's objectives.

No packaging machinery should be built or purchased until everything is clearly described on paper or in a computer display. Never accept a notion derived solely from a person's mind or memory. The hard-copy program allows for feedback and corrective action. Only after all of this information is correctly collated can contingent orders be placed and construction begin.

Each piece of packaging equipment should be built to predetermined specification, with approval required for every deviation. Each unit should be thoroughly tested by the supplier in its ability to meet specifications. As many units as possible should be brought together for test to make sure that the pieces fit both the objective and each other. Tests

FIG. 6.10. Vertical form-fill-seal machine widely used for flexible packaging materials.

should be designed to prove the system components or to provide data needed to change them. Installation in the plant should be initiated only after all feasible tests have been performed, the results have been determined and assessed, and all indicated changes have been made. Installation for production packaging purposes should not be undertaken until the equipment has been thoroughly tested and proved, nor should production packaging be initiated until everything functions to produce the specified package. It should not be assumed that because each of the components operates properly the entire system will function. Further, it should not be assumed that corrections can be made once production has started, for the easiest "solution" to line problems for production personnel is to allow marginally acceptable or even defective products to be placed in distribution.

Control over the output should rest with someone with responsbility to the market, not with someone who is accountable for costs and packaging equipment. The primary rule to be observed is that the objective of the program is to produce packaged products that will be purchased and repurchased.

DEVELOPMENT OF EQUIPMENT

Although it is to be hoped that all components and connecting links can be purchased from suppliers' stock equipment, there will undoubtedly be situations in which no commercially available machinery will be suitable to the task. Packaging equipment design and development should be undertaken only as a last resort. Such a prospect is inevitably expensive, and the cost *always* turns out to be significantly higher than original estimates. Before initiating a development program, precise calculations and projections should be made to determine whether a successful program would return a profit. Simple beliefs, hopes, and wishes are not business-oriented bases for decisions.

The sequence for a packaging machine development would be the same as for a packaging system development:

1. Define the objectives.
2. Define the product and its requirements.
3. Describe the package (form, size, materials).
4. Determine speed.
5. Determine capacity.
6. Determine inputs.
7. Determine expected outputs.
8. Determine expected machine function.
9. Determine cost.

This sequence should circumscribe the machine specifications.

These specifications should be taken to one or more reputable package-machine engineering firms that are experienced in the general area—i.e., flexible containers, paperboard containers, corrugated fiberboard containers, glass bottles, etc. In conjunction with the package-machinery engineering firms, cost estimates and a timetable for the development should be made, and *only then* should the decision be made to proceed. The sequence of development should be laid out in advance, with points for review and decisions to proceed, change direction, slow, accelerate, or stop. It is often better to terminate a poor program than to attempt to rescue it with patchwork corrections.

Obviously, intensive testing is mandatory with equipment that has never been used commercially before.

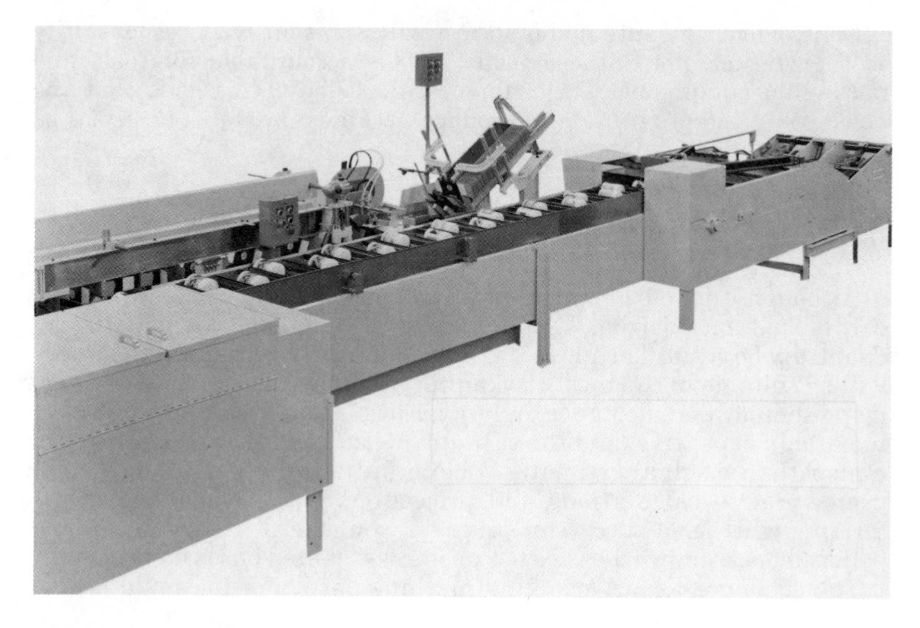

FIG. 6.11. Automatic cartoner.
Courtesy of Superior Packaging Equipment Corp.

The keys to equipment development, then, are

1. Have a clearly determined need.
2. Have a clearly stated objective for the equipment.
3. Obtain statements of feasibility from several disinterested and *qualified* persons.
4. Obtain cost estimates from reputable and *experienced* experts.
5. Determine the economic return if cost estimates are met.
6. Expect a cost overrun.
7. Develop a complete plan with checkpoints and with goals to be met at each checkpoint.
8. Employ engineers with a successful record of packaging equipment development experience in the general problem area.
9. Follow the plan.
10. Do not be afraid to abort the program even if considerable money has been spent.
11. Allow changes only when they do not in any way alter the objective.
12. Thoroughly test.

13. Do not undertake an equipment development program unless it is a business necessity.
14. Expect time overruns.
15. Expect lesser performance than was initially specified.

A number of excellent packaging machines have been designed and built on a custom-need basis. A few of these have resulted from the efforts of the packagers themselves, and, of course, the fruits of their labors have been proprietary, which is their right. Many good machines are built on a custom basis by packaging equipment firms, but probably most are based on some principles with which the firm has experience. And, of course, some merchant equipment has evolved directly from special designs for one packager.

Despite the successes, many of which are publicized, the boneyards and scrap heaps abound with packaging machines that could not be commercially installed or were less efficient than their predecessors. Progress is built on the steps of opportunity and risk, and no one makes advances without taking the chance of failure. But the intelligent packaging development person knows that his budget is generally not unlimited, his position is advisory and not line, and his authority is limited. The risks of failure in packaging equipment development are high, but the tasks are indispensable and the benefits of success are great.

BIBLIOGRAPHY

ALMARKER, C. A., and WALLENBERG, E. 1970. Aspects of the developments of packaging machines in the seventies. Verpack Rundsch. *21*, 1614, 1618. (German)

ANON. 1969A. Your computer gives you instant line management—Four ways. Package Eng. *14* (11), 78–83

ANON. 1969B. Form-fill-seal unit can package four liquids at once. Packaging News *16* (10), 28–29.

ANON. 1969C. Fluidic controls—What they can do for you. Mod. Mater. Handling *24* (5), 78–81.

ANON. 1969D. Code marks you can't see on the package! Mod. Mater. Handling *24* (9), 60–63.

ANON. 1969E. At Tunis—Airfilm conveyors channel a multiproduct flow. Mod Mater. Handling *24* (10), 80–81.

ANON. 1969F. Machinery in motion. Mod. Packaging *42* (10), 124.

ANON. 1969G. Wave of the furture takes shape. Mod. Packaging *42* (10), 154–157.

ANON. 1969H. Successful cartoning machinery installation. Bottler Packer *43* (10), 96.

ANON. 1970A. Packaging standardization and rationalization. Packaging Technol. *16* (110), xxxvii–xl.

ANON. 1970C. Breakthrough for automated processes. Mech. Handling *57* (1), 5.

ANON. 1970D. The End of the packaging line—Equipment to Keep the Flow Tidy. Mech. Handling *57* (1), 47–50, 53–54.

ANON. 1970E. Filling—A major packaging operation. Packaging Rev. *90* (3), 36–37.

ANON. 1970F. Automated packaging is paying high return. Candy Ind. *134* (7), 3, 41.

ANON. 1971A. Printing and packaging. Packag. Rev. *91* (6), 35–37, 39, 41–44, 46, 76.

ANON. 1971B. Filling and weighing systems. Packag. Rev. *91* (8), 17–18, 20, 23–26, 28, 30, 36.

ANON. 1971C. Labelling, printing, and decorating. Mod. Packag. *44* (4), 128, 129, 134, 136.

ANON. 1973A. Focus on: Filling and crowning. Int. Bottler Packer *47* (5), 86, 88, 90–92, 94.

ANON. 1973B. Injection-blow molding finds the way to bigger parts. Mod. Plastics Int. *3* (4), 14–17.

ANON. 1973C. Blowmolded containers—Trends and developments. Packaging *44* (520), 27–28, 30.

ANON. 1973D. Thermoforming and its packaging applications. Packaging *44* (520), 52–53.

ANON. 1974A. A technical review of folding box-gluers. Folding Carton Ind. *1* (3), 17–18, 20, 22, 24, 26, 28, 30, 32.

ANON. 1974B. Thermoforming—A new molding technique. Mod. Plastics Int. *4* (7), 45.

ANON. 1975. Bag packaging system. Can. Packag. *28* (11), 45.

ANON. 1976A. Wrapping, cartoning and casing. Innovations. Food Process. Ind. *45* (531), 19–41.

ANON. 1976B. Blow molds. Plast. Technol. *22* (7), 83.

ANON. 1976C. Heat-seals new and old: Same troubles, novel methods. Paper Film Foil Converter *50* (3), 38–40.

ANON. 1976D. Inks for flexible packaging. Print Trades J. (1067), 32.

ANON. 1976E. Bottling, packing, and shrinkwrapping. Int. Bottler Packer *50* (2), 59–60, 62–63.

ANON. 1978A. Package printing. Can. Packag. *31* (12), 18–21.

ANON. 1978B. Thermoforming: More than ever a distinctive package. Package Eng. *23* (5), 45–48.

ANON. 1979A. Cutting, creasing, stripping, and separating. Folding Carton Ind. *6* (1), 27–28, 30, 32, 35–36, 38, 40, 42.

ANON. 1979B. Recent ice cream developments. Dairy Ind. *44* (1), 13–15.

ANON. 1980A. Loading and marking. Packaging *51* (604), 22–29, 32, 33, 36.

ANON. 1980B. Tinplate, tin printing, container, making, canning, packaging. Tin Int. *53* (5), 202, 204.

ANON. 1982A. Injection molding. Mod. Plastics Int. *12* (4), 48–50.

ANON. 1982B. Extrusion. Mod. Plastics Inc. *12* (4), 51–54.

ANON. 1982C. Filling, capping, and labelling developments. Int. Bottler Packer 56 (8), 34, 36, 38, 40, 42, 44, 48, 50–56.

BRADSHAW, J. H. 1982. U.H.T. and Aseptic packaging. Cult. Dairy Prod. J. *17* (1), 13–16.

BRODY, A. L. 1970. Flexible Packaging of Food. CRC Press, Cleveland, Ohio.

BRODY, A. L., and MILGROM, J. 1974. Packaging in Perspective. Ad Hoc Committee on Packaging, New York.

BRODY, A. L. 1980. New Trends in Food Packaging. Institute of Food Technologists, Eastern Editorial Conference, Savannah, Georgia.

BRODY, A.L. 1982. Storage of Cereal Grains and Their Products. American Association of Cereal Chemists, St. Paul, Minnesota.

BUCHMAN, H. 1969. How to mechanize a line for a problem package. Package Eng. *14* (12), 80–83.

BUFFA, E. S., 1963. Modern Production Management. John Wiley & Sons, New York.

BURNS, J. P. 1976. Extrusion blow and injection blow: A role for both in rigid container processing. Plast. Des. Process. *16* (5), 10–13.

BYRD, B. E., 1968. Cleaning glass for food and beverage use. Proc. Seminar Food Packaging Systems Using Glass, Univ. of California.

CARR, J. H., 1969. Strength and usage performance of corrugated containers. Packaging Technol. *15* (109), 11–12.
CROWE, T. H. 1969. Gas-atmosphere preservation. Mod. Packaging *42* (10), 124.
DAVIS, C. G. 1880. Coding, marking, inprinting: A re-review of methods. Package Dev. Syst. *10* (2), 28–33.
DEAN, D. A. 1970. Considerations in the use of plastics containers for packaging liquid cosmetics. Packaging Technol. *16* (111), 10–17.
ECKERT, G. 1970A. Starting an integrated packaging line for liquids, I. Verpackung. *11* (1), 8–11 (German)
ECKERT, G. 1970B. Starting an integrated packaging line for liquids, II. Verpakung. *11* (2), 47–51 (German)
FAULKNER, H. G. 1970. Logical network planning. Environmental Eng. (47), 10–18.
FLATMAN, D. J. 1970. Shrinkpackaging—The method and materials. Packaging *41* (479), S3–S6, S9–S10, S13–S14.
FOSTER, T. V. 1973. The future of forming. Glass Technol. *14* (6), 157–162.
FRIEDMAN, W., and KIPNEES, J. 1977. Distribution Packaging. Krieger Publishing, Huntington, New York.
GOFF, J. W. 1968. A Study of the Potential for In-Plant Package Manufacturing Machinery, H. N. Brooks and J. R. Hendee (editors) Mich. State University.
HAAK, F. J. 1970. The importance of check-weighing for the industrial large package. Verpack. Rundsch. *21*, 506–512. (German)
HALLIFAX, J. 1970. Problems of labelling at speed. Int. Brewers J. *106* (1252), 72–77.
HANLON, J. F. 1984. Handbook of Package Engineering. McGraw-Hill Book Co., New York.
HARPER, C. A. 1976. What you should know about plastic processing. Chem. Eng. *83* (10), 100–114.
HARTLEY, P. 1973. Stapling and strapping for packaging applications. Packag. Technol. *19* (129), 8, 10, 12, 14.
HEISS, R. 1970. Principles of Food Packaging. An International Guide. P. Keppler Verlag, KG., Heusenstamm, Germany.
JACKAMAN, E. J. 1976. Guide to blister packaging and vacuum forming. Print Buyer *10* (7), 46–52.
JASPERT, W. P. 1969. Systems Approach to Packaging Production. Flexography *14* (12), 16, 45.
JOHNSTON, G., and GROOM, E. J. 1970. A thousand dozen flagons. Engineering *209*, 338–341.
KAREL, M. 1973A. Recent research and development in the field of low-moisture and intermediate-moisture foods. CRC Crit. Rev. Food Technol. *3*, 329.
KAREL, M. 1973B. Quantitative analysis of food packaging and storage problems. AIChE Symp. Ser. *69* (132), 107.
KELSEY, R. J. 1978. Packaging in Today's Society. St. Regis Paper Co. New York.
KOCH, R. H. 1970. Bottling line efficiency—A myth? Brewers Dig. *45* (1), 54–56.
LOPEZ, A. 1981. A Complete Course in Canning. The Canning Trade, Baltimore, Maryland.
LOVETT, W. 1969. Old plant revitalized by conveyor system. Boxboard Container *77* (4), 605–615, 685, 705, 725, 765.
MAKINO, H., and BERRY, R. S. 1973. Consumer Goods. A Thermodynamic Analysis of Packaging, Transport and Storage. Illinois Institute for Environmental Quality, Urbana, Illinois.
McDONALD, D. P. 1972. Developments in the filling of liquid and powder products. Manuf. Chem. Aerosol News *43* (7), 31–32, 35–36.
McGILLAN, D., and NEACY, T. 1964A. Choosing machinery for flexible packaging. Part I. Vertical and horizontal form, fill and seal controlled atmosphere, and preformed bags and pouches. Package Eng. *9* (1), 69–76.
McGILLAN, D., and NEACY, T. 1964B. Choosing machinery for flexible packaging. Part II. Overwrapping, direct or intimate wrapping and vacuum form, fill and seal. Package Eng. *9* (2), 82–88.

McGILLAN, D., and NEACY, T. 1964C. Choosing machinery for flexible packaging. Part III. Conclusion, package closing, heat sealing and choosing the right machine. Package Eng. *9* (3), 74–79.
MILLER, W. 1977. Technology of plastics processing. Pop. Plast. *21* (9), 30–31.
MITCHELL, E. L. 1968. Proc. Seminar Filling, Closing and Thermal Procedures and Machines. Metal Cans for Food Packaging, Univ. of California.
MOODY, B. E. 1963. Packaging in Glass. Hutchinson, London.
PAINE, F. A. 1963. Fundamentals of Packaging. Blackie & Sons, London.
PAINE, F. A. 1967. Packaging Materials and Containers. Blackie & Sons, London.
PAYNE, H. 1969. Conveyors solve producton problems on the finishing line. Boxboard Container *77* (4), 789, 809, 825.
PIKE, E. H. 1976. Progress in thermoforming. Packag. Technol. *22* (141), 5–6.
REGNIER, W. D. 1968. Proper conveyor, machine, and machine component design for abuse-free handling of the glass container on the production line. Proc. Seminar Food Packaging Systems Using Glass, Univ. of California.
RUSSO, J. R. 1969. Better system for bag-in-box packaging. Food Eng. *41* (10), 84–86.
SACHAROW, S. 1970A. The candy manufacturer can choose his shrink film from several types. Candy Snack Ind. *134* (7), 9, 18.
SACHAROW, S. 1970B. Heat sealing of plastic films. Adhes. Age *13* (9), 8–9.
SACHAROW, S. 1970C. Heat sealing of plastic films—Part II. Adhes. Age *13* (10), 8.
SACHAROW, S. 1970D. Heat sealing of plastic films—Part III. Adhes. Age *13* (11), 8, 10.
SACHAROW, S. 1971. Packaging system, what type for your product? Food Eng. *43* (3), 169, 171–172, 175–176.
SACHAROW, S. 1972. Methods for proper heat sealing of plastic films. Flexography *17* (4), 14–16, 41–43.
SACHAROW, S. 1976. Handbook of Package Materials. AVI Publishing Co., Westport, Connecticut.
SACHAROW, S. 1976. Packaging Regulations. AVI Publishing Co., Westport, Connecticut.
SACHAROW, S. 1978. A Packaging Primer. Magazines for Industry, New York.
SACHAROW, S. 1980. A Guide to Packaging Machinery. Magazines for Industry, New York.
SACHAROW, S. 1980. Filling—Key to profitable packaging. Austr. Packag. *28* (5), 107–109.
SACHAROW, S. and GRIFFIN, R. 1973. Basic Guide to Plastics in Packaging. Cahners, Boston.
SACHAROW, S., and GRIFFIN, R. C., Jr. 1981. Food Packaging. AVI Publishing Co., Westport, Connecticut.
SCHAEFER, A. R. 1968. The Labeling Machine. Proc. Seminar Food Packaging Systems Using Glass, Univ. of California.
SCHEINER, L. L. 1978. Thermoforming: A hot bed of activity. Mod. Packag. *51* (8), 41–45.
SCHULTZ, G. A. 1970. Analyze process flow in packaging department for space and labour economies. Candy Ind. *134* (6), 9–10.
SIMMS, W. 1981. Modern Packaging Encyclopedia. McGraw-Hill Book Co., New York.
VALYI, E. 1971. Injection blow molding—a review. SPE J. *27* (1), 44–48.
WEISS, G. 1980. Labelling: Simpler, safer, and faster. Can. Packag. *33* (2), 11–14.
WHITAKER, W. C. 1971. Processing flexible pouches. Mod. Packag. *44* (2), 83–85, 88.
WILSON, B. 1982. Labels and labelling. Packaging *53* (629), 16–30.

7

Packaging Equipment

What do you think the person found? When he got up and stared around? The poor old chaise in a heap or a mound.
Oliver Wendell Holmes, "The One-Horse Shay"

This chapter covers a broad range of packaging equipment either employed or available in the United States. Because of the wide range of equipment available from suppliers as well as the numerous customized specialty machines, it is not possible to cover them all. We will describe sufficient categories to illustrate the majority.

Inclusion in this chapter is not an endorsement any more than exclusion is a criticism. Neither is inclusion or exclusion a criterion of quality nor popularity among users. This chapter serves principally to provide a framework that the packaging developer can use to build on. Since this book's focus is on packaging development, the reader is directed to several reference volumes, including Sacharow's book, *A Guide to Packaging Machinery,* an expanded structure of classification and description of packaging machinery. Packaging's *Annual Encyclopedia* (formerly *Modern Packaging Encyclopedia*) provides general descriptive information on a number of packaging equipment classes, with minimum reference to specific manufacturers.

However, large, the packaging equipment universe is finite and can be comprehended. An intensive study of the range of needs must be made, with in-depth discussions with suppliers, users, and observers. A healthy skepticism is a good attitude when examining packaging equipment. It is better to invest in research than to spend on packaging machinery that is not useful, does not fit, or is a mismatch. No experienced packaging engineer or developer has avoided the pitfall of acquiring packaging equipment that was quickly relegated to a boneyard.

On the other hand, no perfect equipment exists, so while the search for the ideal is important, expecting to achieve it is wrong. Experience and judgment dictate when to study and when to cease. Too much research can be as wasteful as too little. A balance must be struck in the development of packaging systems, with this chapter serving as one of a number of inputs.

CAN-PACKAGING EQUIPMENT

Metal cans are almost always fabricated in a factory other than the actual packaging plant. Metal cans may be three- or two-piece, and composite cans are three-piece. Three-piece cans are fabricated with

FIG. 7.1. Examples of well-designed packaging for food and drink products in metal cans. *Courtesy of The Council of Industrial Design.*

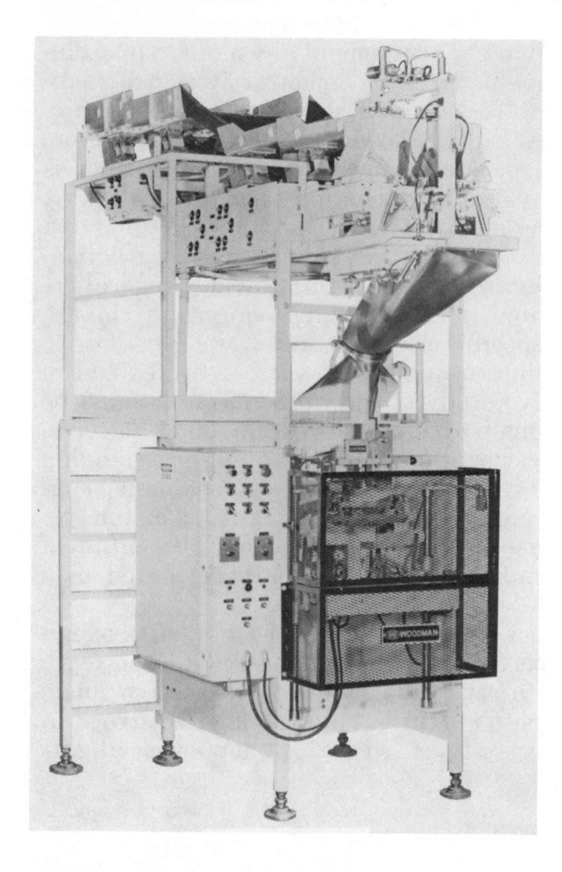

FIG. 7.2. Vertical form, fill, seal machine is used for packaging bran cereal in 10- or 15-oz. high-density polytethylene pouches. This Woodman vertical form, fill, seal packaging machine with weigh-scale filler has a rated speed of 28 to 32 packages/min. *Courtesy of The Woodman Co.*

one end affixed at the fabrication operation. On three-piece cans with convenience or easy-opening ends, the convenience ends are usually attached by the maker, leaving the flat end for the packager to attach after filling. Two-piece steel or aluminum cans have the convenience end affixed after filling.

If the packager is a large user, cans may be formed at a nearby plant—i.e., a through-the-wall operation. This is often found with brewers and in geographic areas in which fruit and vegetable processing occurs, such as in Florida. The can fabricator, whether captive or dedicated, moves cans directly to the packager from a nearby plant on an in-stream conveyor. In such a situation, the canner is generally restricted to one or two styles of can, because long runs are economical. The can maker may reserve the right to supply an alternative can at no increase in price—e.g., the can maker may switch from two-piece aluminum to two-piece steel. Cans may be delivered to the packager's inventory and drawn from that inventory as needed. Direct coupling of two operations often proves difficult, even with computer measurement and control, because the output of a can-making plant usually does not match the requirements of a canning operation.

Most medium and small packagers and many larger users purchase empty cans in bulk as open-top cans from the can makers. Empty cans are wrapped in kraft paper or paperboard and packed in rail cars, in trucks, or on pallets for transportation to the user's plant. Unloading may be accomplished by manually, fork-lifting the cans onto a line, or tiers of cans may be placed onto a moving conveyor unscrambler, which aligns the open-top cans into single or multiple files.

Since cans are not intentionally sterilized at the can maker's plant, and since cans have been handled with open tops, dirt and debris usually enter the can interiors. Major foreign matter is removed by inverting the empty cans and spraying with air, steam, or water.

Following cleaning, the cans are moved to a filler, which may be manual, mechanically aided semiautomatic, or fully automatic. Although the largest volume of products is packed on automatic lines, there are probably a greater number of manual filling lines because of the many different low-volume products being canned. In the United States, relatively few wholly manual operations exist in which the can is used as a scoop to fill the product or in which the operator pours the product into the can to bring it to weight by scaling. Such procedures are, however, valid and acceptable for materials that are not easily handled on mechanical equipment, or where the run is too short to warrant installation of such machinery. Contract packaging is an alternative, but the contract packager is then faced with the same problem. Obviously, in other countries, wholly manual filling is not uncommon.

Fully automatic filling equipment is used for high-volume items, such as beer, beverages, juices, some fruits, canned meats, and powdered products. Each can is fed into a pocket of a multiproduct rotary

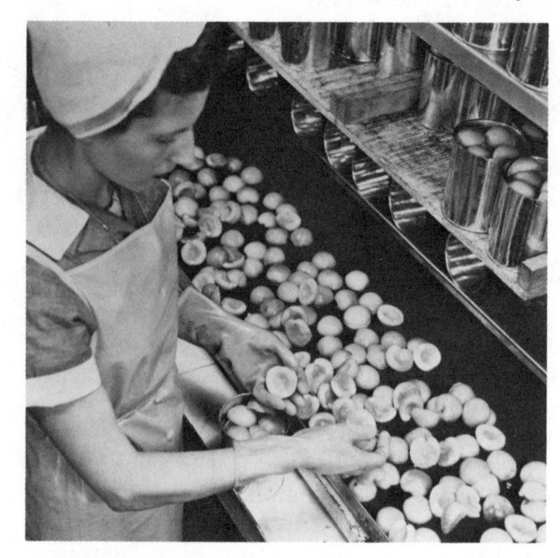

FIG. 7.3. Peach canning line.
Courtesy of National Canners Association.

filler by screw drive or star wheel. The can is elevated to the filling head, and contents are poured from above by gravity or by differential pressure. Liquids can be poured by lowering a multiorificed unit into the bottom of the can and withdrawing it as the can is filled from the multiple openings, with air displaced as the can fills. Rotary fillers make it possible for a large number of filling heads to be placed in a relatively small floor space. As many as 100 pockets in a large rotary liquid filler allow relatively high speeds of up to 1400–1600 cans/min. Such machines generally fill by timing.

Rotary machines are generally continuous so that the can does not have to start and stop and thus withstand the rigors of rapid acceleration and deceleration, with subsequent spillage. Some in-line machines are intermittent-motion or stop-and-start types. The can moves into position and stops. The filling head is lowered over the top of the can, the product is filled by means of conventional mechanisms, and the filling head is removed. The can then accelerates and leaves the filling area. High speeds cannot be achieved because of the intermittent pauses. A few in-line continuous-motion machines use moving filling heads.

Filling can be accomplished by gravity or by positive displacement, depending on the speed required and the viscosity of the product. One significant limitation of high-speed filling equipment for some products (such as baby foods or even beer, on occasion) is the inability of the processing plant to supply enough product to keep pace with the filling machine. The pace is set by the intermittent surges of supply of the product, which should be packed as soon as it is processed to prevent deterioration.

Vegetable-canning Equipment

Because vegetables are particulate and usually irregular in shape and size, vegetable packaging is often either manual or semiautomatic. Some fully automatic equipment is used for baby food, soups, and other reasonably homogeneous products. Vegetables are generally canned with a liquid brine containing seasoning such as salt. Filling is often a two-step process. The solid product is placed in the container by hand, volume, tumble, or even gentle pumping, with drained weight achieved by volumetric means, assuming fairly constant product density. Net weight is obtained by volumetrically adding liquid of known density in the next sequence. The liquid's most important functions are flavoring, improving heat transfer within the container after sealing, and protection of the solid products' form.

The contents are packed hot; the solids having been blanched or heated to set the color, destroy enzymes, reduce the microbiological load, drive air out of the tissues to help with subsequent vacuumization, and reduce the thermal energy required for sterilization. Vacuum is employed to retard oxidative changes and prevent the growth of aerobic (or air-loving) microorganisms. Vacuum is usually achieved by displacement by the hot products of nearly all the air present and by injecting a jet of steam just prior to sealing, which condenses to water, leaving a partial vacuum in the can.

Filling speeds for homogeneous viscous products, such as baby foods, could reach in excess of 900 jars/minute using positive displacement units typified by Pfaudler equipment. Most vegetable packaging occurs at a much lower speed—e.g., 60–120 containers/minute, using Solbern tumble fillers, FMC volumetric fillers, or their commercial analogues.

Liquids are packed hot by volumetric means.

After the product is packed, the can leaves the filler and enters an automatic machine called a double seamer in which a lid is affixed to the top (or bottom) and a double-seam is locked onto the body of the can. Double seaming is a locking mechanism whereby the cylindrical body metal (or paperboard lamination) hooks around the circumference of the end closure. Around the interior rim of the end is a soft, plastic-like gasketing compound which compresses to effect a hermetic seal. Double seaming is accomplished by rotating the can on its vertical axis with the two rims locked between chucks which close during spinning, thus locking body and end together and compressing the sealing compound. Steam in jets or a mechanical vacuum in the double seamer create the vacuum within the can. The sealed can is ejected from the unit onto a takeaway conveyor.

Although a number of firms, including the major can companies, have supplied double seamers, the most popular equipment is manufactured by Angelus Sanitary Can Machine Co. Double seamers perform their action more rapidly than do fillers because they merely

FIG. 7.4. Shrink film pack for fruit juice is currently used for multiple sales of soft drinks. *Courtesy of Reynolds Metals Co.*

rotate the can and do not wait for filling, and so fewer pockets and less total floor space are required to maintain output speed and pace the fillers.

Following filling, the can is heated to sterilize the contents—i.e., to destroy the microorganisms present. For vegetables that are generally considered low acid (i.e., pH above 4.5), temperatures of up to 260°F are required so that the center of the can will reach temperatures above 212°F. To achieve such temperatures, the cans are subjected to steam pressures above atmospheric pressure in autoclaves, retorts, or hydrostatic or pressure cookers. High-acid (i.e., pH below 4.5) vegetables such as tomatoes and sauerkraut may be effectively sterilized by the heat of filling followed by a short cooking period at boiling-water or similar temperatures.

Cans may be physically strapped off in regular rows or tiers onto elevating tables which are lowered into retort baskets. In many processing plants, cans are jumble dumped into the basket, which often dents can bodies and ends. Baskets are then lifted into the retorts. A few canners convey cans through air locks into continuous retorts. Hydrostatic pressure cookers employ the principle of a barometric water leg entry and exit to permit cans to continuously enter and leave a pressure chamber. Regardless of the type of cooker, cans are subjected to high-pressure, high-temperature steam. Continuous agitation during cooking is used in some retorts to increase convection currents of contents within the cans.

After removal from the pressure cooker, cans are cooled. Once the thermal sterilization point is passed, further heating only deteriorates the contents. Cooling is generally performed by spraying or immersing in cold water. Following cooling (usually down to only 100°F), cans are dried to prevent rusting.

Filling and closing operations are generally considered part of packaging, and cooking and cooling are part of the processing operation. The significance of careful integration of packaging and processing should be quite clear in vegetable canning. Obviously, the can is an integral part of the process, contributing to the thermal process by functioning as a pressure vessel.

Aseptic canning has been employed commercially in the United States for more than 30 years, initially for vegetable soup, but more recently for milk-based products such as puddings. Developmental work is underway to aseptically package low-acid products in bulk, as is currently done for high-acid products. The Dole system is used for aseptic canning. The presterilized, cooled product is put into presterilized cans in a sterile environment. Cans, usually of drawn aluminum in single-portion sizes, are sterilized by exposure to superheated steam at 550°C or above. After filling, cans are double-seam closed in a superheated steam environment with closures that have also been sterilized with superheated steam.

Retort pouch packaging is not a recent development, having been originated in the 1940s, adapted by the U.S. Army in the 1950s, and commercialized in Western Europe in the 1960s. The retort pouch is a heat-sterilization resistant flexible package originally designed to be a soft pack for military field rations. By virtue of its large, flat profile, heat penetration is facilitated, thus, in theory, reducing thermal damage to the contents.

In practice, two major and a number of lesser systems are employed. In general, the pouch material is a lamination of polyester/aluminum foil/polypropylene interior. Preformed pouches are picked off a magazine, opened, filled, evacuated, and heat sealed. Pouches from roll stock are formed into a "V," heat sealed on three sides, cut apart, opened, filled, evacuated, and heat sealed. The filled pouches are manually inspected and placed in retort baskets for pressure cooking and cooling under counterpressure. Following sterilization, pouches are manually inspected and packed into paperboard cartons for protection during distribution.

In recent years, attempts have been made to pack up to 10 lb of product in flexible pouches and thermally sterilize for the hotel, restaurant, and institutional markets as a replacement for the no. 10 can.

In many traditional vegetable packaging operations, cans are packed in tertiary containers such as corrugated fiberboard shipping cases prior to labeling in an operation called "bright packing." Alternatively, the filled, unlabelled cans may be bulk palletized to await labeling, casing, and shipping. Paper labels are applied on order from customers virtually on a job-lot basis. Casing may be manual or automatic. Shipping containers are usually corrugated fiberboard loaded from the top, end, or side; they may also be formed around the product. Regardless of the method of bright packing, the tertiary package case must be opened later so that the "bright" cans may be removed for label application. Some national-brand vegetable cans are now lithographed at the can maker's plant so that no new labels are required, but, of course, care must be exercised in packaging and processing to minimize scratching and scuffing. With most vegetable cans, however, a lithographed paper label is applied by rolling the can through a labeling machine, such as a Burt, which places a line of adhesive on the can and on the edge of the die-cut label so that as the can is rolled on its side over the label, the label adheres to the can and then to itself.

After labeling, cans are ready for final casing. Knocked-down corrugated fiberboard case shippers may be manually set up and loaded by gravity from the top. Loading equipment aligns the cans in rows and then drops them. After filling, the corrugated fiberboard cases may be manually sealed or conveyed to an automatic case sealer, which applies adhesive to the major flaps, closes the flaps, and maintains compression until the adhesive partly sets; as an alternative, the sealer may use tape or even metal stitches. Equipment companies such as Standard-Knapp, ABC, and Elliott manufacture such casing equipment. Corrugated fiberboard case packing equipment may also be obtained from case manufacturers such as Container Corporation of America.

Higher volume canners often employ side- or end-loading corrugated fiberboard cases, which may also be received in knocked-down form. Cases are snapped open manually and placed over a horn that keeps them open. Cans are aligned and pushed into the opening, which is then closed and sealed, again using compression to hold the flaps in place. Knocked-down corrugated fiberboard cases may also be removed mechanically from magazines or stacks and automatically erected for loading as integral components of case-packing equipment.

A significant number of canners employ wraparound casing. Flat, die-cut corrugated fiberboard blanks are mechanically formed around a mandrel that is shaped by a block of cans. The equipment accumulates and aligns the cans, draws a corrugated fiberboard blank, wraps the blank around the cans, and seals the flaps. This process takes a fraction of the canning speed, because multiples of 12, 24, and 48 are cased. Typical machinery for this operation is supplied by Thiele.

Fruit-canning Equipment

The relatively high acid content of most fruits allows heat sterilization at 212°F or below. Fruit products are generally packed hot (e.g., 190°F), with a hot sugar syrup. As with vegetables, filling is a two-step process, with the solid product being packed volumetrically or manually. Syrup is added automatically as the second step through a gravity or positive-displacement filler. Vacuum is achieved by means of the hot fill or steam jet followed by steam condensation on cooling.

Purees such as baby foods and applesauce are packed in the same manner as vegetable purees, but the cans are sterilized in hot water, often by continuous conveying through water baths.

Most juices and drinks are relatively acidic (i.e., pH 4.5 or below).

Because acid-type beverages do not require high-temperature sterilization, and because product transfer can be predictably and uniformly performed with liquids, thermal sterilization can be accomplished outside of the can. Hot filling then sterilizes the interior of the can, which is inverted to sterilize the inside of the canner's end.

Tomato juice and its vegetable analogues are processed in a two-stage operation in which the juice is poured at one station and salt is added at a second station. Typical of the fillers are the Pfaudler positive-displacement rotary multiturret units.

An increasing volume and types of juices are currently being packaged by aseptic methods in paperboard/polyethylene/aluminum foil lamination materials in integrated packaging systems. The most commonly used is BrikPak, an import from Western Europe, which employs roll-stock, pre-scored laminate. The material is sterilized by passage through 35% hydrogen peroxide solution and counterflow hot air to heat the H_2O_2 to sterilization temperatures and vaporize the residual sterilant. The material is formed into a continuous tube over a mandrel using an overlap heat seal and a tape seal to close the raw edge of the lamination. Product is poured, and cross-seals are made by pressure and induction heating. The finished pouch is cut off and forced into a block-like shape before exiting the machine. By sealing through the product, air is excluded from the package.

The principal competitor in the United States as of this writing is BlocPak, also an import from Western Europe. Presealed sleeves of paperboard/polyethylene/aluminum foil lamination are erected on a mandrel and discharged into a presterilized chamber. Then 35% aqueous hydrogen peroxide is sprayed into the open top container and vaporized to effect sterility. The container is filled and then closed by fin heat sealing a gusseted top.

Equipment for Canning of Beer and Soft Drinks

Because much beer is distributed under cold, wet conditions, can exteriors are prelithographed. Beer-can appearance is an important marketing tool, and so the lithography should remain intact despite any scratching, shipping, and denting the can might undergo during processing and any impacts and rubbing the can might experience during distribution. Although beer may be pasteurized outside of the can, in most of the larger American breweries, beer is pasteurized after being poured into consumer-size cans. A few brewers employ microfiltration followed by refrigeration and tight control over the distribution process in place of pasteurization. The concept of maintaining cold throughout distribution life obviously affects the packaging process. A similar process is used widely by American brewers for barreled beer and has been introduced recently for beer in plastic bulk packaging.

Because the number of cans used by any one "bottle shop" or packaging line is very high, manual and semiautomatic equipment is almost unknown in the United States.

Many breweries are directly or indirectly connected to can-making lines. Almost all breweries have either direct conveyor links to can making or automatic can unloaders and unscramblers. Beer cans are conveyed through cleaners at high speed to 72–100 head rotary filling lines; at one time, many of these were made by George J. Meyer Co. and Crown Cork and Seal, but now they are increasingly being imported from Western Europe. Beer and other carbonated liquids require a counterpressure to keep the liquid, which is under pressure, from spraying up and to reduce liquid losses. Speeds of filling can commonly reach 1400 cans/min for beer and up to 1200 cans/min for carbonated soft drinks, with the principal limitation being demand and not necessarily technical obstacles within the equipment. Duration of the entire filling cycle is a few seconds to allow flow of liquid into the can.

Immediately following filling, the open-top cans are transferred in continous motion to an adjacent double seamer for end application. The brewer's or canner's end is usually flat, tin-free steel, although with two-piece aluminum cans, the aluminum convenience end is attached in the double seamer. All aluminum beverage cans have single, double, or even triple necked-in tops so that the body wall and double seam are flush or tapered and cans do not bump against one another and dent the relatively soft aluminum in multipacking. Following closure, beer is heated for pasteurization. Warm-water sprays are used to bring the temperature up (usually to less than 150°F), followed by cooling with water sprays. Carbonated beverages are at about 40°F when packed, and the cans may require warming after closing to reduce surface moisture condensation. Warm-water showers are commonly employed to warm the cans.

Most beer cans and a very significant portion of carbonated beverage cans are distributed and marketed in multipacks, such as six-, eight-, and twelve-packs. The main example of beverage can six-packing is the Conex Hi Cone polyethylene ring carrier that fits over the double-seam chimes at the top of the can. The Hi Cone carrier represents a total machinery–material system commercially developed by the material supplier. Reels of die-cut polyethylene are threaded onto continuous-motion machines which align the cans in two parallel rows, stretch the plastic over the cans, and cut off at three or four pair, depending on whether a six- or an eight-pack is being made. In the early 1960s, paperboard carriers held a significant portion of the beer multipack business, but the economics and efficiencies of the plastic ring carrier virtually erased paperboard six-pack multipacks in the United States, except for Coors. On the other hand, 12 can packs have become significant. All of the paperboard six-packers are similar in operation: a flat die-cut paperboard blank is drawn down, over, and around the bottom of two rows of cans that have been aligned. The cut edges are locked beneath the cans after tightening. Coors employs adhesive closure. All paperboard six-packers represent total machinery-material system developments, with the stimulus provided by the material supplier.

In the early 1970s, brewers and, later, carbonated beverage packers introduced higher-unit-count packages. A significant advantage of many paperboard 12-, 18-, or 24-packs in this country today is that they may be distributed without tertiary packaging.

Although many brewers ship 12-can packs with no further tertiary packaging (which regulations permit), there is an increasing trend toward the use of corrugated fiberboard trays to help protect the 12-pack. Therefore, the paperboard 12-, 18-, or 24-pack for cans may be both a secondary and a tertiary package—an advantage in attempting to achieve cost-effective packaging/distribution systems.

In Western Europe, there has been rapid growth in the use of 18- and 24-count glass packs in paperboard cartons, E-flute corrugated fiberboard cartons, corrugated fiberboard trays plus shrink film, and what Americans would call cases.

Both preglued paperboard sleeves and flat paperboard blanks are employed to make 12-, 18-, and 24-packs for both cans and glass in the United States.

In the sleeve system for 12- or 24-pack cans, cans are collated. The paperboard sleeve, preglued, is removed from a magazine and snapped open into the shape of a tube. In continuous and parallel motion, cans are pushed into the open-ended tube either from one side, as on Pemco, Pearson, and Superior equipment, or from two sides (half from each side), as on Jones, Mead Packaging, and Manville equipment. The ends are closed by folding them down and either gluing or mechanically locking them in place. A 12-packing system using paperboard sleeves loads cans into sleeves at speeds of up to 100 12-packs/min in 3 × 4 or 50 24-packs/min in 4 × 6 arrangements.

A few systems use flat paperboard blanks to wrap 12 cans into the blank. End panels are glued into place after wrapping. Speeds of up to 130 12-packs/min are possible on such equipment.

Regardless of type, multipacks are packed in corrugated fiberboard cases or trays for shipping. For trays, a flat die-cut corrugated fiberboard blank is stripped from a magazine and two rows of multipacks are conveyed on the blank. The ends and sides are folded up and glued in place, with the multipacks acting as a mandrel. R. A. Jones & Co. is a typical supplier of this type of equipment, which runs at up to 50 multipacks/min to keep pace with the remainder of the line.

Tertiary packaging in corrugated fiberboard cases and in corrugated fiberboard plus shrink film is described below under glass packaging.

Canning Equipment for Dry Products

Metal cans are also used for granular dry foods, such as roasted and ground coffee. This product is subject to oxidation and so must be protected against exposure to air. Roasting causes considerable evolution of carbon dioxide gas, and so the coffee is sometimes permitted to wait to release the excess carbon dioxide, which might otherwise cause excessive pressure within the can. After grinding, the coffee is packed into wide-mouth cans by a combination of auger feed and gravity. The auger, a rotating helical device within a conical stator, forces precise quantities of coffee through a bottom orifice, from which the coffee's descent continues by gravity. With uniform-density, free-flowing dry products such as coffee, the number of turns of the auger can be related to the weight of the product, and so auger fillers provide excellent weight control. The most widely used auger fillers in the United States are manufactured by Mateer-Burt.

However, coffee-filling equipment, has also been made by B. F. Gump, which also provides mechanical vacuum equipment and some machinery for instant coffee packaging in glass jars. Mechanical vacuumizing is performed by literally applying a vacuum to the open-top filled can prior to double seaming. A series of filters exclude coffee dust from the vacuum lines.

A considerable number of dry granular products, such as beverage mixes, cocoa powder, coffee whiteners, and gravy mixes, are packed in composite paperboard cans. Again, auger or volumetric fillers, either rotary or in line, are used. Following filling, the cans are shaken to eliminate bridging and to level the top. Cans are closed by double seaming with an easy-opening metal end or by means of a plug lid. Vacuum or vacuum plus backflush with nitrogen gas may be employed prior to final closure to help retard product oxidation. In recent years, injection-molded, high-density polyethylene cans have been used in increasing quantities to package dry powders.

Other Canned Products

Many other products are also packed in metal cans, such as paints and other coatings, motor oil, shortening, and waxes. Except for motor oil, most nonfood can applications are low volume and low speed. The openings on paint and motor oil cans are so large that high-speed filling is employed, although larger quantities are filled per container than are common with most foods. Motor oil cans today are largely made of spiral-wound paperboard, although this is changing to plastic.

Metal cans represent a rugged, dependable, fairly uniform package capable of (1) withstanding temperature and pressure extremes, (2) containing liquids and solids, and (3) barring the passage of moisture and gas. Metal cans may be conveyed at high speeds, stopped quickly with little damage or breakage or denting, and stacked many tiers high. On the other hand, metal cans are relatively heavy and expensive. The most economical shape is cylindrical, which limits display value and increases shipping volume.

Steel can makers, recognizing the potential intrusions of containers of other materials, have reduced the weight of metal (simultaneously strengthening the metal in the process) and have improved coatings, opening convenience, and decoration. Aluminum fabricators have introduced two-piece drawn-and-ironed aluminum cans that offer wrap-around decoration and lighter weight. The "traditional" can companies have countered with tin-free steel, using draw-and-redraw fabrication to produce two-piece cans.

Equipment for packaging-product metal cans was generally designed for heavy-base-weight tin-plated steel. Lighter-weight tinplates are not wholly interchangeable with the heavier tinplates on this equipment, but major changes are no longer required. Similarly, the use of aluminum cans, two-piece cans, or spiral-wound composite paperboard cans does not require major packaging equipment changes. Weight, frictional properties, minor differences in shape, and other factors all influence the design and/or modification of canning equipment. But packages that perform like a can in laboratory tests sometimes do not perform as well on existing equipment without binding, added wear on parts, buildups on guide rails, etc. It is best to test package-equipment compatibility by making actual off-line machine runs. Unfortunately, new packages or machines are frequently tested for only a few minutes of dry-running time before being placed on the production floor. Real and near failures have resulted from this risk-laden procedure.

GLASS PACKAGING EQUIPMENT

Compared to the number of metal cans used as packages in the food industry, only a little glass is employed. Of course, large quantities of glass bottles and jars are used for beer, carbonated beverages, and baby foods, with lesser amounts used for edible oils, wine, condiments, drugs, toiletries, cosmetics, household chemicals, etc. The use of glass bottles has in recent years been adversely affected by blow-molded and injection blow-molded plastic bottles, especially for health and beauty aids. The major use of glass in packaging is for beer and carbonated beverages. The total volume of glass used for packaging has been increasing, but each year sees more conversions to other materials that may be cheaper or easier to handle or both.

Except in large-volume operations, empty glass packages are usually received at the packager's plant in preshipper corrugated fiberboard cases with paperboard or corrugated fiberboard partitions to separate each jar from its neighbors, as required by railroad and common-carrier truck regulations. Large users receive glass on bulk pallets, often unitized by shrink or stretch film with slip sheets between tiers.

When received in partitioned corrugated fiberboard cases, glass may be removed manually, because the cases are then reused for shipping after the jars or bottles have been filled. Most larger users, except for those using bulk shipment, employ semiautomatic or continuous-

FIG. 7.5. Bottle filling at the start of the bottling line, Cincinnati plant, National Distillers Products Co.
Courtesy of National Distillers and Chemical Corp.

motion equipment to separate the glass from the partitioned shipping case. Because the case has not been sealed, it is relatively easy to open the flaps and separate the glass from the case by sliding the case up an inclined plane away from the glass. The glass may also be removed by vacuum cups lifting the bottles vertically out of the shipper.

Because of the presence of fibers in the shipping case, the glass must be cleaned and washed after removal. Cleaning is sometimes performed by air blast. With products such as nonreturnable bottles for carbonated beverages, a water rinse has proved satisfactory. Glass bottles or jars are unscrambled, aligned, and conveyed to a machine that clamps a nozzle over the open end, sprays in water, and inverts the containers to drain. Returnable bottles undergo a complete detergent wash, rinse, and drain cycle on large soaker-washers such as are made by Archie-Ladewig or Barry Wehmiller. Glass jars for certain food products, such as baby foods, are also completely washed prior to filling.

Bulk-packed glass bottles or jars are removed by pusher bars on automatic or mechanized depalletizers.

Because of the relative fragility of glass, most glass packaging lines operate at slower speeds than do equivalent can lines. Beverage bottle lines operate at lower speeds than can lines mainly because of the smaller neck opening on bottles, which allows only small-diameter filling nozzles to enter and fill. The general top speed for commercial lines is about 1200 containers/min, but usually glass bottling lines operate at about half the speed of lines metal cans—and generate more packaging material waste because of breakage. The highest speed commercial glass packaging lines, except those for beer and carbonated beverages, are probably the lines for baby food because of the smaller quantity of contents and the wide mouth of the container, which significantly shortens the filling time.

On high-speed glass lines, the speed of conveyors is increased by applying a continuous coating of liquid soap lubricant so that the glass slides more readily and spins on its vertical axis. This rotation reduces the probability of direct point-to-point glass impact contact, which may scratch and thus seriously weaken the container.

To reduce breakage during stops and starts on the line, glass containers are generally externally coated, a treatment that reduces but does not eliminate the potential for scratching and breakage. Hot products such as vegetables, fruit juices, baby foods, and catsup are packed into preheated jars to avoid shock.

A few products with irregular shapes, such as asparagus spears and place-pack olives, are packed manually. At the other end of the spectrum, products such as baby foods, catsup, and beer are packed at very high speeds, on fully automatic rotary fillers. As with metal cans, jars or bottles enter the filler from a screw-worm spacer plus plastic star wheel. Rails are plastic coated to reduce scratching. George J. Meyer and Crown Cork & Seal, as well as several Western European firms, make bottom-up fillers for placing carbonated beverages in glass; in these fillers, a nozzle extends down into the bottle to minimize foaming and is withdrawn as filling proceeds to a predetermined height within the bottle. Pfaudler manufactures positive-displacement pressure-volumetric fillers used for viscous fluid materials. Pneumatic Scale Corp. makes a wide line of filling equipment which employs vacuum assist to draw still liquid into the glass container. Shutoff of a flow of low-pressure air by the rising fill level stops the product flow.

Nonliquid products may be packed using semiautomatic methods. Glass jars are conveyed and placed automatically under pockets through which mechanical devices or operators manually push solid or particulate foods. Fruits and vegetables that require an added syrup or brine are packed in a two-step operation, with the liquid added volumetrically.

Dry powders, such as instant coffee and powdered coffee creamer, can be handled by a combination of gravity and vacuum, often with an auger assist. As with liquids, the wide-mouth jar is brought under a single station of a rotary turret. The head clamped over the glass finish has an inlet and an outlet, with vacuum being drawn on the latter while the product falls by gravity through the inlet. The vacuum controls the rate of flow. Powders that tend to stick or clump are forced into the gravity drop by an auger feed in the hopper. B. F. Gump manufactures equipment for instant coffee. Pneumatic Scale Corp., Nalbach, and Mateer-Burt make powder-filling equipment used in the United States. (Note that packaging machinery firms often regarded as complete manufacturers of their own machinery may use functional components made by others.)

Glass packages are closed and sealed in such a way as to exclude the environment, to contain the contents (including carbonation), to prevent loss and pilferage, to be attractive, and, above all, to allow the container to be opened. Because glass packages have only one opening, and that opening is used for adding the contents, filling equipment is *not* used for closing with currently known closures. A separate machine in line with and/or near the filler is employed for closing. On automatic and semiautomatic lines, jars are conveyed to a rotary turret unit with relatively fewer stations than on the filler because it takes less time to close than to fill the jars.

A number of different types of closure methods have been designed, with the method of choice depending on the contents and customer and consumer needs. These methods may be basically divided into normal, pressure, and vacuum. All are based on the principle of seating a resilient material, formerly mainly cork and now mainly film or foam plastic, against the surface finish of the glass jar or bottle opening, compressing the resilient material to seal the periphery, and holding the closure and glass in this relative position. The closure is held in position by locking onto a protrusion molded into the glass. Four basic types of closures are in common use: screw, push, crimp, and roll or spin-on. Screw closures include the threadedtype, which requires one or two complete turns for application and removal, and the lugtype, which requires less than one turn. Both closures are prefabricated and turned into the glass. Roll-on closures are threaded by forcing the soft aluminum side-wall body into the screw threads on the glass finish. Crimp closures are used for vacuum and pressure; the crown is mounted on the glass and mechanically bent into the finish.

Screw caps are sealed to the glass by mechanically placing the cap on loosely, holding the glass package stationary, and spinning the cap. Lug and continuous-thread caps are sealed in the same manner. Screw caps with a glassine wax lamination liner, as used for instant coffee, are affixed in the same manner, except that the screw cap already contains the glassine liner, and adhesive is applied to the glass lip before sealing. Screw caps can be used for vacuum sealing, with a vacuum apparatus added to the equipment and cap liners that can retain vacuum. Among the many firms supplying screw-cap closing equipment are Pneumatic Scale Corp., Consolidated Packaging Machinery, Resina Automatic Machinery Co.

Crowns, formerly the principal means of sealing internally pressurized glass, are loosely seated on top of the glass bottle opening. The glass is elevated into a tapered opening that exerts vertical pressure downward to compress the liner and exerts pressure inward around the circumference at many points to grip the glass ring. Crown Cork & Seal is a major supplier of both closures and equipment for applying the closures at speeds that match those of beer and carbonated beverage filling lines.

Roll-on closures are fabricated from aluminum, which is soft enough to be formed by pressure. Pressure is exerted downward to compress the inner resilient sealing material, and the cap and bottle are spun on the vertical axis while the edges are formed into a thread by pressing them against the molded glass thread finish with knife-like blades. Roll-on closures are a relatively inexpensive means of providing reclosure, and so their use for beer and carbonated beverages grew at a phenomenal rate in the 1970s. Equipment for attaching these closures is supplied by Alcoa, which also supplies the roll-on aluminum caps.

In beer, fruit and vegetable glass packaging, the product undergoes further processing in its container. Most beer is pasteurized after bottling, using hot-water showers. Carbonated beverage bottles are warmed with hot water to reduce surface moisture condensation. High-acid foods in glass are conveyed into a hot-water bath for sterilization. Low-acid products, such as baby food meats and vegetables, may be lifted and placed on an elevated platform in a retort basket. Jumble fill of retort baskets, as is used for metal cans, may not be used for glass. If the closure is steel, jars may be removed by electromagnets that handle the filled containers gently. If the closure is aluminum, jars are

FIG. 7.6. Screw closure on Coca-Cola bottle.
Courtesy of Reynolds Metals Co.

strapped or pulled off onto the platform. The basket is then placed in the pressure cooker, where air pressure must be applied to keep the lids from being forced off by internal steam pressure as the temperature rises during cooking. Air and steam pressure are adjusted so that the total pressure remains constant on the caps as the internal temperature of the contents changes. Cooling is performed similarly; air pressure is retained in the retort to make sure that the closures are not distorted or popped.

After cooling, jars are dried and labeled (unless the glass has a permanent applied decoration, a relatively expensive graphic procedure not often used today except with returnable glass). Most beer and catsup bottles use spot and neck labelers, and many other glass-packed products use only spot labels. With very few exceptions, glass labeling is automatic, requiring only operator attendance to refill label magazines, glue pots, etc. Some labeling is performed using pressure-sensitive labels.

A number of single-head labelers are used in sequence with multi-head rotary fillers and closers. Batteries of such labelers in tandem are still found in beer and carbonated beverage plants, which otherwise contain a single-bottle filler and a single-bottle closer. Labeling requires removal of the label from the magazine, application of adhesive, wiping the label on the usually cylindrical bottle, and applying some pressure. Labeling may also be accomplished on continuous-motion rotary units, which run at higher speeds than the intermittent-motion units now so commonly used. Equipment for beer and carbonated soft drink labeling is supplied by George J. Meyer in the United States and Krones and Jagenberg in Western Europe. Labeling equipment for other industries is supplied by firms such as NJM Corp. and Pneumatic Scale Corp.

After labeling, beer and carbonated beverage containers are most often multipacked, using paperboard systems.

Preformed basket carriers fabricated from virgin or recycled paperboard are used for glass bottles. Those for brewers are designated as meeting compliance because they are fabricated with paperboard separation to permit distribution by common carriers. Because they usually travel on company-owned trucks, carbonated beverage bottles do not normally require compliance carriers.

Prefabricated paperboard basket carriers are used for returnable bottles by the carbonated beverage industry, providing a unit in which to transport the full glass bottles from the bottler to the consumer and that facilitates return of the empty glass bottles to the bottler. Prefabricated paperboard basket carriers are also used for nonreturnable glass bottles, particularly for premium and super-premium beer.

Frequently, knocked-down paperboard basket carriers are delivered to glass bottle manufacturers, where they are mechanically erected and placed in corrugated fiberboard shipping cases to be filled with empty glass and delivered to the bottling plant.

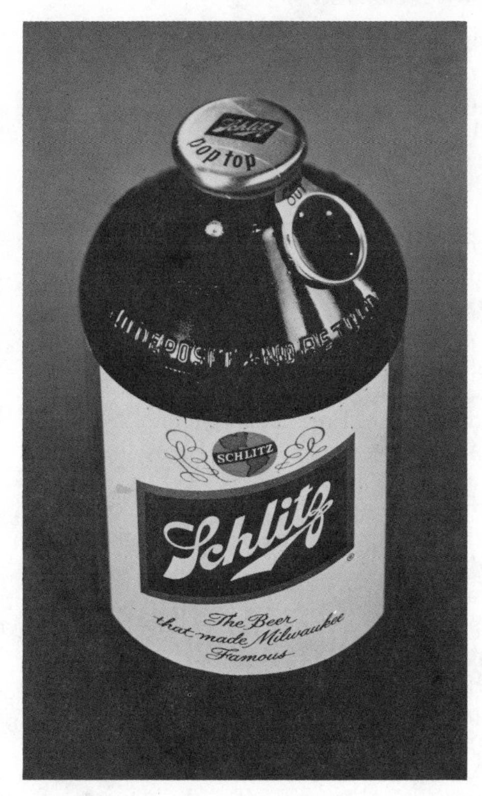

FIG. 7.7. "Pop-top" beer bottle.
Courtesy of American Flange and Manufacturing Co.

Prefabricated paperboard basket carriers thus may arrive knocked down at the bottler's plant or in corrugated fiberboard cases filled with empty glass bottles. At the bottling plant, the glass bottles are removed from the carriers; go through cleaning, filling, and closing operations, as described above; and return filled to the open paperboard basket carriers in corrugated fiberboard cases for distribution.

Preglued knocked-down paperboard carriers erected by a snap-open mechanism have preapplied partitions in compliance with transport Rules 41 and 222. The filled bottles are drop-loaded in after collation above the carrier.

In a few parts of the United States and in Canada, rigid, injection-molded, high-density polyethylene basket carriers are used for returnable bottles. Some reusable plastic units have solid-wall construction and thus contain graphics, or they may be skeletal.

In the Gulf States region of the United States, where relative humidity is high, paperboard carriers can lose strength. Plastic carriers are waterproof and may be used indefinitely. In other areas with high returnable-glass-bottle volume, even with an initial cost of ten times the paperboard carrier, the plastic basket could represent a good investment for the bottler.

Wrap-around paperboard secondary packaging begins with a flat blank, which has a lower initial cost to the packager than does a preglued basket or sleeve. The elements of a wrap-around paperboard secondary packaging machine include blank feed, product in-feed, folding, and locking or gluing. The bottles are collated into a grouping such as 2 × 3 or 2 × 4 on a continuous-motion machine. The flat paperboard blank is removed from a stack by vacuum or mechanical lugs and positioned above and moving in parallel with the collated bottles beneath.

In position, the paperboard blank is broken at score lines and wrapped around the bottles. The edges are brought together and over each other and the bottles are tightened in place. Paperboard locks are then brought together or the overlapping edges are adhered in a manner similar to that described for metal cans.

Among the wrap-around multiple packaging machines in common use for multiple packaging are those supplied by Mead Packaging, Manville, Kliklok (Certipak), Federal, and others with model speeds in the 100 and 200 multipacks/min range.

Structural designs for paperboard wrap-around secondary packaging can provide complete light protection for six sides, with paperboard dividers between bottles to ensure compliance with Rule 41. The wrap-around design also permits the paperboard to band and unitize bottles.

More like a knocked-down paperboard carton is the paperboard sleeve used for full-enclosure multiple packaging by several American brewers to protect beer in clear glass bottles against light. Over-the-top paperboard sleeves have a gable-top shape to conform to the tapered shapes of the bottle tops. Further, die-cuts in the base are projected upward to act as glass-to-glass separations and thus comply with Rule 41 requirements. The paperboard sleeve is printed and die-cut as a flat blank and then glued into a sleeve form in the converter's plant. Knocked-down sleeves are delivered in stacks to the bottler.

The multiple packaging machine removes the sleeve from the magazine and snaps it open. Collated bottles conveyed in parallel with the moving opened sleeve are pushed into the sleeve while fingers from beneath push the separation tabs up from the sleeve base. Glue is applied to the end flaps, which are mechanically folded down and adhered to form a six-sided carrier.

Glass-bottle 12-packing is increasing in the United States. Because glass bottles have nonstandard design, sloping shoulders, and long necks and because of regulatory limitations of glass-to-glass contact, 12-packing of glass containers in the United States lagged behind 12-packing of cans. Several systems have been developed and are now being commercially used in the American brewing industry. These systems are contoured at the top to fit snugly over the tapering necks of bottles and thus hold them firmly in place. Since the weight of glass

FIG. 7.8. This aluminum foil/paperboard folding box is designed to hold chips for picnics, outings, and home entertaining.
Courtesy of Reynolds Metals Co.

bottles is significantly greater than that of metal cans, the carrying handle is often reinforced. Paperboard partitions are inserted between bottles as they are being collated and before packing to meet regulations.

Container Corporation of America's GlassWrap system collates the 12 bottles into a 4 × 3 configuration, inserting the paperboard partition between the bottles to comply with regulatory provisions. A flat paperboard blank is moved into position with the collated bottles and wrapped fully around them. Hot-melt adhesive is employed to effect a longitudinal seal. Hot-melt adhesive is applied to the end flaps to close the carton on the ends. Speeds of up to 200 12-packs min may be achieved with the GlassWrap system.

In other systems, glass bottles, are collated into two groups of six at speeds of up to 1200 bottles/min. In continuous motion, the groupings are pushed into the two open ends of the preglued sleeves from both sides and the ends are folded over and closed with hot-melt adhesives.

Foam-plastic-clad glass bottles provide some protection against shattering and have large label areas. These bottles may be distributed with no added separation and comply with Rule 41. These bottles have been used for a variety of glass bottle applications, including prepackaged cocktails, toiletries, carbonated beverages, and fruit juices. Part of the success of this package has been the result of integration with a multiple package which fits over the closures of six foam-plastic-clad bottles to form a single sales unit. In addition, 0.5-liter plastic bottles, have been introduced for carbonated beverages; these bottles require no separations and can be multipacked via the same system used for foam-plastic-clad bottles.

The multiple packaging material is thermoformed from extruded polyethylene sheet. The sheet is die-cut with openings to retain the bottles at their necks. Although shrouding the necks and shoulders of bottles, the packs are not printed. Further, bottles connected only by their necks do not stack easily in stack display in retail supermarkets or in pallet loads.

Hartness and Dacam and Owens-Illinois, one of the bottle makers, manufacture equipment to apply denested carriers—i.e., the multipackers—to bottles held in corrugated fiberboard trays beneath. Carriers are removed from the nest and aligned above the bottles. As the cased bottles are conveyed past, the multipacker is settled over the top. Pressure bars complete the process of locking the necks of the bottles to the plastic units.

Injection-molded plastic units which fit over the necks of either plastic or plastic-clad bottles are clipped either manually or by automatic equipment. Container Corporation of America's Versaclip carrier is an injection-molded plastic structure with a printed paperboard field that combines multipacking function with graphic decoration. This clip may be applied manually or by Hartness equipment in the corrugated fiberboard shipping tray.

In geographic regions in which Rule 41 does not apply (for example, Western Europe), paperboard clips such as Waddington's Neckline have been used to unit two or more narrow-neck bottles. With the introduction of PlastiShield plastic-clad glass and all plastic bottles in the United States, paperboard constructions have been used for multiple packaging. Preglued structures are snapped open above bottles held in corrugated fiberboard trays and pushed down over the bottle tops. With plastic bottles, the plastic retaining rings are used to hold the clips. With plastic-clad glass bottles, the paperboard structure is often doubled to minimize possible movement of the bottle necks through the die-cut openings.

A more recent development has been a paperboard clip with a skirt that provides a much larger decoration area than does the earlier over-the-neck version. All clips are placed on bottles on equipment especially made and provided by the systems supplier.

A further example of current interest in plastic multiple packaging is manifested in the implementation of shrink-film secondary packaging of cans, smaller-size polystyrene-foam-clad glass bottles, and plastic bottles. A thin polyolefin shrink film should represent an inexpensive means to unitize cans or bottles. Shrink-film secondary packaging systems are commercial today in the United States and to a limited extent in Western Europe and in South Africa for both cans and glass bottles.

A few West German Bauer machines are used in the United States to multipack composite paperboard cans of refrigerated biscuits. In Western Europe, however, Bauer machines coupled with bottle collators and tray erectors are widely used to shrink wrap groupings of beer bottles and, in food plants, to shrink wrap groups of cans and jars. Bauer is no longer an active company.

Kisters equipment, also from West Germany, has recently introduced equipment that can register printed shrink film in a double-line configuration that permits an output of up to 200 six-packs/min. High-speed Kisters shrink-film multiple packaging equipment is being commercially used to multipack bottles.

Tertiary packaging may consist of corrugated fiberboard cases derived from preglued knocked-down units, corrugated fiberboard wrap-around packaging, or corrugated fiberboard trays plus shrink film. There are three basic methods of tertiary packaging: horizontal, vertical, and continuous-motion.

Horizontal packing is used for cartons or cans. The equipment forms the product into the required case-load pattern and with a horizontal pushing motion transfers the load into the corrugated fiberboard shipping case. Speeds of up to 30 cases/min are limited by intermittent motion. Vertical packing is used for packing glass bottles or cans. The product is either lowered or dropped vertically into the case. Vertical packing is performed on either intermittent-motion or continuous-motion machines. Intermittent-motion case packing attains speeds of up to 60 cases/min. Continuous-motion machines operate at up to 75 units/min. Continuous-motion packing is used for high-speed bottle and can lines that require no changeover. Continuous-motion machines reach speeds of up to 75 cases/min.

Horizontal packers are commonly used for packing paperboard cartons and cans. Most are semiautomatic, requiring an operator to place the empty case.

Many horizontal packers are available, but only two basic principles are used. In the single-level packer, the package pattern is a single tier. The product is accumlated as it is fed into the packer and pushed at a right angle into the case.

In a multilevel unit, product is fed into the packer. When a desired fill is achieved, the contents are lifted vertically up through a set of gates. The gates open and close as the product is pushed through. The elevator then lowers, leaving the product supported by the gates.

This operation is repeated to achieve the desired number of tiers. The horizontal pusher advances the load through a funnel. The sequence repeats until the desired number of tiers is attained. The complete load is then pushed into the case.

Semiautomatic packers require an operator to erect the knocked-down corrugated fiberboard case and place it onto the horizontal form-

FIG. 7.9. This form fill machine is used to fabricate filled cartons from blanks on an automatic cartoning line.
Courtesy of Reynolds Metals Co.

ing horn. Once the corrugated fiberboard sleeve is on the horn, the machine holds it in position, packs it, and lowers the filled case to a take-away conveyor. Its speed is about 10 to 12 units/min.

In a typical automatic case-packing operation, the basic packer is integrated with an opener and positioner and the discharge conveyor is linked to a sealer. With this system, the knocked-down corrugated fiberboard case is erected automatically, placed on the horn, and, when filled, discharged to a sealer at speeds of up to 25 units/minute.

In another automatic packer, an integral machine performs all the functions of case set-up, positioning, and gluing at speeds of up to 25 units/min.

Contents normally packed on vertical machines are rigid cylindrical or nonround containers. Primary packages can be handled either loose or in secondary packages.

In intermittent-motion machines, primary packages are distributed into a number of lanes with a reciprocating or vibrating divider to establish one dimension of the case load pattern. The laned product is transferred to a grid assembly that releases the product down into the opened corrugated fiberboard case. The packer-conveyor terminates at the entrance to the grid assembly, where line pressure pushes the product across a narrow metal band to a positive stop. The other dimension of the load pattern is established by the length of primary packages supported by the strip. When the riding strips are filled and a corrugated fiberboard case has been conveyed into position for loading, the strips are shunted laterally and the primary packages drop through fingers that guide them into the open-top case.

The corrugated fiberboard case position is handled by the case feed conveyor, which restrains the incoming case supply, indexes cases for loading, opens all flaps to the horizontal, elevates the empty case, lowers the filled case, and discharges it to the out-feed conveyor.

Apparent output limitations of the intermittent-motion vertical packer are overcome by simultaneously handling multiple cases during each cycle to attain speeds of up to 60 cases/min.

Continuous-motion packers for vertical loading have also been developed. Both primary packages and corrugated fiberboard cases remain in motion throughout the cycle. Primary packages are conveyed on a conveyor above the corrugated fiberboard case feed conveyor. A gripper assembly moving with the primary packages lowers and grips a caseload of product. A case on the lower level is transported at the same speed as the gripper assembly and directly below it. The product conveyor ends and the gripper assembly gently lowers the load of primary packages into the case. The assembly then rises and returns to repeat the cycle. Speeds of up to 75 cases/min have been attained.

Machines are available to set up corrugated fiberboard cases automatically. A magazine holds the knocked-down sleeves, which are pulled out of the magazine and erected either mechanically or with vacuum. Once opened, the case flaps at one end are sealed with hot-melt adhesive and closed at speeds of up to 60 cases/min.

Glue, tape, and staples can be used for final sealing, with the most common being water-based cold glue. Hot-melt adhesives provide instant bonding. Speeds of gluers are in excess of 50 cases/min.

Tertiary packaging for carbonated beverage and beer cans was formerly confined to corrugated fiberboard cases and trays. Since the mid-1970s in the United States, cases have been increasingly replaced by corrugated fiberboard trays with full and partial shrink-film overwraps to reduce tertiary packaging material costs. Although these systems have been largely confined to use with carbonated beverages and beer, they may penetrate other tertiary packaging applications in the 1980s.

Most shrink-wrapping equipment, regardless of purpose, consists of reciprocating-motion shutter machines. Shutter-type shrink-wrapping machines use two rolls of shrinkable polyolefin film, one above and one below, which bond together at a heat-weld line on each cycle. The corrugated fiberboard case or tray filled with product travels through a web of the film formed from the rolls past a reciprocating heat-seal bar. The web of film forms a floor beneath the case or tray while at the same time draping loosely over the top. After the tertiary package has passed, a bar containing the heat sealer descends to form a film-to-film heat seal behind the case or tray and simultaneously cuts the film. The bar then retracts up like a window shade. In this manner, the continuous web of film is maintained while the filled case or tray is passing through the shutter bar. Reciprocating mechanisms generally limit the speed of the equipment to about 25 cycles/min for case-sized configurations, although by incorporating horizontal reciprocating motion to the bars or cases, speeds of up to 40 units/min can be achieved. After leaving the wrapping area, the trays pass through a forced-air heat tunnel which balloons the loosely wrapped film sleeve, heats it, and causes the film to shrink into a tight bundle. Conventional shutter-mechanism shrink-film wrappers have been imitated many times in Western Europe.

Kisters shrink-film-wrapping equipment from West Germany, marketed in the United States by Simplimatic, is based on a principle of cutting a sheet of film from a web beneath the conveyor, transporting it to the moving, filled cased tray, and wrapping it around this tray with a continuously moving cantilevered arm. Being continuous in motion, speeds of up to 70 cycles/min have been attained in commercial use, with speeds of up to 100 cycles/min in surges. The system overlaps the film edges beneath the tray, which bond in a forced hot air shrink tunnel rather than a heat sealer. System costs of shrink-film tertiary packaging can be significantly lower than those of all corrugated fiberboard case tertiary packaging systems.

In the United States, about 40 companies produce shrink packaging machines, the overwhelming majority of which are of the low-speed shutter type. Shrink-wrapping equipment introduced by such firms as Weldotron, Great Lakes, and Battle Creek in the early 1970s is still serviceable but has been supplanted in higher volume applications by newer designs with higher speeds.

Using shrink film plus corrugated fiberboard tray in place of corrugated fiberboard cases could reduce the cost of tertiary packaging by reducing the amount of packaging material required. A shrink-film–wrapped tray is easier to open and allows easier identification of contents. A roll of shrink film occupies significantly less volume in a food plant than knocked-down corrugated fiberboard case blanks.

The tertiary package serves almost exclusively in a bundling or unitizing capacity, and so a packaging material combination that performs this role at lower cost than corrugated fiberboard alone could constitute the basis for a better system. Low-density polyethylene shrink film has sufficient tensile energy to unitize a grouping of primary or secondary packages, such as cans, cartons, or pouches indefinitely and tightly.

The Uniform Freight Classification Board of Railroads permits a variety of shrink-film tertiary packages. The National Classification Board for Motor Vehicles has issued regulations permitting selected shrink-wrapped trays of primary packages.

Converting from corrugated fiberboard cases to corrugated fiberboard trays plus shrink-film overwrap can be followed by elimination of the corrugated fiberboard tray altogether, replacing it with a flat corrugated fiberboard or even paperboard pad plus the film overwrap. Advantages of the flat pad plus film include even higher visibility, much less disposable waste, and faster shelf stocking.

The speed of packing on commercial equipment, including tray and film, can be as little as two-thirds the speed of packing with all-corrugated-fiberboard case construction. Primary package damage could be lower than with fully enclosed corrugated fiberboard cases because the primary packages are more tightly bound and move as a unit rather than as individual cans. Packaging material scrap at the retail level should be significantly reduced.

Two basic corrugated-fiberboard/shrink-film tertiary packaging concepts are available: one involves a total shrink-film overwrap and the other a partial-film overcap on a tray. In the latter case, the film covers the tops and walls of the primary or secondary packages, adheres to the sides of the corrugated fiberboard tray, and is shrunk.

Continuous-motion shrink-packaging equipment, typified by that made by Anderson, employs heat-seal bars on a moving overhead chain. Corrugated fiberboard trays filled with product are conveyed through the two-film web-wrapping machine. A heat-seal bar is lowered behind the tray, carrying a top web of shrink film with it. A heat seal of the top to a bottom web is made behind the case by an impulse sealer on the bar. The loosely draped tray is conveyed to a forced-hot-air shrink tunnel. Multiple-impulse heat-seal bars on the chain permit operation at up to 75 cases/min.

The VisiCase 920 machine was developed for high-speed, continuous operation in beverage-canning plants. In a single, fully integrated machine, the VisiCase System collates primary packages, forms a corrugated fiberboard tray, mates the tray with the primary or secondary packages, and wraps them in shrink film. Operating speed is in excess of 70 packs/min. The principle of wrapping is to use multiple-impulse heat-seal bars on a continuous overhead chain to seal two webs of shrink film and drape them over the top of and beneath the trays on a synchronously moving conveyor belt. The wrapped trays exit through a forced-air heat tunnel.

The Wrapcap and PacCap systems developed by Huntington Industries (no longer existent) make use of partial rather than full shrink-film wrapping. Rather than fully enclosing the tray, the Wrapcap and PacCap systems affix the web of shrink film to the walls of the corrugated fiberboard tray and thus form an overcap on the primary packages in the tray. By using only a fraction of the shrink film employed in a full overwrap, significant savings in shrink-film use can be effected. The shrink-film overcap is easy to remove by the retailer. The bottom of the tray is corrugated fiberboard, thus eliminating conveying problems that might arise in automated grocery-distribution warehouses. Shrink film is affixed to the walls of corrugated fiberboard trays either by applying hot-melt adhesive film to the corrugated fiberboard prior to applying monolayer film or by welding coextruded film to the walls using heat-seal bars. In either case, continuous-motion overhead constant heat-seal bars are used, thus permitting high-speed operation. Wrapcap and PacCap equipment uses infrared radiant heat tunnels to shrink the film gently over the tops of the primary packages. Because less film is used, no need exists for a forced-air ballooning of film as with fully wrapped corrugated fiberboard trays.

In the Wrapcap system, prepacked trays are overcapped with heat-shrinkable film sealed to two or four sides and to all four corners of the tray at up to 75 packs/min.

PacCap is an integrated system which collates and groups primary packages, forms a corrugated fiberboard tray around the grouping, applies shrink film over the product, and adheres it to the sidewall. The package then passes through the heat tunnel, where the film is shrunk tightly around the product.

OTHER PRIMARY PACKAGES

Metal canning equipment has been adapted to spiral-wound composite paperboard cans, and glass equipment to plastic bottles, but direct replacement has not always been possible because of the inherent differences in the packages, such as in compressibility and impact strength.

Blow-molded and injection-blow-molded plastic bottles can be received scrambled and can be aligned without danger of weakening by scratching, shipping, etc. The coefficient of friction between two plastics is such that binding can occur. To attain speed, the bottles must be held or else they will literally fly. Excessive compressive pressures during filling or capping can cause collapse, and so retaining collars are designed into the bottles. High temperatures and large pressure differentials in vacuum filling can cause bottle collapse. Thus, while the basic principles of glass bottle handling can be applied, significant modifications are engineered in to accommodate packages fabricated from the plastics. Experience with conversion from glass to rigid plastic demonstrates that, despite careful testing of product and package compatability, enormous problems in packaging-machine line operation can be encountered in direct conversions.

PAPERBOARD CARTONING

The original paperboard carton, still used today, was the set-up box, which consists of a relatively inexpensive grade of chipboard to which printed paper may be adhered for decoration and for some structural support. Corners may be glued or taped together. Because the paperboard used is usually recycled, there is a relatively high degree of fiber orientation. Thus set-up boxes can be distorted by physical damage, moisture absorption, etc. Lithographed paper stock adhered to the chipboard can become wrinkled and scuffed during the fabrication process.

Most firms using set-up boxes are small and have only short and intermittent needs. Consequently, the paperboard set-up box is formed at one location and stored just prior to use. The set-up box requires a large storage space.

The set-up box is still used for box chocolates, jewelry, and specialty products. It is a relatively expensive carton to manufacture (consisting of two separate structures) and use (it does not nest or stack well and so usually must be placed manually on packaging lines.) The set-up box thus offers relatively little advantage over the folding carton. Set-up boxes may be made in-plant or even in line with packaging with a machine such as the FMC Corp. unit that makes taped-corner cartons from flat die-cut blanks plus roll stock tape cut off by the machine to proper length.

Fortunately for both the products and the consumers, folding paperboard cartons have been available and widely used for a century. Most folding paperboard cartons fall into one of four categories: preglued flat blank, tubular, or tray.

Paperboard cartons that are preglued or preadhered by heat-sealing, convertor-applied coatings are generally delivered in knocked-down form to the packager for erection immediately prior to filling and closing. Flat blanks are die cut by the converter and delivered flat to the packager for fabrication before or as part of use. Flat blanks are often found around the product they are to contain.

Tray-style cartons are characterized by single-piece paperboard bases with side and end members hinged to the central panels. Each side and end panel is connected to the adjacent wall member by a glue flap, a mechanical locking tab, a friction lock, or a similar device.

Tray-type cartons are frequently called open-top cartons because the opening is on the largest surface, which could normally be the top. The cover is the main difference between tray-type cartons and open paperboard trays.

Tube styles are made by folding the paperboard into a square or rectangular tube and adhering the overlapping flap. End panels are hinged to the tube in a variety of closing constructions, including tuck, sealed, and lock ends. Two basic designs for tube-type cartons are standard and deep draw. A standard tubular carton has an open top and a bottom closed by folding flaps and tabs. A deep-draw paperboard carton is formed with a solid bottom and flaps and tabs on the top. Tubular or open-end cartons are formed from flat paperboard blanks by folding them to form a tube and gluing or heat sealing the long side seam. Both ends are closed by overlapping flaps and tabs. The tubular or open-end carton, so called because the opening is on one of the smaller surfaces, may be turned for filling so that the opening is facing upward or to the side as it passes the filling station on the packaging machine. The bottom end of the carton may be closed before, during, or after the product is filled.

Some wrap-around cartoners use the product as a mandrel around which the carton is formed; other packaging equipment shapes the cartons in dies. In some applications, the wrap-around carton is very similar to a wrapper, except that it is made of paperboard rather than paper or plastic film.

Paperboard cartons are made in a variety of special sizes and shapes to meet the individual requirements of the products to be contained.

Cartons are closed by one of three techniques (or combinations thereof): overlapping and gluing, tucking into the opening, and locking. Glue is usually employed to adhere the tabs or panels together to form the carton. Adhesive may also be used to seal the top and/or bottom after the product is filled. Adhesive may be used to hold the corners of the tray-type carton together and to seal the side seam and both ends of tubular-type cartons. Either hot-melt adhesives or cold glues may be used. Glued-bottom cartons that can be erected automatically or with ease by manual means may be formed by the paperboard convertor, which die cuts and prints the blanks and adheres the long side seams. Adhesive is applied to the flaps near the center of the closure; the carton is folded so that it folds inside and snaps into place as the carton is opened. This technique provides a relatively rigid carton and eliminates the need for in-line gluing.

In tuck-flap closing, the open ends of the carton may be closed by folding a panel over the opening and tucking the attached flap into the opening. The bottom may be adhered with glue and the top tucked, or both openings may be closed by tucked flaps. Carton corners may be fastened by end locks in which part of one flap is passed through a slit or die-cut opening in the other flap.

Some products require more protection than can be obtained from either a paperboard carton or flexible packaging alone and thus may be packaged in a lined carton or double package. The addition of a lining combines the benefits of a flexible package with the rigidity, protection, ease of handling, and display characteristics of the paperboard carton.

The most popular liners for cartons include inner-bag liners, double packages, pouch-in-a-carton liners, and laminations. The inner-bag liner is formed outside the carton and inserted into an already erected carton. The flexible material is shaped around a mandrel to form an inner bag with sealed bottom and sides but with an open top. The mandrel and liner are pushed into an open carton, the block is withdrawn, and the liner is left in the carton. After the liner has been filled with the product, the top of the liner is sealed and the carton is closed. Inner liners may be made from glassine, laminations, or coextrusions.

A double package is similar to a paperboard carton with an inner-bag liner except that it is produced in a different manner. Liners are cut from a roll and wrapped around a mandrel and the seams are sealed. A printed carton blank is scored and formed around the liner while it is still on the mandrel. Glue spots are often added to fasten the liner to the carton near the opening. Most double-package makers are products of Pneumatic Scale Corp. Products such as cereals, snacks, cookies, crackers, and cake mixes are often packaged in pouches on vertical form, fill, seal machines, and the filled packages are inserted into cartons on equipment such as that made by R. A. Jones.

The barrier properties of some cartons are enhanced by laminating a moisture- or gas-barrier material to the inside surfaces of the carton.

Working from these very basic types, the numbers of both patented and public-domain paperboard carton constructions have mushroomed.

Normal folding carton styles include such variations as the tuck end, the reverse or straight-tuck glue end, economy flaps or full flaps, the tuck-and-glue end, trays, sleeves, and shadow box cartons.

Cartoning-machine types include semiautomatic machines in which the operator fills the carton manually. The machine feeds the carton from the magazine, forms it, and closes its ends after manual filling. In fully automatic equipment, the product is automatically filled. In vertical-loading machines, the product is loaded from the top as the opened carton moves through the machine in a vertical attitude. With horizontal loading, the carton moves through the machine in horizontal attitude, and product is loaded through one or both ends.

Semiautomatic cartoners are used where many sizes and changeovers are required. Speeds are 30 and 150 cartons/min. Product is usually filled manually.

Fully automatic cartoners have higher speeds but less size range than semiautomatic cartoners.

A horizontal machine can work directly with filling machines or it can be equipped with automatic fillers. Fully automatic horizontal machines have a speed of up to 600 cartons/min for certain products. Typical horizontal cartoners are made by R. A. Jones, CECO, Adco, and Hayes.

Fully automatic machines can be equipped with one of several types of automatic fillers. Bottles or jars, for example, are usually brought into the cartoner upright. Bottles move into buckets on the cartoner conveyor by means of a continuous-motion transfer wheel. Collapsible tubes are transferred automatically and directly into cartoner buckets. To increase speed, multiple fillers are placed in a line along the conveyor.

Free-flowing products such as candy or detergent can be filled on vertical equipment by means of a volumetric cup system synchronized with the carton as it moves under the filling nozzles.

Flat paperboard blanks are printed, die cut, scored, glued, and stacked for use in feed magazines on packager equipment. The flat blank is formed into a package ready for filling at the packager's plant, either by locking or by adhering die-cut flaps together, leaving an opening for filling. Equipment employed for this purpose includes various models of Kliklok machines, which lock four corners in place to form a tray or hinge-top tray, depending on the original blank; the Sprinter from Sweden, which performs much the same operation; and U.S. Automatic Box Co. equipment, which usually allows for gluing the flaps. Kliklok also manufactures a Head Seal machine for glue closure of the flaps.

Corner locks are generally intricate and difficult to form manually with either precision or speed. Slightly less paperboard is required for glued corners, but the differences are relatively small and should not warrant serious consideration in most applications.

Although many paperboard cartons are formed from uncoated board, many are made from one- or two-sided polyethylene extrusion-coated or even wax/plastic-coated boards for use in grease-resistant containers for baked goods, confectionery snacks, and ice cream; for moisture resistance with frozen foods; and for decorative value and scuff resistance with cosmetics and toiletries. With moisture-sensitive foods, hot-melt adhesive and even web corners with glue are required.

Paperboard coatings can interfere with the use of inexpensive water-based adhesives. Adhesive for carton set-up and sealing requires good tack, because the time in the carton set-up unit is short. Most set-up units are single-station units that rely on glue tack to hold the flaps in place until a set is effected. Polyethylene-coated or wax/ethylene vinyl acetate copolymer-coated paperboard for frozen foods or ice cream requires the use of hot melts if the carton is adhesive sealed. Polyethylene extrusion-coated milk cartons represent a special case, with fusion heat sealing of polyethylene coating to polyethylene coating to form and close the carton. Some wax and polyethylene extrusion-coated cartons for frozen foods are sealed by melting the wax coating and bonding. Conventional methods of applying heat to the exterior of the board are not satisfactory, because paperboard is a heat insulator and could be burned in the effort to melt the interior wax. Further, the exterior wax would be melted off. Hot-air blasting on only the surface to be melted allows coating-to-coating seals without resorting to addition of hot melts.

Frequently, frozen food cartons are made from polyethylene-coated solid bleached sulfate paperboard that is printed, cut, scored, and adhered after extrusion coating.

Although about 40% of paperboard packaging is fabricated from recycled board—expecially where the primary package is being contained, as in toothpaste tubes, inner cereal and snack liners, and dehydrated soup pouches—most foods and related products in direct contact with the paperboard are now in solid bleached virgin Fourdrinier paperboard. This grade is so commonly employed that it is often called food board.

Because of the irregular shapes and sizes of many food products contained in paperboard cartons, filling is often manual. For example, prefrozen fish sticks and chicken pieces are usually hand packed. Closure may then also be manual, since an operator is already performing a task with the carton. For more efficient use of filling labor, however, closing may be automatic.

A large quantity of packages is packed semiautomatically or automatically into end-loading cartons. For example, pies and dinner platters are automatically pushed into paperboard sleeves which have been snapped into position at one location and conveyed to a continuous-motion double belt which pushes the product into the open end and then folds the ends shut, as on the R. A. Jones machine.

The Pneumatic Scale Corp. double package maker is probably the carton machine most frequently employed for free-flowing products, such as ready-to-eat cereals, cookies, crackers, and cake mixes. The package is essentially a flexible film liner used to impart both grease and moisture resistance in paperboard carton. The film is cut from roll stock onto a mandrel, where the long seam and a bottom seal are made. A preglued paperboard sleeve is snapped open and pushed over the film liner. The combination is ejected to a conveyor, which transports the lined carton to the filling station. Alternatively, the sleeve may be formed from a flat paperboard blank, thus saving one converting cost.

Free-flowing products are volumetrically packaged if they are relatively inexpensive or gravimetrically packaged if they are relatively expensive. As has been indicated, the type of filler can be dictated by the developer on the basis of specific need. Such products as cookies, snacks, and dry cereals, all of which tend to settle after packaging, are packed by weight. Compact, free-flowing foods are usually filled volumetrically. A hopper fills a cup of predetermined volume which is then doctored clean at the top and moved into position over the package, at which point the contents are released into the open package.

Through the use of rotary turret fillers, relatively high speeds (up to 300 units/min) are possible.

Following filling, the inner liner, if present, is folded over or heat-sealed shut, depending on the package requirements. The external paperboard carton is closed at the next station and either locked in place or glued, using wet glue if the paperboard is uncoated or hot-melt adhesive with coated paperboard. Hot-melt adhesive may be preapplied to the paperboard package, as has been done with ice cream packages. Preapplied hot-melt adhesive requires application of hot air or another heat source directly to the hot-melt adhesive to activate the adhesive and allow the closure to be made. Hot-melt adhesive is active for only a brief period (a second or less), and so the tack must be made immediately following activation of the hot-melt adhesive.

Rather than using the relatively elaborate procedures demanded to prevent burning, ink smearing, and coating deterioration, most packagers apply hot-melt adhesive to paperboard carton flap interiors with Nordsen adhesive-applying equipment in conjunction with the closing operation.

Secondary Cartons

Paperboard cartons are employed as secondary packages, especially with cosmetics and toiletries. Cartons add decoration and protection. Several flexible packaging machines include paperboard cartoning elements as a synchronized part of the machine. For example, vertical form, fill, and seal machines are being used in conjunction with continuous-motion horizontal cartoners to package snack crackers. A pouch is made and conveyed to a moving line running parallel to another line containing an opened paperboard sleeve. The two lines converge, and the pouch is transferred into the carton by pushing.

Bartelt horizontal form, fill, and seal flexible-packaging equipment can include cartoning devices that operate in much the same manner.

Secondary Packaging

Beer and carbonated beverage multipacking, described in detail earlier, is the principal example of paperboard multiple packaging. However, paperboard cartons are employed for secondary or multiple packaging of many other products to unitize and to assist in distribution to retailers. Examples include candies, proprietary pharmaceuticals, and health and beauty aids. Candy bars in six packs are often in paperboard, with wrap-around cartoners among the more novel commercially employed machines. Most cartoners, however, are top-manual or side-automatic loading.

A considerable amount of unitizing in secondary paperboard cartons is performed manually. This method helps to provide an inspection station and, more important, reflects some of the difficulties of accumulating and aligning primary packages by mechanical means. Although equipment that produces the desired action has been engineered and installed, the reliability of manual methods sometimes has proven superior.

EQUIPMENT FOR FLEXIBLE PACKAGING

Bags

Bags are usually prefabricated at a converter's plant geographically removed from the packaging line, filled, and closed. Probably the largest-scale use of preformed bags is for bread packaging, with low-density polyethylene film the principal packaging material used. Printed bread bags are usually received stacked on wickets or posts. Most of the equipment used by the bread-baking industry is made by AMF, which conveys the finished loaves to a station where a prefabricated bag is blown open and brought forward to meet the bread being pushed into the open end. After insertion, the ponytail end of the

FIG. 7.10. AMF Mark 70 Bagger.
Courtesy of AMF Corp.

loose bag is twisted shut and a plastic-coated tie is clamped on the narrow portion of the twist. Although the reciprocating-action bagger is slower than an overwrap machine and the package costs more than film from web stock, consumer acceptance of the reclosure feature has forced adoption of the less efficient equipment.

Bread packaging is the largest single use, but preformed low-density polyethylene bags are also widely used when production quantities do not dictate use of automatic equipment, when runs are short, and when relatively low-cost packaging is desired. Preformed, low-density polyethylene bags are highly adaptable to slow-speed operations, which have many changeovers from one product or package to another. They are widely used for hardware parts, soft goods, multipacking of candy, special deal packs, etc.

Because polyethylene is a limp material with electrostatic attraction to itself, some simple equipment is usually required to package in preformed polyethylene bags. As with bread, bags may be received stacked on wickets. Each is individually opened by an air blast, filled, torn from the wicket, and then closed. A large number of relatively inexpensive machines for this application are commercially available. Typical are those manufactured by Doboy and Tele-Sonic.

Soft goods are often packaged in a polyolefin film in pillow-form-centerfold or J-fold roll-stock film. By using an L-sealer such as that made by Weldotron Corp., a two-side open package can be formed. The product can be slipped into the opening and the L-sealer brought down again to heat seal and cut off a three- or four-side seal package.

Pouches

Pouches are usually, but not always, formed simultaneously with packaging the product. Some preformed pouches are commercially used today by packagers. Most pouches, however, are formed from flat roll stock on two types of equipment: vertical or horizontal. Vertical machinery is most applicable to free-flowing dry products or fluids that can be pumped. Gravity feed is almost always the method of entry into the vertical package after other mechanical means, such as an auger or pump, have brought the product to the pouch opening. The only exception is liquid aseptic packaging, in which the fluid is present above the level of the closure.

Vertical Form, Fill, and Seal Pouches. Vertical form, fill, and seal equipment is usually intermittent-motion, requiring completion of one cycle before another cycle or package can be formed. A number of units have been designed with continuous motion using multiple sealing jaws on a continuously moving chain. Several types of vertical machines are in commercial use, including those that form a three-side or pillow-pouch package and those that form a four-side face-to-face seal pouch. Although dry, flowing products may be packaged on either type of machine, fluids and moisture-sensitive products are generally packaged with the four-side seal because face-to-face fusion seals are more positive and have less tendency to leak or otherwise fail. Pillow-type construction, on the other hand, affords greater capacity per unit area of packaging material.

In addition to the variations possible as a result of basic design differences among machines, and as a result of differences in collars, tubes, and sealing jaws, fillers should be selected to fit the product-package requirements. In this way, the versatility of basic vertical form, fill, and seal machines can thus be enhanced.

Speeds on reciprocating machines are limited by the ability of the jaws to reciprocate mechanically and also by the speed at which products fall by gravity into the pouch.

Four-side seal pouches may be formed from two webs of material, as on Circle or Prodo Pak machines used for liquids or fluids.

The pillow pouch is generally constructed with two end seals and a back or side seal running the length of the pouch. The end seals are face to face, but the long seam may be overlap or face to face (usually called "fin").

In almost all commercial equipment, vertical form, fill, and seal three-side seal pouches are formed by drawing flat, flexible packaging material from a roll. Pouch paper, polypropylene laminations, polyethylene, and lightweight flexible aluminum foil laminations are common vertical form, fill, and seal packaging materials.

The flexible material is drawn over and around a collar, which is usually above a cylindrical tube. The long seam is formed by heat sealing, usually against the tube. A horizontal draw bar then pulls the material the length of a pouch, simultaneously forming the top end-seal of the succeeding pouch. Cut-off may be made at this point by the action of a knife in the heat-seal bars or may be made by a separate cutter beneath the heat-sealing site. The draw bar opens and returns to the top of the pouch, thus undergoing a reciprocating action. After the bottom end seal has been formed and while the draw bar is moving up, product drops from the filler into the newly formed open-top pouch.

A large number of vertical form, fill, and seal machines are on the commercial market. Among the more widely known are those manufactured by Package Machinery Co. (Transwrap, a name that has become somewhat generic for the machine type), FMC Corp., Hamac Hansella (West Germany), Triangle, Woodman, General, and Hayssen.

Numerous variations differentiate the equipment and make the package developer's task more interesting, since these variations are not always widely known. For example, Hayssen machinery can employ power-driven rolls to assist the draw bar in pulling flexible materials from the roll and to assist film movement. Rovema machines also employ rubber wheels against the tube to drive the formed pouch past a fixed sealing and cut-off bar. Rovema machines also allow for gusseting the sides to increase the cubic capacity of the pouch. The MiraPak machine, although no longer available new, incorporates a filling cup that moves down to meet the pouch, thus reducing the distance of the gravity free-fall. As a result, momentum is reduced and the fall is more gentle, so that fragile products such as potato chips are more effectively handled. Several models of the Woodman machine are slanted rather than vertical to allow the product to slide rather than fall. Since it also allows for more gentle falls, Woodman equipment has also found considerable use for fragile products such as potato chips and pasta. General equipment has auger fillers for coffee, etc.

Tethrahedral Packages. Tetrahedral packages can be formed by turning two horizontal end-seal bars at a 90° angle to each other in parallel horizontal planes. This configuration allows for reduction of the amount of flexible material needed to contain a given weight of contents, particularly for small packages. TetraPak manufactures vertical tetrahedral packaging machinery which is being supplanted by other equipment producing packages of more conventional shape. By

FIG. 7.11. Twin-tube vertical form, fill, seal machine. Weigh-scale fillers are above each of the two tubes on this Triangle Pulsamatic Machine, which makes pouches from roll-stock flexible materials. Speed is 80 pouches/min. Output is transferred to a horizontal cartoning machine to create a bag-in-box.
Courtesy of Triangle Package Machinery Co.

nature, tetrahedral packages have noninterlocking shapes and so occupy large volumes and are difficult to align and stack. Although tetrahedral packaging might save in terms of flexible materials, the saving could be offset by cubic loss or increase in secondary packaging materials. Tetrahedral packaging has found some applications for liquid coffee whitener packaging, for which large numbers of packages or large quantities of material are required. Tetrahedral packaging is also used for a number of limited specialty-packaging applications where the units can be aligned in a pattern, but these are mostly outside the United States.

Horizontal Pouch Makers. Horizontal flexible-packaging equipment forms the package while the flexible material travels horizontally. Product may be filled from either a horizontal or vertical direction. Because face-to-face fusion seals are made on all four sides, both liquid and moisture-sensitive products (such as dry soups and beverage

FIG. 7.12. Gravimetric form, fill, seal machine; vertical form, fill, seal machine with multiple-weigh-scale filler for pouch packaging of snacks, crackers, cookies, and cereals. This is a single-tube machine. *Courtesy of Package Machinery Co.*

mixes) can be enclosed. Free-flowing products are generally filled from the vertical axis into a vertically positioned pouch that has had three sides formed, with the product flowing through the open top. Horizontal form, fill, seal systems are employed for packaging retortable foods, which are vacuumized after filling and retorted after sealing. More often, however, retort pouches are preformed and pulled from a stack magazine prior to opening, filling, vacuumizing, and sealing. Retorting is carried out under counterpressure to reduce the possibility of rupturing heat seals, which are weak at elevated temperatures.

By moving the sealing and filling mechanisms in parallel with the pouch, continuous-motion filling can be effected.

The best-known horizontal form, fill, and seal machines are Rexham (Bartelt) and the Canadian Delamere and Williams. Since the flexible materials are not stressed during forming, high-barrier materials such as aluminum foil laminations can be and are used to provide product protection. Horizontal form, fill, and seal intermittent-motion equipment can operate at speeds of up to 120 packages/min, and continuous-motion equipment can operate at speeds of up to 300 packages/min. Contents may be filled under inert gas or even vacuum because of the positive seals.

FIG. 7.13. In the bag-in-box systems, the SIG machine for packaging ready-to-eat breakfast cereals. Systems make a carton from a flat blank and a bag from roll-stock film. Product is added and both bag and carton are closed.
Courtesy of Raymond Automation Co.

Horizontal equipment may also form the flexible material into a pouch oriented in the horizontal plan. Product may enter from the side, between two sheets of flexible material, or from the top, with a second web coming down on the top to form a four-side seal around the package, as on a Rotowrap machine. Speed is limited only by the filler.

Yet another application of horizontal form, fill, and seal packaging is found on Hayssen RT equipment, in which a single web of flexible packaging material is folded and sealed into a three-side fin-seal pouch around the product in a continuous-motion action. By connecting the machine in series with a conveyor, relatively high-speed packaging can be achieved. With gas flush, this type of equipment has found considerable use for processed meats and cheese.

Units that form pouch-style wraps from roll stock in a horizontal plan include those made by FMC Corp., Doboy, Rose Forgrove in the United Kingdom, and SIG in Switzerland. The biscuit, cracker, baking, and candy industries use this type of equipment extensively both for wrapping of individual portions and for overwrapping. The equipment forms a tight, moisture-proof, and even gas-proof face-to-face seal wrap around products without using the product as a backing for sealing or cut-off. Thus relatively fragile products are generally not in danger of damage from the packaging operation. Further, less packaging material is employed for each unit of volume wrapped. In such equipment, the web is drawn from roll stock in a continuous motion

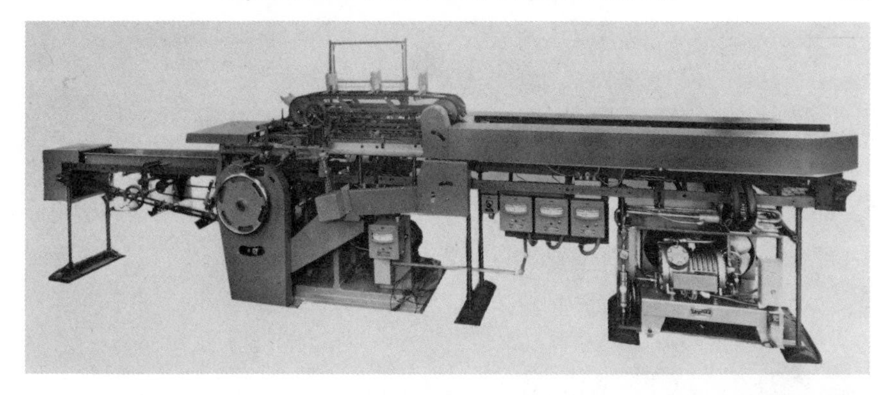

FIG. 7.14. Multiple package overwrapping machine collates six individual portion paperboard cartons of breakfast cereals and overwraps with printed polyethylene film. Feed is from left and exit is at right.
Courtesy of Package Machinery Co.

and folded around product being conveyed beneath the moving packaging material. The bottom can be overlapped (in which case the product is the heat-seal anvil) or fin sealed. Because of the continuous motion, speeds as high as 600 portion packages/min can be attained, with speeds decreasing as the size of the package increases. The end seals are formed by a rotating head anvil, which simultaneously seals two ends and cuts off the package.

Wrap King overwrapping machine and its analogues are used to portion-package individual sweet, soft-baked cakes. Since these products are in distribution channels for only a few days because of staling, high moisture protection is not required. Grease resistance is needed, however, and this is provided by unsupported aluminum foil. These units make a formed wrap or bunched fold, with the product being pushed up through the web, which is folded and tucked under. The product is soft and compressible and requires packaging to maintain its integrity and to keep it apart from adjacent pieces in multipack units. Working from roll stock, the equipment cuts off the appropriate length and wraps the product so that it is not damaged by machine action. Dead-fold foil is used to obviate the need for heat and pressure for sealing.

Pouch Packaging for Fluid Milk. Packaging of fluid milk in film pouches started in some limited areas of France and leaped to success in Canada. Pouch packaging of milk was successful because of a classical mating of packaging material to packaging machine. The equipment used for this purpose is similar in principle to conventional vertical form, fill, and seal machines, but the volume of milk filled (a quart) is higher than the volumes of products normally handled on

FIG. 7.15. Package Machinery Company's "Transwrap" is an excellent example of a vertical form, fill, seal machine.
Courtesy of Package Machinery Co.

FIG. 7.16. Thermoform seal machine is widely used in the cured meat industry for the packaging of frankfurters and luncheon meats.
Courtesy of Mahaffy and Harder Engineering Co.

conventional machinery. Liquids are also not normally packaged in this manner. Further, because dairies are involved everything must be sanitary and cleanable. Three-mil DuPont Sclair linear low-density polyethylene film, a material with an excellent heat-sealing range and tensile strength, is employed, and speed is relatively slow—about 25 strokes/min. The two major machines used in Canada are PrePac and Thimmonier, both originally French. The latter is marketed under the Twin Pak name.

THERMOFORMED PACKAGES

In recent years, several European companies and their American affiliates, as well as some American companies, have introduced thermoform, fill, and seal equipment. Thermoform, fill, and seal machines, built by the packagers of portion-pack jams, jellies, and condiments, have been used in the United States for many years but have not been available on the commercial market.

These machines heat a web stock consisting of a plastic such as impact polystyrene and then form the sheet with pressure into a female cavity mold. The cavities are moved under a filling head, where the contents are filled. Contents may be fluid (e.g., cream, yogurt, juice), viscous (e.g., jam), or solid (e.g., roasted peanuts). After filling, a film or laminated flexible membrane is heat sealed to the flat flanges formed between the successive cavities at another station. A die-cutting device separates the individual cavities.

Almost all thermoform, fill, and seal machines in the world are made in Western Europe—e.g., ERCA in France, being marketed by Continental Group as of this writing; Form Seal in France; Gasti in West Germany; H&K, made by Robert Bosch Group in West Germany; and Hassia in West Germany. Systems for pharmaceutical tablet or capsule individual packaging are made by firms such as Hassia and Package Machinery Corp. (United States).

In recent years, solid-phase pressure forming (SPPF), introduced by Shell and by Moore, has been used in the United States to permit use of polypropylene sheet to make preformed cups.

The ERCA, H&K, Thermoforming, and Benco machines are being marketed in the United States in what the marketers call aseptic packaging modes.

Several American and Western European firms market machines that thermoform a barrier web "in line" with an area for filling and either vacuum packing or gas flushing, and heat sealing a printed flexible web. In the United States, Mahaffy & Harder and now Hayssen produce and market such equipment for processed meat, cheese, pharmaceuticals, etc. From Western Europe, Multivac equipment is marketed by Koch Supplies, Tiromat by Kutter, and DixieVac by American Can, all principally for processed meats, but with other applications as well.

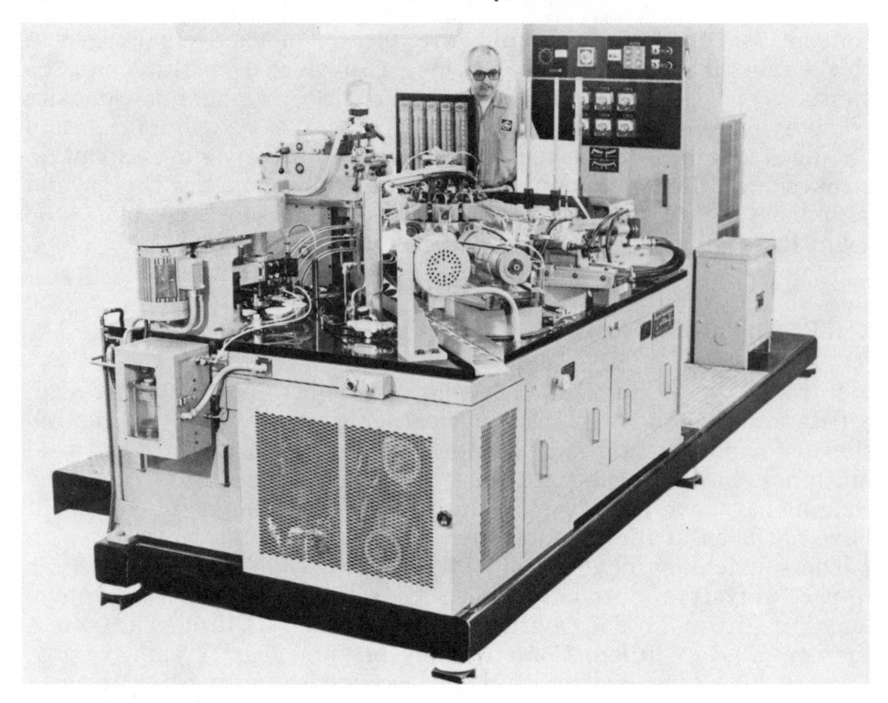

FIG. 7.17. Paperboard can body making machine.
Courtesy of Container Corp. of America.

FIG. 7.18. Paperboard can body making machine suitable for fabricating two-side polyethylene extrusion-coated paperboard.
Courtesy of Container Corp. of America.

FIG. 7.19. Single-wound paperboard cans in a variety of shapes and sizes are fabricated in line with filling and closing lines.
Courtesy of Container Corp of America.

These brief descriptions do not begin to exhaust the subject of package machinery available to packagers. Just a listing of this machinery could occupy this entire volume. A critical review useful to packaging developers would be more complete than the treatment given here and would, of course, include far more equipment. This brief glimpse, however, should give evidence that not all packaging machinery is universally applicable to all products. No one has yet attempted to provide a neat, well–cross-referenced, comprehensive listing of the stock equipment available, its limitations, and its application—a work that would be most valuable to all package developers.

BIBLIOGRAPHY

ANON. 1975. The Folding Carton, Paperboard Packaging Council, Washington, D.C.

ANON. 1969A. International Encyclopedia on Packaging Machines, 2nd Edition. B. Behr's Verlag, Hamburg, West Germany.

ANON. 1980. The 1980–81 PMMI Packaging Machinery Directory. Packaging Machinery Manufacturers Inst., Washington, D.C.

ANON. 1968. Guide to Labeling Equipment and Practices, Modern Packaging Encyclopedia, McGraw-Hill, New York.

ANON. 1970A. Automatic thermoforming machine for package production. Packaging *41* (480), 58–59.

ANON. 1970B. Two new developments in liquid packaging. Packaging *41* (489), 50–51.

ANON. 1970C. Pouch formed-filled-sealed-and cartoned-at up to 800 a minute. Packag. News *17* (6), 3–4.

ANON. 1970D. U.S. form-fill-seal unit takes both powder and liquids. Packag. News *17* (9), 6–7.

ANON. 1970E. Filling machine for bag-in-box or rigid containers. Packag. Rev. *90* (3), 47.

ANON. 1970F. High speed can labelling with printed tape. Packag. Rev. *90* (10), 70.

ANON. 1971A. New range of shrinkwrap machines Converter *8* (6), 44–45.

ANON. 1971C. Shrink-wrapping machines. Print. World *189* (2), 132.

ANON. 1972A. Wrapping machines—A review of the scene and some types available. Packag. Rev. *92* (5), 34–35, 38, 40, 42, 44, 50.

ANON. 1972B. Development trends in blow moulding machinery. Packaging *43* (508), 32–35.

ANON. 1975A. A technical review of folding box-gluers. Folding Carton Ind. *2* (4), 14, 16, 18, 20–22, 24–26.

ANON. 1975B. Focus on: Stitching machines. Int. Paperboard Inc. *18*, No. 9, 38, 40, 42, 44, 46–47, 49.

ANON. 1976A. Wrapping machine. Food Drug Packag. *34* (3), 20.

ANON. 1976B. Cartoning machine. Food Drug Packag. *34* (5), 24.

ANON. 1978A. Machinery breakthrough slashes cost of single-use package. Package Eng. *23* (3), 41–43.

ANON. 1978B. Hot meat application equipment: A survey of manufacturers. Adhes. Age *21* (8), 29–32.

ANON. 1978. Flexible Packaging and Associated Machinery Markets, Frost & Sullivan, New York.

ANON. 1979. Glossary of Packaging Terms, The Packaging Institute, U.S.A., New York.

ANON. 1979. Machinery '79: What's new in technology? What's the processing outlook? What's the sales growth picture? Mod. Plastics Int. *9* (4), 43–53.

ANON. 1980A. New developments in extrusion blow moulding machines and controls. Mod. Plastics Int. *10* (1), 25–27.

ANON. 1980B. Recent trends in packaging machines Packag. J. *1* (1), 73–74, 76.

ANON. 1980C. Machinery now and in the future. Food Drug Packag. *43* (6), 67–73.

ANON. 1980. Machinery now and in the future; Food & Drug Packag. 43, 6–7.

ANON. 1981. Marketing Guide to the Packaging Industries. Charles Kline & Assoc., Fairfield, New Jersey.

ANON. 1982. Packaging Digest Machinery/Materials Guide. Delta Publishing, Chicago, Illinois.

ANON. 1982A. Marking and coding equipment. Packaging *53* (622), 21, 23, 24, 27, 28, 30.

ANON. 1982B. New folding carton equipment. Paperboard Package *67* (4), 64, 69, 70.

ANON. 1982C. Machinery. Aerosol Age. *27* (7), 32–42.

ANON. 1984. Fibre Box Handbook, Fibre Box Association, Chicago, Illinois.

BABB, A. C. 1981. Coors packaging methods. Modern Brewery Age (33), 32–33.

BARKER, H. R. 1970. Current developments in automatic weighing and filling. Food Process. Ind. *39* (467), 29–31.

BRODY, A. L. 1970. Flexible Packaging of Foods, CRC Press, Cleveland, Ohio.

BRODY, A. L. 1971. Food canning in rigid and flexible packages Crit. Rev. Food Technol. *2* (2), 187–243.

BRODY, A. L. 1975. The last frontier for cost reduction: Physical distribution packaging, The Packaging Institute, U.S.A., New York.

BRODY, A. L. 1982. Double package maker. Cereal Foods World, January, p. 21.

BRODY, A. L. 1982. If I were a carton. Cereal Foods World, October, 1982.

BRODY, A. L. 1972. Flexible Packaging of Foods, CRC Press, Cleveland, Ohio.

BRODY, A. L. and MILGROM, J. 1974. Packaging in Perspective, Ad Hoc Committee on Packaging, New York.

BRODY, A. L. 1981. Shrink films for cases Food Eng. 53 (6), 62–64.

BRODY, A. L. 1981. Shrinking the costs of tertiary packaging, Can. Packag. 34 (4), 20–21.

BRODY, A. L. 1982. Packaging. In Storage of Cereal Grains and Their Products, American Assoc. of Cereal Chemists, St. Paul, Minnesota.

BRUINS, P. F. 1974. Packaging with Plastics, Gordon & Breach Science Publ. New York.

BUDD, S. M., and MOODY, B. E. 1972. Glass containers—technical advances. Packag. Rev. *92* (6), 21–22, 26.

CARON, P. E, 1967. Encyclopedie Internationale du Conditionnement de Liquides, Compagnie Francais d/Editions, Paris.
CAUNT, A. 1979. Machinery design for the 80's. Packag. Today *1* (1), 38–40.
CHASE, D. 1980. Packaging moves into the eighties, Milk Industry (U.K.) *82*, 30–31.
CLEMEN, J. D. 1967. Vacuum packaging machinery, Proc. Seminar Food Packaging with Flexible Laminates, Univ. California, Los Angles.
DAVIS, C. G. 1978. Filling dry products: A review of method. Package Dev. Syst. *8* (3), 30–33.
DAVIS, C. G. 1978. Filling liquid and semi-liquid products: A review of methods, Package Dev. 8 (1), 1–7.
EWEN, T. 1978. Filling and closing glass and plastic containers. Packaging *49* (577), 11–14.
FRIEDMAN, W. F. and KIPNEES, J. J. 1977. Distribution Packaging. Krieger, Huntington, New York.
GILLIES, T. 1973. Automates retortable pouch packaging. Food Drug. Packag. *28* (3), 4, 14.
GRIFF, A. 1975. The Plastic Can, Edison Technical Services, Bethesda, Maryland.
GRIFF, A. 1981. Plastic Containers for Soft Drinks, Beer and Liquor, Edison Technical Services, Bethesda, Maryland.
HANLON, J. F. 1971. Handbook of Package Engineering. McGraw-Hill Book Co., New York.
HEALEY, G. 1967. Atmospheric machines. Proc. Seminar Food Packaging with Flexible Laminates. University of California, Los Angeles.
HEATHER, R. P. 1978. The advance of automatic glass container forming machines and production technology. Glass *55* (9), 66, 68–69.
JOHNS, R. W. 1973. Machinery and equipment: past, present, and future. Austl. Plast. Rubb. *24* (10), 39, 41.
KEARNEY, A. T. 1966. The Search for a Thousand Million Dollars, National Association of Food Chains, Washington, D.C.
KELLER, R. C., and MARGIO, J. 1980. Packaging line information system increases production efficiency. Packaging Technol. 10 (5), 45–49.
KELSEY, R. F. 1981. Better packaging control with electronics, Food Drug Packag. 44 (6), 16–17.
LAMPI, R. A. 1981. Retort pouch. J. Food Proc. Eng. 4 (1), 1.
LEONARD, E. A. 1971. Economics of Packaging. McGraw-Hill Book Co., New York.
LEONARD, E. A. 1971. Packaging Specifications Purchasing and Quality Control. McGraw Hill Book Co., New York.
LEONARD, E. A. 1977. Managing the Packaging Side of the Business. American Management Association, New York.
LOPEZ, A. 1981. A Complete Course in Canning. Canning Trade, Baltimore Maryland.
LUCIANO, R. A. 1977. Today's machinery: Breakthroughs in versatility and speed. Mod. Packag. *50* (2), 18–21.
MOODY, B. E. 1963. Packaging in Glass, Hutchinson, London.
MORETON, D. A. 1970. Shrink-packaging systems, machinery and equipment. Packaging *41* (479), S21–S23, S26, S28, S30–S36.
PAINE, F. 1977. The Packaging Media Blackie & Son, London.
PALMER, D., and RUSSO, J. R. 1970. Computerized packaging. Food Eng. *42*, 74–76.
PARK, W. R. R. 1969. Plastics Film Technology Van Nostrand Rheinhold, New York.
PICKERING, G. Buying a blow moulder. Many factors shape the choice. Plastic Technol. *24* (7), 35–38.
RAYMOND, C. 1981. Automated bar wrapping and packaging line, Manuf. Confect. 61 (6), 75–76.
REDDY, D. 1976. Plant and equipment guide. Aerosol Age *21* (7), 32, 34–35.
RUSSO, J. 1981. Aseptic packaging, Food Eng. 53 (3), 20.
SACHAROW, S. 1971. Packaging machines. What type for your product? Food Eng. *43* (3), 70–74, 77–78, 81.

SACHAROW, S. 1976. Handbook of Packaging Materials. AVI Publishing Co. Westport, Connecticut.
SACHAROW, S. 1979A. Packaging machinery bags and bagging. Austr. Packag. *27* (7), 30–31,
SACHAROW, S. 1979B. Know your packaging machinery. Wrappers and overwrappers. Aust. Packag. *27* (10), 28–29.
SACHAROW, S. 1980. A Guide to Packaging Machinery, Books for Industry, New York.
SACHAROW, S. and GRIFFIN, R. 1981. Food Packaging, AVI Publishing Co., Westport, Connecticut.
SHANNON, D. 1979. Stretch machinery has it wrapped up. Mod. Packag. *52* (3), 33–37.
SHORTEN, D. W., and BOUTELL, N. J. 1974. The basics of vacuum forming. Print. Trades J. (1048), 67–68, 70–72.
SIMMS, W. C. 1976. Equipment: Trends in the '70s. Mod. Packag. *49* (1), 39–41.
SIMMS, W. 1982. Packaging Encyclopedia, Cahners, Chicago, Illinois.
STYLES, M. E. K., HINE, D. J., and PAINE, F. A. 1971. The machine/material interface in flexible package production. Packag. Technol. *17* (118), 26–30.
TOENSMEIER, P. A. 1961. Paperboard Packaging Handbook, Board Products Publishing, Chicago, Illinois.
THOMAS, R. H. 1961. Cosmetic Packaging Technology. Columbia Univ. Press, New York.
THORPE, T. 1971. Form, fill and seal machinery developments. Packag. Technol. *17* (118), 22–23, 32.
TRIETER, R. A. 1974. Selecting injection blow machine-the in-depth approach. Plast. Technol. *20* (5), 27–32.
WHITE, E. N. 1972. Improving maintenance and utilization of critical packaging machines. Chem. Process. *18* (10), 75–78.
WOODROOF, J. G., and PHILLIPS, G. F. 1981. Beverages: Carbonated and Non-Carbonated, AVI Publishing Co., Westport, Connecticut.

8

The Relationship of Packaging to Marketing

A good business is one which also makes money.

Camillo Olivetti (1925)

Marketing is a relatively modern invention, and it has become intricately involved with the packaging function. In early times, products were produced principally by hand for very limited market areas and readily found purchasers through the simplest of advertising techniques, word of mouth, and the sign over the door identifying the shop. With the invention of printing, it became easier to put the manufacturer's identification or "brand" on the product, thus reminding the customer where he purchased the product. For the most part, however, the customer sought out the producer in order to buy.

The change to a marketing philosophy whereby the producer seeks out the customer occurred as a result of increased production via machinery invented during the late seventeenth, eighteenth, and nineteenth centuries. Higher production rates at lower costs expanded marketing areas. Manufacturers no longer sold directly to the buyer, but instead shipped their goods to retail shops over a wider area, and the shopkeeper resold the goods to the consumer.

Except for the previously noted labeling, packaging was utilized only to contain and to protect products during storage and shipment. Indeed, many products were shipped only in bulk containers up to the 1890s. The customer carried his purchase home in a wrapper or bag or even in a container which he himself furnished. The shopkeeper was the salesman. Traveling "drummers" sold wholesale lots to the shopkeepers.

During the 1890s and early 1900s, production of goods increased to such a level that it outstripped demand. The purchaser now had a choice among competitive products. It became necessary for the manufacturer to seek methods for influencing that choice. Thus, simultaneously with the development of mass production and mass distribution of goods came the development of what is now known as the marketing function. Although packaging was used in this new function, it was not accepted as an integral member of the selling and marketing team until the late 1930s. The package was regarded strictly as a necessary evil that added cost to the product.

One of the first innovations of the marketing function was the combined introduction of the unit package, brand identification, and advertising. Advertising served to promote customer loyalty to the identifiable product. The unit package carried the identification and provided better quality and convenience. Advertising explained why the product was better than a competitor's. Brand identification called attention to the product-package and implied a warrantee of the advertised quality.[1]

The historic first in this regard was the Uneeda biscuit package introduced in 1899 by the National Biscuit Company (Fig. 8.1). This package protected the product better by means of inner wrappers and an outer overwrap. It also advertised the product brand name via the printed overwrap. It was instrumental in driving the old-fashioned cracker barrel out of the marketplace and into museums.

During the Depression years, the money for advertising campaigns used so lavishly in the 1920s dried up. Management discovered, however, that the package could do an excellent job of advertising its contents. The era of use of styling and design in packaging to motivate purchase had begun. The package had at long last been recognized as a marketing tool. With the advent of the self-service store, which minimized or eliminated the sales clerk, it became imperative that the package serve as a silent salesman.

During the following decades it became apparent that the package could also apply functional conveniences for both the consumer and wholesalers. Easy-open devices, reclosure devices, and mere product visibility aided the consumer. Ease in stacking, price marking, shelving, and identification aided the wholesalers. Marketing people discovered that in an affluent society, customers buy on impulse products they do not really need. Customer motivation at point of purchase became an important function of the package as a marketing tool.

Since all of these package-marketing functions are important to the success of the package-product combination, it is vital that the package development scientist be aware of them and how they relate to his efforts in creating a functional package.

PRODUCT LIFE CYCLE

One of the primary responsibilities of the marketing function is to identify and describe the parameters of the product life cycle. All products, like living creatures, go through a life cycle. The cycle comprises a number of definite sectors.

[1]The power of the brand name is illustrated in the Tylenol incident of 1982. In the middle of the scare about the product being contaminated with cyanide, people were willing to buy under-the-counter bottles of Tylenol even when the product was taken out of distribution by both Johnson & Johnson and the government!

FIG. 8.1. Early Uneeda biscuit package is from the turn of the century. Its successor, a folding carton, was the origin of the first unit package.
Courtesy of Walter Landor and Assoc.

Conception: The creation of the product idea.

Gestation: Determining whether the product should be developed and creating the product.

Birth: Launching of the product into the marketplace.

Infancy: A period of struggling growth of sales, with heavy promotion and little return.

Childhood: Sales begin to grow and to support the cost of promotion.

Adolescence: Sales begin a steep climb and maximum profits are now realized. Advertising is still needed but it can begin to taper off.

Young adulthood: Sales continue to grow but at a lesser rate of increase. Profits are healthy, with modest advertising.

Middle age: Sales start to level off. Profits level off or begin to decline. Advertising is necessary to prop up sales.

Old age: Sales begin to decline. Profits are marginal and growing smaller and smaller. The product can no longer support advertising.

Death: Sales have declined to the point that profits are gone and there may even be a loss.

Burial: Management kills the product and discontinues its manufacture.

It is essential that the corporate marketing department be sensitive to the position of each of their products in relation to its life cycle. Marketing must participate heavily in the decision to create a product and in the development program that brings it to the launching point. Marketing has maximum responsibility in sustaining the product via promotion through its launching, infancy, and childhood. They alone must decide whether the product is showing healthy progress or whether it should be given extra help or killed. In like manner, it is extremely important that marketing be aware of when a product reaches the point of old age. Much money can be wasted trying to keep a corpse alive. However, during "middle age," a product can often be given a new lease on life through a product or package modification. Marketing must be able to sense the need and to have the change ready in time to realize the most good. If a package change is aimed solely at reducing package costs to keep an old product alive, there is a danger that the expense of developing the new package will never be recovered. Marketing must, therefore, keep a proper balance in the product mix, so that there will always be a crop of infants ready to replace the dead and the dying.

PRICING POLICY

It is a marketing responsibility to establish pricing policies with respect to each product in the marketplace. To this end, marketing personnel must be keenly aware of the prices of competitive products as well as the costs of making and selling their own. In launching a new product or attempting to capture a new customer or a new market with an old product, pricing may be made enticingly low, even at a loss of profits. For those products that are approaching maturity, pricing can be as high as the market will bear in order to maximize profits. An old product may be kept alive merely to help carry its share of overhead while a young product struggles to maturity or because its purchaser also buys a number of healthy products. Through the skillful application of experience, instinct, and known facts, marketing can establish a product pricing mix that maintains a desired profit level.

DISTRIBUTION POLICY

It is a marketing responsibility to decide where a product should be marketed. The scope may be anything from local to international export. The decision is based on knowledge of supply and demand; of pricing; of the costs of shipping, packaging, and promotion; and of product shelf-life. Obviously a product that has only a 30-day shelf-life cannot be distributed to a market that will bring it to the consumer with deteriorated quality. A good example of this is the distribution of potato chips. Because of limited product freshness over time, most potato chips have been marketed in local or limited regional areas. National distribution would require a longer shelf-life and better, more expensive protective packaging. Such packages are now being introduced, and there is a growing trend toward wider distribution. It

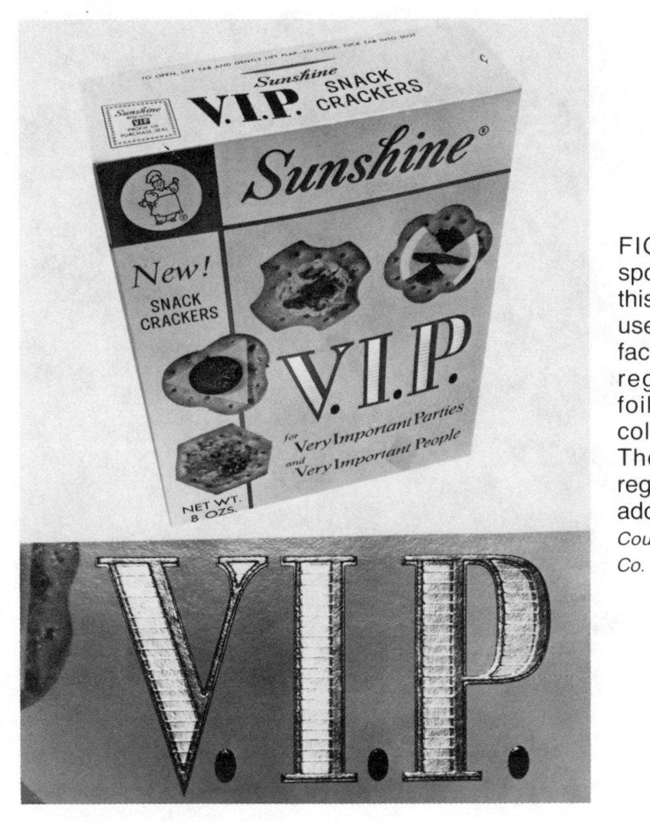

FIG. 8.2. Helping to spotlight new crackers, this represents the first use by a biscuit manufacturer of a distinctive registered embossed foil carton with eight-color gravure printing. The initials "VIP" are registered embossed for added eye appeal. *Courtesy of Reynolds Metals Co.*

is vital that packaging development be aware of the intended distribution cycle for a new product, as this will influence decisions on what materials or package forms to use.

MARKET RESEARCH

It is a marketing responsibility to establish through market research the need for a new product-package and what the potential market volume might be. This effort can also give some indication of what the market will bear in terms of price. During and after the creation of the new product-package, market research is needed to measure consumer response to the new concept, to encourage or discourage further developmental effort, and to guide the development in the best direction. Market research helps plan, execute, and evaluate a test market to determine whether consumers actually will buy and repurchase the product. Data so obtained will also help marketing in establishing pricing, distribution, and promotional policies.

FIG. 8.3. This array of Reynolds' aluminum packaging products illustrates the metal's extensive use by the drug industry.
Courtesy of Reynolds Metals Co.

Concept Testing

A new package-product concept can be concept-tested by showing sketches or mock-up samples to a panel of consumers and asking for their opinions. Later, a consumer placement test can be used to develop still further reaction to the concept. In the latter instance, working samples of the concept are placed with a number of consumers in their homes together with alternative or unidentified competitive concepts. Each consumer is asked to try them all, and then, in a follow-up interview, his or her reactions are solicited.

It is important to note that even "concept testing" may lead to failure because of the unreliability of results. A new package should not be tested and developed solely via consumer conceptual testing. In our present-day affluent society, people buy psychological pleasures and not biological necessities. It is important that psychological devices be integrated into a concept test. An additional problem inherent in concept testing is that only creative people deal with concepts. Most customers must relate to discrete "things" and not broadly defined concepts. For these reasons, concept testing is often unreliable.

It is interesting to note a few examples of the unreliability of concept testing. About 25 yr ago, consumers were asked whether they would want to purchase yellow margarine. Over 90% said that yellow margarine was not needed. Yet it is still on the shelves. When consumers were asked whether they would use instant coffee, they said it was an unnecessary product. Yet it is highly successful.

Test Marketing

New package concepts should be test marketed prior to consumer introduction. The full-scale test marketing of a package is an attempt to evaluate the nature and degree of consumer acceptance by actually putting the product on the market in selected areas. It may utilize a single full-scale marketing program within the test area, or it may use different combinations of marketing factors—i.e., different promotional appeals, different prices, or different types and intensities of distribution.

Test marketing should be regarded by management as a research operation seeking out areas of potential profit and as an effort that may minimize the losses on new and hazardous ventures.

Since product or package faults make up more than half of the factors responsible for new-product failure, a suitable test-market program is essential for package success. When vacuum-packaged bacon was initially introduced to the U.S. housewife, it was test marketed in certain geographical areas selected for population, class profile, urban density, trade structure, delivery ability, and other special factors. It was found that the flexible plastic film package used was not suitable for supermarket display. Because of a too-high slip value, the packages did not stack well in the supermarket display cabinets and constantly slipped when stacked. Since many housewifes overturn a stack of bacon packages in a never-ending search for lean bacon, the stack of bacon became a mess and a constant problem for supermarket personnel. After this fact was learned via the test market, the films were changed to ones with higher coefficients of friction.

The following precautions are necessary in all test market programs.

1. Do not allow the sales force to be overly enthusiastic and overpromote the concept.
2. Do not use nonrepresentative special products or packaging.
3. Do not let the test run into seasonal imbalance.
4. Beware of a competitive thrust into the market area.

PRODUCT-PACKAGE PROMOTION

Marketing is responsible for establishing promotional policy with respect to each product-package combination. This includes primary direct advertising via mass media, secondary direct advertising via point-of-purchase displays and package styling and design, and indirect advertising via the establishment of a desired corporate image through public relations. This last factor also plays a part in the selection of the other advertising forms.

Advertising

Advertising is essential for the success of a packaged product. It is an integral part of a firm's promotional program. Advertising serves to create the desire in the consumer to buy the specific package and to instill a favorable feeling and image. Total communication is mandatory, and the advertising agency must be brought into intimate contact with the package development. Not only must the package fit into the advertising campaign, but the campaign must be compatible with the package.

In planning a promotional venture for a new package, several factors must be considered: (1) advertising budget, (2) message content, (3) media selection, (4) use of an outside agency, and (5) evaluating effectiveness.

Advertising Budget. The method by which management reviews expenditures for advertising must be carefully controlled in a planned budget. Media advertising, i.e., in newspapers and magazines and on radio and TV, is the most critical. Depending on the product selected, a choice is made as to the medium to be used. Although it is difficult to promote a package per se, the positive advantages of a product in a specific package can and often are promoted.

Message Content. Advertising copywriting is a grueling vocation demanding orginality. Good copy can make or break a new packaged product. The content of the copy must convey the message demanded by the company. Reynolds' aluminum can reclamation program, a very successful pioneer in ecology, was a public relations message carried by most media.

Media Selection. This is the method used to transmit information from the company to people they hope will be either buyers or influencers of those who actually do the buying.

There are a host of both general and trade magazines that sell advertising space regularly. It is their main money maker, and a smart advertising director can wisely select the proper source. Private advertising agencies are great aids to a company's advertising department. They are expert in copy and media selection. Fees are determined by either a commission to the buyer or service charge from the media selected by the buyer. The effectiveness of advertising is measured by the extent to which it attains the objectives set for it.

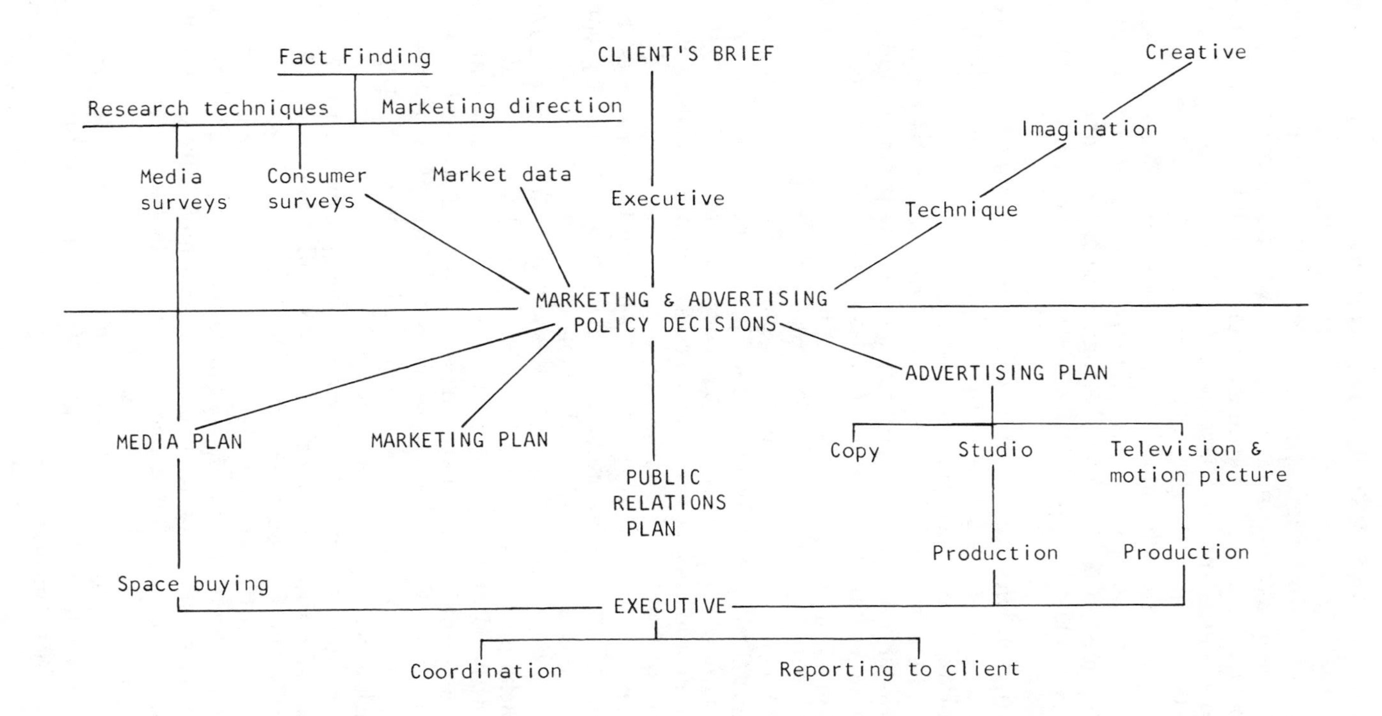

FIG. 8.4. Organization of a typical advertising agency.

A MARKETING CASE HISTORY: DOY-N-PACK VS. BRIK PAK

In 1979, Shasta Beverages made its first tentative step outside the carbonated beverage business with a new product most observers thought looked more like something astronauts might take into space than a key to future corporate profits.

The product, a juice-based drink called Capri Sun, was strange by U.S. standards. Although the U.S. fruit drink business was dominated by 46-oz. cans, Capri Sun was available in single-serve portions in an unusual flexible foil pouch. And although most fruit drinks contained preservatives, additives, and artificial colors, Capri Sun was all natural.

Beyond that, it was expensive; its price was twice that of its nearest competitors, Coca-Cola Foods' Hi-C and R.J. Reynolds Industries' Hawaiian Punch in large cans.

Shasta did encounter initial problems with Capri Sun—leakage, a limited number of flavors, difficulty in opening the package, and company confusion as to what were the most marketable features of the product. But company executives are optimistic that most of the big worries are behind them.

They may not be, however. Shasta's big plans face what is probably their biggest test—the threat of aseptic packaging, a new packaging form that is now widely found on the supermarket shelf.

Capri Sun could very possibly be overtaken by market forces before it even gets into national distribution.

Capri Sun is now in approximately 70% of the United States. It is doing such a brisk business that the company is building new production lines to increase its output threefold.

But getting there will not be easy. While Capri Sun's unusual Doy-N-Pack packaging was unique in 1979, when the product was introduced in the United States (in Atlanta and Buffalo test markets), today it faces formidable competition not only from Hi-C and Hawaiian Punch, but from Borden and Ocean Spray; all have introduced drinks—the last two pure juices as opposed to juicebased drinks—in aseptic packaging.

Aseptic packaging, which has for many years been used in Europe for beverages ranging from juice and milk to coffee and wine, was first introduced in the United States in 1982, after the Food and Drug Administration authorized its use.

Its developer, Switzerland's Tetra Pak, already has a U.S. subsidiary called Brik Pak, Inc., in full swing. And the marketers of 100 pure juices, which until now have had to refrigerate their products, are taking to the new Brik Pak with great speed.

The Doy-N-Pack, one of six kinds of foil pouches on the market in Europe, is the exclusive property of Wild Group of Companies, based in Heidelberg, West Germany.

Four years ago, Shasta's parent reached an agreement with Wild for the exclusive rights to market the Doy-N-Pack in the United States. The product is marketed in other countries through arrangements with various companies.

Capri Sun's Doy-N-Pack came into the United States offering distinct advantages over traditional cans and bottles—it is portable, lightweight, and unbreakable, has no sharp edges and a longer shelf-life, can be frozen, and has the ability to keep the drink inside significantly cooler than the outside environment.

But the Capri Sun package has few attributes to set it apart from the new aseptic packages.

With the exception of the Doy-N-Pack's eye-catching, flashy graphics and ability to be frozen, drinks in aseptic packaging offer almost the same advantages as the Doy-N-Pack at a similar per-ounce–basis price. Retailers are generally more receptive to aseptic packaging because it is said to be easier to stack on shelves and less apt to leak.

Capri Sun's share increases have come mostly at the expense of smaller regional brands, and in some cases they have resulted from a reorganization of grocery store shelves by retailers who have gotten rid of marginal juice drinks to take on Capri Sun.

PACKAGE STYLING AND DESIGN

Prior to 1920, package design was considered to be the responsibility of the professional artist. The acceptance of the need for design as a sales stimulator occurred in the mid-1920s along with the growth of mass advertising and mass production. One of the pioneers in design work was a man named Lucien Bernhard, a German designer newly arrived in the United States. He developed a certain poster-type style and used it quite extensively in the early 1930s. Packaging design firms made their debut in American industry during the 1930s. An interesting description of present-day packages follows:

> Wanted: An Extraordinary Salesman: One who can dress up in the company uniform, call on major distributors, attractive enough to appear in our magazine and trade journal ads, and stay with our product up to the point of purchase. Beyond that, we are looking for a salesman who will stay with the consumer after the purchase, and keep selling our product at the point of consumption. No pay, no vacation.
>
> Right now we need approximately 10,000 such salesmen, expect to need over a million within the next six months.

A housewife must weekly choose from a wide range of packages, products, and sizes. In just 1/5 sec, the package must be able to convey

FIG. 8.5. Early package design on a tea can bears the mark of the professional artist. Note the ornate illustrations. *Courtesy of Walter Landor and Assoc.*

the message "buy me." No advertisement is ever read as often as a package.

Package design involves more than the surface aesthetics of a package. A final choice is influenced by the entire marketing program, including the package-product combination, the corporate symbol, the distribution and pricing policy, and the promotional effort.

Elements of Good Packaging Design

In designing a package, the designer must aim for several goals in his final design. These are (1) attract the buyer, (2) communicate the message to the buyer, (3) create a desire for the product, and (4) sell the product.

Attraction of the Buyer. The package must be able to stand out in the amazing bazaar of packages now found on supermarket shelves. This can be accomplished by the use of color, shape, copy, trademark, logo, or additional features. How to accomplish this effect is a major problem for package designers.

A package which draws attention to itself is more than half sold. Since the average time a shopper has to look at a package in the supermarket is less than 1 sec, attention is won by exciting the eye with something about the package that is new and interesting. Since most people tend to notice only pleasing objects and ignore disagreeable objects, the attention provoker must be pleasing, whether it be intellectual or sensuous. Packages designed to look like books may appeal to college professors, while other purchasers may be attracted by the use of sex appeal in the form of pictures of beautiful girls. In both cases, the requirement is to evoke attention. If attention is not obtained, the consumer will simply select a competitive product.

Sometimes long-used designs must not be changed, since brand loyalty has been built up in relation to the specific package design. Lavoris mouthwash offers a good example of a seemingly outdated design that sells quite well. The Hershey chocolate bar has retained the same color for many years. Heinz Ketchup still is sold in a bottle that is a poor pourer. Yet, when the bottle shape was changed, sales dropped.

Communication to the Buyer. The package must be able to communicate its message to the buyer. All the necessary information must be clearly visible or implied through the factors of color or design.

Communication involves both direct and subtle communication. Direct communication depicts the product and describes what it is and how to use it. It may also describe its advantages. Indirect or subtle communication involves the use of color, shape, design forms, and tex-

FIG. 8.6. This label design, winner of the Design Centre's show (London), is an example of modern communication. The use of three oranges and lemons tells the consumer that the product is triple concentrated.
Courtesy of Council of Industrial Design, London.

ture to imply certain properties, such as purity, strength, value, delicacy, feminity, masculinity, fun, and dignity.

An interesting example of a planned program of subtle communication is a Swedish milk package. Before the TetraPak unit was introduced to the Swedish market in competion with glass bottles, the new package design was carefully evaluated with Swedish consumers. Favre (1969) reports this case in great detail.

Color was considered to be the main aspect involved in consumer communication. Tests revealed that in order to denote cleanliness and purity, light colors, a white background, and modern design motifs should be used. Consumers associated red, orange, and brown with fat. In order to play this down and suppress the idea that milk is fattening and high in cholesterol, these colors were avoided in favor of light colors. To communicate that milk is a thirst quencher, various shades of blue with fresh, light colors were found to be best. For communicating the fact that milk does not have an unpleasant taste, vivid color contrasts were found to be best. To imply good value, excellent technical design and multicolor printing detail were employed. Childish designs were not used in order not to limit the age appeal through infantile connotations. Similarly, to overcome the feeling of many men that milk drinking is effeminate and childish, strong, masculine, vigorous colors were used with linear and angular designs. Red is considered a masculine color. For appeal to a broad range of social economic levels, a clean graphic design was used, with the avoidance of cheap colors.

The result was a modern-looking package with blue, white, and red colors. It met with excellent success in the marketplace.

Creating a Desire for the Package. Desire to purchase a package can be created by proper design. The package can help the consumer recall a theme that other advertisements have promoted. The package may convince the purchaser that the product can fill a need or satisfy an inner desire for luxury or prestige or some other psychological want.

In alcoholic beverage packaging, many liquor firms market small glass miniatures of their product, not only to airlines, but on the retail market. Even though the product profit on these packages is small, the package is a direct replica of the larger bottle. This is done to create a desire by the consumer to buy the larger bottle and establish brand loyalty.

The aerosol can spurred the development of a totally new market for women's—and now men's—hairsprays. Prior to the use of aerosols, no such product existed. Yet there was a definite need for an easily applied and functional method of hair care. Through the use of a then-new package concept, an industry was born. In this case, a desire was fulfilled by the package and purchasers jumped on the bandwagon.

Other convenience packaging follows much the same vein. Through convenience packaging, consumers were introduced to easy-to-prepare pot pies and other complete dinners. In many cases, these items are more expensive than if they had been prepared at home, but the desire to save time stimulates the consumers' desire to purchase.

"Selling" the Product. The package must not only sell the product, but create the desire for repetitive purchases. The package must overcome consumer reticence by being a useful, functional, or satisfying addition to the consumer's purchases.

Re-use features, special giveaways, and easy-dispensing devices are all instrumental in repeat sales. The introduction of milk packaged in flexible polyethylene bags is an example of a package design stimulating repeat sales. To facilitate use of the package, specially designed molded polypropylene pitchers are sold with the bags. The consumer need only purchase one pitcher, but every time he is out of milk, there stands an empty pitcher just "begging" to be refilled. The consumer has no alternative but to buy the same package again.

Color. In order to achieve the goals of good packaging design, the artist has at his command a selection of materials which may be limited by the technical and functional requirements of the package. He also may alter form and shape. He may utilize texture. His greatest weapon, however, still lies in the use of color and line through the graphic arts.

Package Form. Size and shape must be both functional and attractive. Obviously, size must be dictated by the quantity of the contents, but shape can be varied extensively. Shape can make a package more convenient to hold or to stack on a shelf. Shape may permit better communication of package advertising. Shape can imply lightness or thinness, delicacy, or strength.

FIG. 8.7. New bag design uses the lightness and transparency of a polyethylene bag and the aesthetic appeal and strength of aluminum foil, thus showing its usefulness as a bread package.
Courtesy of Walter Landor and Assoc.

Graphic Design

The package must convey messages through graphic design. Printed copy describes the contents and how to use them. Illustrations or symbolic designs convey direct or indirect messages about the product and its quality and value. The arrangement of these elements is extremely important to the need for attracting attention to the package and communicating the desired information. Copy should be simple, legible, complete, and attractively arranged to harmonize with the overall package design.

It is best to use a minimum of copy and allow the package to depend on color imagery, design, and symbols to get the sales message across. The lettering used should supplement the package's color plan. Color has a specific effect on legibility. Black lettering on a yellow background is very legible; however, yellow on white is poor. In addition, copy is less legible in capital letters than small letters, and a word is more legible if the space between the letters is larger than the type thickness. For the part of the message containing ingredients, weights, etc., it is extremely important to use good, legible colors. The purchaser must not think that the label is designed to conceal these facts.

Copy can also be included to convey a certain message. A homegrown quality can be conveyed by the use of fancy, curly lettering. To modernize a product, graphics can be given straight, simple lines.

Color in Graphic Design. By definition, color is a quality by which objects can be differentiated with the eye independent of their form. It is a property of light rather than of the object, although the molecular composition of the object determines its color by means of the light vibrations returning to the eye. In perfect darkness, objects have no color. White light is a mixture of all spectral wavelengths, and it can be divided into its components by prisms or water droplets, as in the familiar rainbow. Black "light" is the absence of all spectral wavelengths. Color as a form of light is a subtractive thing. We achieve a yellow light by subtracting all other wavelengths. This can be done by use of certain filters or absorptive materials. Most materials have this property and appear to the eye to have a color. A small quantity of a "yellow" pigment absorbs all other wavelengths and permits only yellow light to be reflected or transmitted to the eye of the observer. Achieving different colors using pigments is an additive process. Adding a red pigment to a yellow pigment results in the additive effect of both the red wavelengths and the yellow wavelengths, which creates a new color, orange. The principal spectral colors are red, orange, yellow, green, blue, indigo, and violet. Red, yellow, and blue are known as primary pigmentary colors because other pigmentary colors are achieved by adding primaries together: red and yellow give oranges. Blue and yellow give greens. Blue and red give purples.

The change in color due to wavelength alterations is known as change in hue.

When white and black are mixed together, a wide variety of gray shades and tints can be achieved. This scale of whiteness or grayness is known as the value scale. If white is considered high-value and black low-value, then other pure colors can be compared on the value scale. Pure yellow is a high-value color, whereas pure red and pure blue are low-value colors. A pure color may be diluted by adding a white, gray, or black to it. If a gray of the same value is added, the color is diluted in chroma and has less purity or less saturation, but it retains its hue and its value. If white or lighter gray is added, the result is a higher value and is called a tint or a lighter tone. If black or darker gray is added, the result is a lower value and is called a shade or a darker tone.

Vivid chroma colors with strong contrasts in value produce a dynamic, forceful impact that can in some cases be actually painful to the eye. Weak-purity colors (pastels and grays) with weak contrasts in value produce a calm, serene, peaceful impact which can become boring or insipid. The careful selection and use of color is of extreme importance in package design.

Package design utilizes color as a vehicle for evoking a consumer response. In selecting an appropriate color for a package, certain considerations are important. Color provokes discrete physiological and psychological reactions. By understanding these properties, a package designer can capitalize on the use of color in design.

Physiological Reactions to Color. In a red room, a person's pulse and respiration rate increase. It has been ascertained that when anemic children are exposed to red and yellow rays, their red blood corpuscles increase in number. Green is a quiet color that evokes a decrease in blood pressure. Although light green provokes indifference, dark green evokes a feeling of calmness.

Psychological Reactions to Color. Color causes people to associate with certain ideas. It evokes specific psychological responses and, if used properly, can become a valuable sales tool. By choosing a color, a special mood can be developed for a package.

Red is an exciting color. It signifies strength and virility and causes people to look at the package. In most cases; the use of red must be carefully controlled. Light red is a cheerful color, but dark or bright red is more likely to induce depression and irritation. Cherry red is sensuous. As red becomes darker, it becomes more serious and depressing.

Orange expresses action and has the ability to communicate. It looks clean and appetizing and has an intimate character.

Yellow is a cheerful color and is the loudest and brightest of all colors. Pale yellow looks dainty, golden yellow is active, green yellow is sickly, and a deep, strong yellow suggests sensuousness.

Pink is suggestive of femininity and deep affection. It lacks vitality and gives an impression of intimacy and gentility. When a bright magenta pink is used, the viewer tends to feel frivolous.

FIG. 8.8. "Twist bag" used for fruits.
Courtesy of Walter Landor and Assoc.

Green is a quiet, refreshing color. It is associated with youth, growth, and hope. Being an undemanding color, it evokes neither passion nor sadness. When darkened to olive, the same color becomes a symbol of decay and rot.

Blue is a cooling and subduing color. It is a tranquil color suggestive of celestial infinity. It differs from green in its tranquility, since green suggests earth-like quiet, whereas blue suggests a heaven-like quiet. If darkened to indigo, it becomes a severely depressing color.

Black is a symbol of death and despair. It is a depressing color that contains an impenetrable character. But, when used as a shiny color, it can convey an impression of nobility and elegance. In the 1980s, black is being used extensively for cosmetic packaging to connote luxury, quality, and expense.

White denotes virginity and purity. It connotes the unexplainable and inaccessible. When used near blue, it has a refreshing and antiseptic effect.

Brown is a utilitarian color and is suggestive of work and compactness. It is a healthy color that is neither brutal nor vulgar. As it becomes darker, it assumes the character of black.

By utilizing two colors in a package, a certain mood can be established by the purchaser. For example, a combination of red (excitement) and yellow (happiness) evokes a feeling of dynamicism. Additional factors important in package color selection are shape and personal and national preferences. Although a triangle is best viewed when yellow, blue is best suited for a circle. The happiness and movement of yellow radiates in a triangle, whereas blue tends to disappear from sight with a circle.

Color Combination. The color wheel enables one to select harmonious colors quickly and easily. By noting the psychological response from the consumer, package visibility and emotion are determined.

The color wheel contains six basic colors—red, orange, yellow, green, blue, and purple—and six intermediate colors—red-orange, yellow-orange, yellow-green, blue-green, blue-purple, and red-purple. Colors that are opposite each other on the 12-color wheel are called complementary colors. Colors that lie side by side on the wheel are called analogous colors. In a two-color scheme, colors that are complementary, such as red and green, or colors that are analagous, such as red and orange, may be used. One three-color scheme utilizes one color plus the colors to the right and left of its complement—for example, red plus yellow-green and blue-green. Another three-color scheme, known as the triadic, uses colors that are at equal distances from each other on the wheel—i.e., red, blue, and yellow.

Color as an Attention Provoker. With increased impulse sales, a package must be able to attract attention and not be subdued by a supermarket's internal decor. A housewife spends about 27 min shopping in a supermarket and selects only about 14 items out of a possible 10000 items. More than 75% of all purchases made are impulse ones. Orange and red attract attention the best. Blue also has high visibility, probably because it is a well-liked color. Yellow is not a strong attention getter, probably because of taste preferences. The four colors leading in attraction are orange, red, blue, and black. Since these are not necessarily the most visible colors, attraction is also attained by the use of contrasts, eccentric colors, and shapes and by using a color that differs from that used with a competitive product.

Color as an Implication of Product Quality or Special Features. The quality of a product must be conveyed through the use of color. An expensive product should look costly and not employ the color scheme of a mass-merchandized item. Yellow is usually regarded as a cheap color. For commonly used products, lush, dark-colored designs tend to upgrade the package and can often cause the product to command an increased price.

Special characteristics of a product should be conveyed on the package itself through the medium of color. A chlorophyll containing product should somehow integrate green in the package design. In the drug and pharmaceutical field, the curative action of the product should be conveyed. For a durable and useful product, brown should be used in the color scheme. Toothpaste should be packaged in blue and white, since these colors suggest cleaning and hygiene. Cosmetic packages connote skin care and softness through the use of pink and blue. A stronger type of cigarette is best conveyed in a black or dark-red pack, whereas weaker ones can be put into a white or pale-gray unit. Poisonous products such as insecticides demand glaring, striking colors which signify a strong and noxious effect. Black and yellow are particularly useful.

It is important to make a food package appear clean, appetizing, and representive of the product. Blue stripes are best used with white eggs, since blue makes the eggs appear more white. Mayonnaise looks yellower and more appetizing when used with blue. White sheets with a blue band appear sparkling and new.

Color as a Means for Establishing Corporate Symbols. The proper use of packaging color can establish corporate symbolism for a broad range of products. Eastman Kodak yellow is a classic example of a corporate type color. This yellow causes people to readily identify with Eastman Kodak. In other cases, colors can be used to differentiate between product variations. Shampoos and electric bulbs are packaged in different-colored packages which identify their product variations.

Color and Symbolism. People of different nationalities relate to various colors in different ways (Table 8.1). The market served by the specific product is of extreme importance in selecting a suitable color.

FIG. 8.9. Potato dumplings or pirogen, an Eastern European product, in foil cartons has been designed to appeal to the nonethnic market. Note the English name with the European translation.
Courtesy of Reynolds Metals Co.

TABLE 8.1. Color and Symbolism in Packaging for Asian markets[a]

Country	Color	Color Connotation	Symbol	Symbol Connotation
China	White	Mourning (avoid)	Tigers, lions, and dragons	Strength (use)
Hong Kong	Blue	Unpopular (avoid)	Tigers, lions, and dragons	Strength (use)
India	Green and orange	Good (use)	Cows	Sacred to Hindus (avoid)
Japan	Gold, silver, white, and purple	Luxury and high quality (use)	Cherry blossom	Beauty (use)
	Black	Use for print only; prefer gay, bright colors	Chrysanthemum	Royalty (avoid)
Malaysia (Population is mixed Malaya, Indian, Chinese)	Yellow	Royalty (avoid)	Cows	Sacred to Hindus (avoid)
	Gold	Longevity (use)	Pigs	Unclean to Moslems (avoid)
	Green	Islamic religion (avoid)		
Pakistan	Green and orange	Good (use)	Pigs	Unclean to Moslems (avoid)
Singapore	Red, red and gold, red and white	Prosperity and happiness (use)	Tortoises	Dirt, evil (avoid)
			Snakes	Poison (avoid)
	Red and yellow, yellow	Communist (avoid)	Pigs and cows	Same as for India and Pakistan (avoid)
Taiwan	Black	Avoid	Elephants	Strength (use)
Thailand			Elephants	National emblem (avoid)
Tahiti	Red, green, gold, silver, and other bright colors	Use		
Arab and Moslem states	White	Avoid	Animals	Avoid
			Pigs	Religious pollution (avoid)
			Star of David	Political (avoid)

[a]Information courtesy of Hygrade Packaging Co. (New Zealand).

In addition, social stratum, rural or city dwelling, age, and race are important factors.

In the United States, red is indicative of cleanliness, but in the United Kingdom, red is the least clean of all colors. In Sweden, blue is masculine, but in Holland, it is a feminine color. Red and blue are the most popular colors in Holland. In Italy, a land with plenty of sunshine, red is a preferable color. The United Kingdom, Sweden, and Holland prefer blue and golden yellow. Blue is serious in Holland, whereas green is a serious color in the United States. Red is very popular in the Balkan nations.

An interesting observation is that in times of economic stress, different package colors sell best than in good economic times. In prosperous times, red and blue sell best. When depressions arise, green-yellow is preferred. Red is a strong color suggestive of strength and vitality. This is indicative of an extravagant economy. As the wheels of progress slow down, people prefer a harmonious, quiet, and serious color, such as yellow-green.

In an evaluation carried out by the Polish government, it was found out that older people prefer to buy children's products in pink and blue packages, whereas younger buyers prefer light green and warm yellow. Thus, an item intended for a grandfather market should be colored differently than packages intended for younger parents.

A basic consideration to follow is that in nations having bright sunlight, colors should be strong. Under the influence of tropical light, pastel shades tend to appear old and musty. This fact can and should be considered when using a color for a symbolic message.

Just as important as color is the symbolism used and the printed message. In Taiwan, the elephant denotes strength and is a desirable symbol for a package. But in nearby Thailand, the elephant is the national emblem and its use would be equal to decorating a U.S. package with the "stars and stripes." Pigs and cows are to be avoided in Islamic and Hindu nations for religious reasons; the Star of David is also to be avoided for religious and political reasons.

Many times, developing nations will use the printed package to convey a certain message. Through the use of gravure-printed labels or flexo-printed polyethylene bags, protein-rich infant food packages can promote birth control in overpopulated nations. Pakistan has tried this approach.

Snowflakes used in a design are meaningful to people who have seen snow but meaningless to tropical people who may never have heard of it. The mushroom is a symbol of good luck in Central Europe but is meaningless elsewhere.

Packaging Design for Ethnic Minority Groups

While many minority groups have been assimilated into the American melting pot, others have encountered a discriminatory and repres-

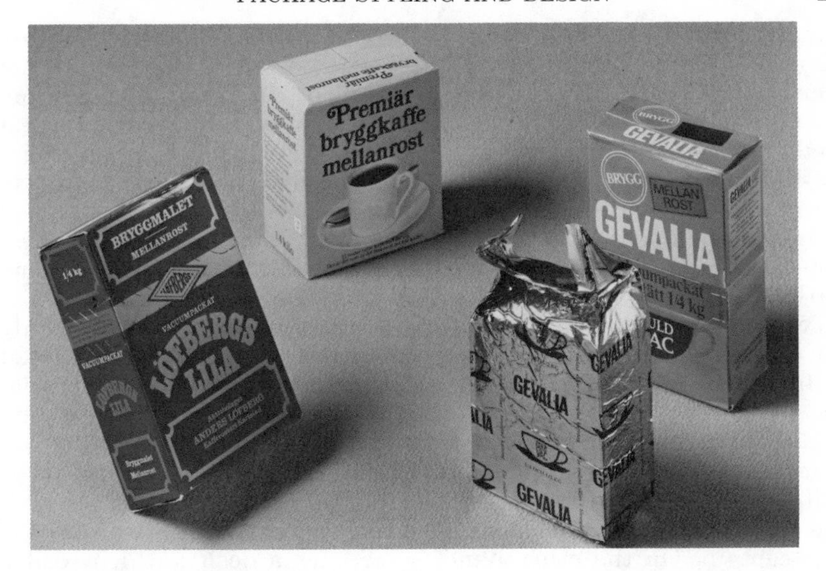

FIG. 8.10. Vacuum-packed coffee in a foil lined carton.
Courtesy of ICI Plastics.

sive attitude on the part of the majority culture. This has led to the formation of concentrations of minority groups in particular geographic areas. In some extreme cases where this has coincided with extreme poverty, these areas have become known as "ghettos." Because of repression, rejection, and segregation, some of these minority groups have developed acute sensitivity to and dislike for certain words, phrases, or symbols. Despite their being at the lower end of the socioeconomic scale, the general affluence in the United States has made these cultural groups an increasingly important factor in the consumer marketplace.

The last two decades have been an era of social unrest and upheaval as these minorities have become aware of their economic power and begun to struggle not for assimilation and loss of identity but for recognition and acceptance of their individuality.

Packaging design aimed at these groups must take notice that symbols once considered acceptable may now appear to be offensive to minority groups if they call attention to repression or discriminatory practices.

In the religious area, symbols denoting a specific religion are not used in mass-market packaging, but they may be used for a specific ethnic market. For example, Hebrew letters appear on Passover products for the Jewish market. A package design incorporating the image of Christ would not sell widely because it might offend the feelings of many non-Christians as well as many Christians who feel that such commercialism would profane their religion.

In the area of race, only recently have U.S. designers become aware of the need for packaging that is specifically designed to appeal and not offend black citizens. Blacks represent 12% of the national population, have an estimated $32 billion in buying power, and are the largest single market (in numbers) for packaged goods in those areas where two-thirds of all such sales are made. The need of American business to tap this latent market has caused both the advertising industry and package design personnel to study methods of appealing to the market and the elimination of prior designs considered offensive by the black market.

Two factors are involved in design for the black market. Not only must offensive packages be eliminated, but the presence of the 24 million black consumers must be recognized in the packaging media. The picture of "Uncle Ben" was eliminated from the carton containing this brand of rice. The reason given by the company was that the firm updated its graphics in light of present-day marketing requirements as well as to bring about increased distribution of items other than food. The smiling Uncle Ben logo has been replaced on TV by a white riverboat captain. On the other hand, sometimes a design which whites might think would be offensive to blacks is actually not offensive to blacks. Quaker Oats' Aunt Jemima pancake mix features a picture of a black Aunt Jemima, suggesting an era long past. Market research by Quaker Oats indicated that black consumers do not feel that the package is offensive and that it is even more widely used by blacks than whites. The General Foods cherry-flavored Kool Aid package features both black and white children in its design. General Foods research indicated that blacks drink more Kool Aid per capita than whites. The selection of an integrated design is thus a logical business decision. Integrated "looks" are now appearing on many other packages, ranging from cookies to children's toys. In some cases, the new designs incorporate a "no-people" approach. An interesting note is that blacks tend to be more attracted by yellow colors and tend to reject brown.

There is a definite need for designers to undertake market tests for their specific products in the actual environment where the package is to be sold. The psychology and rationale of the black market then becomes apparent. Many black areas of large cities contain small grocery stores, and these do not lend themselves well to massive promotional efforts. The design of a package intended for this market must carry an appropriate sales message. In addition, since incomes are smaller in ghetto areas, smaller packages sell better. The need for status is often expressed by increased sales of well-known brands as opposed to cheaper private labels. Although specific designs intended for blacks are not needed or even requested, sensitivities must be understood in the light of reality.

Future cosmetic package designs will probably incorporate a greater number of beautiful black women. In fact, at least one new line of cosmetics has been designed specifically for black women.

BIBLIOGRAPHY

ANON. 1969. Planning for tomorrow's packaging realities. American Management Association, New York.

ANON. 1971. Marketing and standardization of packaged imported goods. Trade Ind. 2,503.

ANON. 1973. Choosing alternatives in a tight market. Mater. Handl. Eng. (Special Issue, Fall), 32–36.

ANON. 1975B. Market data 1975 for basics, it's wait 'till next year. Can. Chem. Process. *59* (9), 19–34.

ANON. 1975C. Plastics' resource impact said equal to other packs. Food Drug Packag. 33 (10), 20, 4, 12.

ANON. 1976A. Economics of materials. Mater. Eng. *81* (1), 20–29.

ANON. 1976B. The economics of packaging: the editor asks the expert. Packag. Rev. *96* (2), 41–42, 46–47, 72, 74.

ANON. 1976C. U.S. Foreign trade in containers and packaging materials 1975. Containers Packag. *29* (2), 5–7.

ANON. 1976D. Flexible packaging: outlook to 1981. Food Eng. *48* (7), 52–54.

ANON. 1977. What users thinks of flexible packaging. Packag. Digest. *14* (4), 20–21.

ANON. 1978A. U.S. plastics sales data: 1977 vs. 1976. Mod. Plastics Int. *8* (1), 51–57.

ANON. 1978B. Cellophane: Down but not out. Mod. Packag. *51* (5), 51–53.

ANON. 1980A. Packaging outlook 1980: 52 Billion dollars. Packag. Dig. *17* (1), 52, 53, 55, 56.

ANON. 1980B. Converted flexible packaging markets. Predicast Inc., Cleveland, Ohio.

ADAMSON, J., and OKON, L. W. 1977. Depletion and cost of containers in a deposit/return system. Package Dev. Sys. *7* (6), 28–32.

BRACK, R. W. 1973. Packaging and the public. Austr. Packag. *21* (5), 104, 107, 109–110.

BRADATCH, A. 1980. Market and technological developments in the corrugated board industry consequences for manufacturer. Int. Paperboard. Inc. *23* (1), 27–32.

BRISTON, J. H., and NEILL, J. 1972. Packaging management. Gower Press Ltd., Epping, Essex, United Kingdom.

CALE, P. 1980. Success on the shelf. Can. Packag. *33* (6), 16–20.

COHEN, S., and HEITZ, W. D. 1977. The technical and marketing aspects of the plastic merchandiser. Tappi Synthetic Paper Conf., pp. 165–172.

JONES, A., and WOLPERT, V. M. 1979. Plastics and metals packaging. International techno-economic survey. Wolpert and Jones Ltd., London.

COLANIGELO, M. 1979. Packages feel oil supply impact. Food Drug Packag. *40* (9), 4, 44–5.

COOPER, A. 1979. Packaging is more than a marketing tool. Marketing, July, pp. 43, 45–46, 48, 50, 53.

DEAN, J. M. 1970. Packaging in the seventies. Packaging *41* (488), 29, 31, 33.

DENSFORD, 1980. Commerce seeks pack growth slowing with economy. Food Drug Packag. *42* (2), 4, 9.

FaVRE, J. P. 1969 Color Sells Your Package. ABC Edition Zurich (Switzerland).

HAZELL, H. L. 1978. The VK economy 1979–81—and its effect on the packaging industry. Paper *190* (10), 20, 620–622.

KELSEY, R. J. 1977. Packaging Marketplace. Food Drug Packag. *36* (8), 21, 4, 29, 32–34.

KENNEDY, J. P. 1975. Changing economy: Its impact on equipment. Chem. Eng. *82* (6), 17, 54–60.

LEONARD, E. A. 1977. Managing the packaging side of the business. American Management Associations, AMACOM Division, New York.

LEONARD, E. A. 1979. Packaging's economic responses to the forces in the world today. Package Dev. Sys. *19* (5), 10–12.

LOUIS, P. J. 1972. Marketing and packaging guide for the common market countries. Eur. Packag. Digest. No. 108, pp. 2–9.

MASON, E. Inflation sparks move to cost-cutting systems. Can. Packag. *34* (10), 9–12.

MILLS, R. 1971. Economic factors in the development of packaging during the 1970's. Packaging *42* (493), 21, 23, 25, 27.

MILLS, R. 1976. Boom and slump psychology in the packaging industries. Packag. Rev. *96* (3), 21, 23–24.

NOBLE, P. 1972. Marketing Guide to the U.S. Packaging Industries, 2nd Edition Charles H. Kline & Co. Inc. Fairfield, New Jersey.

RAPHAEL, H. J., and OLSSON, D. L. 1976. Package Production Management. AVI Publishing Co., Westport, Connecticut.

RAUCH, J. A. 1977. The Kline Guide to the Packaging Industry. Industrial Marketing Guide, 3rd Rev. Edition Charles H. Kline and Co., Fairfield, New Jersey.

ROSATO, D. V. 1976. Packaging market: more coverage for plastics. Plast. World *34* (2), 28–33.

SHARRING, F. 1979. Industry economic review. Paperboard Packag. *64* (8), 49–83.

TINMOUTH, T., D'ARCY QUINN, F., and MORRISON, J. 1982. Packaging for profit. Can. Packag. *35* (6), 21–23, 25.

WOOD, R. 1970. Thin-wall disposable containers: Manufacture and markets. Plast. Rubber Text. *1* (5), 207–208, 210.

ZANES, G. W. 1980. The market makes the product—not the designer. Paperboard Packag. *65* (9), 90, 92–94, 96.

9

Governmental Regulations Affecting Packaging

The problem is that the United States has not learned to use science as a regulator, although no other country has either, and the solutions have been based on laws passed in reaction to crises rather than on science.

The New York Times, March 15, 1977

Regulations regarding food and, indirectly, packaging go back to antiquity. Archaeological discoveries from the Hittite period (3500–1200 BC) have proved that even then there was a fairly comprehensive system that covered foods. Egyptian pictographs have also been unearthed showing various officials measuring and weighing grain. But perhaps the earliest civilizations to which food regulations can be directly traced were the Roman and Greek.

In its prime, the Roman Empire included almost all of modern-day Western Europe. Intrinsic to ancient Roman values was not only cleanliness but the value of good commercial practice. For the former, the Romans built baths and concerned themselves with the purity of the body as well as of food. And for the latter, an elaborate system of weights and measures was devised. Every Roman market had a *Libripen* (master of the public weights) whose function was to observe the weight of the commodities sold and payment received in various transactions.

In the ancient Greek colony of Cyrene (Cyrenaica) in North Africa, archaeologists have uncovered mosaics illustrating King Arcesilas supervising the weighing and storing of silphium (an ancient herb.).

In later centuries, Islamic civilization (600–1100 AD) contained regulations regarding a weights and measure system which certified various containers as to their proper fill.

As early as 1266 in England, a law called *statutum de pistoribus* regulated the size of bread loaves, and a later regulation required bakers to stamp their mark in each loaf to certify compliance with the law.

But these very early regulations were not directed toward providing a safe and wholesome supply of food for the common man. Rather, they were intended solely to protect each merchant from the other and were

FIG. 9.1. Woodblock print of a scale maker. He produces many different kinds of balances for many uses, including merchants' scales for weighing gold and grocers' scales for spices.

From The Book of Trades, Standebuch, 1568, (Dover Reprint, 1973).

used for tax purposes by the ruler. This was carried right through into the Middle Ages, when Europe was fragmented into small fiefs or city-states. Each charged a tax on goods sold and each had its own version of food regulations. Bread was stamped in medieval Europe to certify that the baker had complied with the law. But in spite of these regulations, food adulteration was widespread. Wine and bread were so commonly adulterated during this time that several specific regulations were promulgated covering these products. In virtually all parts of Europe, manufacturers of food products considered dangerous to health were imprisoned or otherwise punished. Yet, severe adulteration of foods continued because of both poor enforcement and widespread ignorance of these laws.

At the end of the Middle Ages and right up into the late 1800s, bread was mixed with alum, beer adulterated with various narcotics, and tea leaves colored with carbonate of copper. The eminent British painter James Hogarth wrote letters to the newspapers of the day complaining about these practices. The situation caused Charles Dickens to write in one of his novels, "If they be like to die, let them do so." And in 1820, Frederick Accum published the first book solely devoted to food adulterations, *A Treatise on Adulterations of Food and Culinary Poisons.*

As early as 1790 in the United States, Thomas Jefferson then secretary of state, proposed federal legislation on trademarks. But there was so little interest in his idea that Congress took almost 100 years to pass the first trademark law, and even that was declared unconstitutional.

In the United States, the climate relative to proper and meaningful food legislation started to change after the Civil War (1860–1865). Spurred on by the conservation and human rights movements, it reached its climax during the Spanish-American War (1898–1901). When the battleship Maine exploded in Havana, Cuba, U.S. soldiers were rushed to the battlefront. And feeding these soldiers was as important as clothing them. One type of provision supplied to the U.S. Army was canned beef. The beef was purchased by the Army, and even before the U.S. soldiers had arrived in Cuba, many had died (in Tampa, Florida) from the tainted product. Investigation showed that the beef was pickled in formaldehyde and had even been canned more than 30 years before, during the Civil War. This so outraged the American public that William Randolph Hearst picked the story up and published it in all his newspapers. Public opinion was forming and a group of concerned citizens led by Dr. Harvey Wiley pressed hard for regulations capable of stopping such practices. They did not succeed in capturing general consent until the publication of Upton Sinclair's book, *The Jungle*. It was then that the unsanitary conditions prevailing in some meat-packing plants first became known to the public. Theodore Roosevelt recognized the need for improving conditions in these meat plants and pushed for the adoption of food and drug laws. All this culminated in the passage of the first federal Food and Drug Law in 1906 (it went into effect in 1907). (Table 9.1)

TABLE 9.1. Catastrophes and Their Regulatory Result

Event	Result
1. Ten children died of diphtheria antitoxin.	Biologics Control Act of 1902
2. Soldiers died as result of eating tainted canned beef.	Food and Drug Law, 1906
3. "Elixier sulfanilimide" event—70 people died.	Federal Food, Drug, and Cosmetic Act, 1938
4. Physicians meeting in Milan, Italy, on cancer.	Food Additives Amendment to the Federal Food, Drug, and Cosmetic Act, 1958
5. "Thalidomide" incident—birth defects caused by new drug.	New Drug Act, 1962 (Public Law 87-871)
6. Over 2000 deaths of children due to accidental poisoning.	Poison Prevention Packaging Act of 1970
7. Over 350 reports annually of allergic reactions to cosmetics.	Cosmetic ingredient labeling—voluntary and mandatory provisions
8. Marketing of useless radiological devices for cancer cure; other useless devices on market (e.g., arthritis cures).	Medical Device Amendment, 1975 (GMP published in late 1977)
9. Tylenol incident—seven deaths caused by drug, 1982.	Legislative activity for tamper-resistant packaging (OTC)

Between 1906 and 1912, the law referred only to statements on composition or identity. The passage of the Sherley Amendment in 1912 corrected this interpretation and made violators open to prosecution, "if the package shall bear or contain . . . any statement . . . which is false or fraudulent." The basic problem with the 1906 law and 1912 amendment revolved around the fact that the burden of proof was on the government. If a preservative was harmful, it remained up to the government to prove its harmfulness. In addition, the Bureau of Chemistry had to show injury to health. In spite of these factors, by 1917, 6000 violators had been brought to trial. More could have been done, but the Supreme Court considered these cases in the realm of business and not public health cases.

Industry responded to this legislation, and many manufacturers voluntarily policed their products. Adulterated food became less common, and a start was made toward providing healthful and safe food.

The McNary-Mapes Amendment of 1930 began the codification of the quantity of contents in packaged foods. Exceptions were usually found in packaging of meat or milk products due to agricultural pressures.

FEDERAL FOOD, DRUG, AND COSMETIC ACT—1938

In 1938, the passage of the Federal Food, Drug, and Cosmetic Act directly affected the food industry. It was the first law that provided the modern fundamentals of food legislation. A new agency was set up called the Food and Drug Administration (FDA), and it was made responsible for all administrative decisions. Food additives were not controlled except where such additives were known to be poisonous substances. Even this was an improvement over the 1906 law, since between 1907 and 1938 classification and regulatory laws regarding toxic additives had been handled piecemeal as problems arose, and the chemicals prohibited for use in edibles or in materials coming in contact with them were mainly those that had a well-established history of toxicity.

The Second World War interfered with the full application of the 1938 act, but several specific details of the act are important. Standards of identity and a reasonable standard of quality and of fill of container for various foods were specified. Traffic in injurious foods was prohibited. More requirements for stating ingredients on labels were specified. Slack filling was brought under review, and the use of deceptive containers was prohibited. Penalties were invoked to control fraud or deliberate intent to violate the act.

THE 1958 FOOD ADDITIVES AMENDMENT TO THE FOOD, DRUG, AND COSMETIC ACT

In 1958, the Food Additives Amendment to the Food, Drug, and Cosmetic Act was passed. The rules and regulations proposed by the FDA under this amendment are of significant concern to the packaging industry.

Other important amendments to the Food, Drug, and Cosmetic Act which have been passed since 1958 include the Color Additives Amendment of 1960, which establishes the rules under which colors can be qualified for use in or on foods, drugs, and cosmetics, and the New Drug Act Amendment of 1962.

The most important amendment ever made to the original law as far as the food packaging industry is concerned is the Food Additives Amendment of 1958. This law deals with food supply in the United States and takes cognizance of all chemical components of this supply, whether these components enter the food by direct addition or by indirect means.

The packaging industry is concerned with the part of the 1958 amendment that deals with indirect food additives—those substances which can become a component of the food by migration from a packaging material.

FIG. 9.2. An FDA inspection team examines conductors that carry a whole-egg mixture to various parts of the plant. They are examining the equipment for cleanliness.
Courtesy of FDA.

When the 1958 Food Additive Amendment was passed, the FDA was faced with a great deal of work. It had to find a way to administer the new law as well as bring under control, by scientific evaluation, the many thousands of substances involved in foods and in packaging materials for foods.

In order to accomplish this, the FDA first set up the rules and regulations for administering the new law. These rules and regulations included subparts A through G. The subparts had the following definitions:

Subpart A: Definitions and Procedural and Interpretative Regulations

Subpart B: Exemption of Certain Food Additives from the Requirement of Tolerances

Subpart C: Food Additives Permitted in Feed and Drinking Water of Animals or for the Treatment of Food-producing Animals

Subpart D: Food Additives Permitted in Food for Human Consumption

Subpart E: Substances for Which Prior Sanctions Have Been Granted

Subpart F: Food Additives Resulting from Contact with Containers or Equipment and Food Additives Otherwise Affecting Food

Subpart G: Radiation and Radiation Sources Intended for Use in the Production, Processing, and Handling of Food

If each subpart is evaluated separately, the full scope of the law is obtained. These subparts are under Part 121 (Food Additives) of Title 21 (Food and Drugs). The prefix of 121 for each section refers to Part 121.

Subpart A contains several sections and relates primarily to administrative matters pertinent to the administration of the new law. This includes the rules and regulations for the filing of petitions, objecting to proposed regulations, public hearings, judicial review, matters of evidence, etc. This subpart makes up the substantive law with which the Food Additives Amendment is administered.

Subpart B is quite short and includes only two sections, but it nevertheless affects the food packaging industry to a great extent. Section 121.101 of this subpart lists all those "substances that are generally recognized as safe." These substances have come to be known as having a generally recognized as safe (GRAS) status. The forerunner of this section was the so-called "White List." The FDA does not add new materials to this list, but it often removes materials from the list when evidence becomes available to show that the "generally recognized as safe" status should not be continued.

The list of GRAS clearances is under careful review by FDA officials. Due to the cyclamate issue of 1969, a reevaluation of GRAS components was instituted. All status-opinion clearances issued prior to the 1958 Food Additives Amendment have been revoked. The FDA intends to reevaluate these former approvals using modern scientific tools. Companies that have received preamendment approval must send a copy of the letter of approval and a copy of the inquiry letter to the FDA. The FDA will then issue a qualified current opinion based on each component's safety. Over 10,000 additives are affected; however, the FDA has permitted interim use of the products in question. Reevaluation is expected to take many years. This is due to the Reagan administration's views on regulation and the high costs of animal feeding tests ($250,000 per test).

Subpart C includes approximately 75 sections and addresses "Food Additives Permitted in Food and Drinking Water of Animals or for the Treatment of Food-producing Animals." It is not of great relevance to packaging suppliers, because here the food additive is a direct additive, a deliberately added food component. Pet food packaging materials would fall under Subpart F as concerns indirect additives.

Subpart D address "Food Additives Permitted in Food for Human Consumption." This subpart includes approximately 186 sections and lists those additives which are made directly into food. It is of value to those concerned directly with food processing; however, packagers must know which substances are permitted in a packaging material.

Subpart E relates to substances for which prior sanctions have been granted. These substances are listed in the only section under Subpart E, i.e., 121.2001, which is entitled "Substances Employed in the Manufacture of Food-packaging Materials." The "prior sanctioned materials" listed in this section are materials that were sanctioned by FDA for use in food-packaging materials prior to 1958. Dr. Lehman of the FDA issued prior sanctions before to 1958. No further additions have been made to the prior sanctions list since 1958. The reevaluation instituted by the FDA for GRAS clearances also affects all additives with prior sanction or approval. The same letters of approval and inquiry must be submitted to FDA officials.

Subpart F concerns "Food Additives Resulting from Contact with Containers or Equipment and Food Additives Otherwise Affecting Food." This is the most relevant subpart for the packaging industry. The subpart contains approximately 97 sections. Within these sections many components of food packages are regulated. Section 121.2501 concerns olefin polymers, which include polyethylene and polypropylene. Section 121.2502 addresses nylon resins. Even though there are regulations for food-packaging materials such as polyethylene and a determination can be made by simply referring to these regulations, it is also necessary to obtain assurance from a particular supplier that the material in question actually meets with provisions of a specific regulation. Polyethylenes are not all the same, and many polyethylenes are not food grade. Therefore, even though there is a regulation under Subpart F pertaining to polyethylenes, permission must be obtained, in writing, from the FDA to offer assurance from suppliers of polyethylene that the grade of polyethylene to be used by a supplier in a good packaging application is "food grade" by virtue of its compliance with Section 121.2501 of regulations under Subpart F. This also holds true for all other packaging components, such as lubricants, plasticizers, coatings, and adhesives.

The Delaney Clause

A most important part of the 1958 Food Additives Amendment is the Delaney Clause. At a 1956 World Congress on Cancer in Milan, Italy,

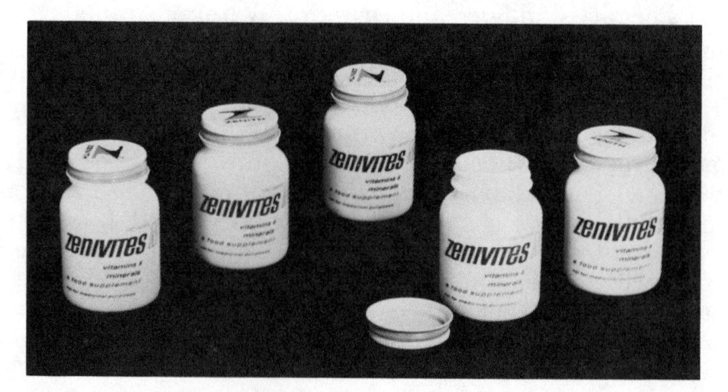

FIG. 9.3. Zenith Laboratories has converted from glass to polyethylene bottles. This conversion of packaging material required FDA approval.
Courtesy of E.I. DuPont de Nemours & Co.

it was concluded that a cause of the world increase in cancer was the increased use of chemical additives in food. These words were heard by Representative Delaney (NY), who immediately moved to ban cancerous substances from foods. He was aided by Sen. Estes Kefauver, a Democrat from Tennessee. The net result was the Delaney Clause—a clause that allows zero tolerance for any additive that is a carcinogen.

This clause is becoming of increasing concern for the packaging industry of the 1980s. At the time of its inclusion into the 1958 amendment, the test methods used to detect carcinogenic substances were valid to about 20 parts per million. Testing advances have now brought these analytical methods to about a few parts per trillion. The industry is now seriously considering how to revoke the Delaney Clause and substitute some type of benefit/risk statement.

Extraction Tests

Several FDA regulations under Subpart F require that extraction tests be run. This is required by some regulations to make sure that the chemicals involved do not migrate to the food beyond the level established by the FDA to be safe.

Extraction data may be complex. All extraction data must contain the amount of a single substance which migrates under standard conditions and the total contamination of the food product. In addition, toxicity data must be available for FDA evaluation.

Migration of additive depends on time, temperature, area, and type of food. While extraction studies may be conducted on a specific food, most tests utilize special extracting solutions. The solution selected is specific for the food to be tested. For particularly sensitive applications, radioactive tracer analysis is used. When tagged carbons are employed, excellent results are obtained. However, radioactive tracer tests are expensive and require extremely intricate equipment. Due to these factors, the process is not in widespread use.

In order to formulate a range of solutions capable of simulating food extraction, several factors are significant. The various solution types that should be included are acid, alkaline, and neutral as well as alcoholic and oil solutions. A representative list suggested by the FDA is (1) distilled water, (2) 3% aqueous NaCl, (3) 3% aqueous $NaHCO_3$, (4) 3% aqueous acetic acid, (5) 3% aqueous lactic acid, (6) 20% aqueous sucrose, and (7) lard or vegetable oil.

The list of solutions recommended by the French authorities is more complex: (1) distilled water, (2) 10% ethyl alcohol, (3) 50% ethyl alcohol, (4) 95% ethyl alcohol, (5) 10% acetic acid, (6) 1% NaCl solution, (7) 5% NaCl solution, (8) 10% sucrose solution, (9) 2% citric acid solution (pH 4.5), (10) lard, and (11) arachide oil.

It is essential that all extractants be used correctly to arrive at an accurate determination of the migrant. Since different migrants often require different extracting solutions, the proper test method for the specific chemical must be used. For example, highly poisonous substances require extremely sensitive test solutions.

In principle, migration studies should actually be conducted for the entire shelf-life of a packaged food. In practice, however, the amount of time involved makes such long-term tests impractical. Most tests are conducted at accelerated temperatures and the results are correlated with actual conditions. U.S. authorities state that the test conditions used must correlate with actual packaging conditions. If a film is autoclaved during a packaging operation, migration tests must be conducted at the autoclaving temperature. Other nations feel that the test conditions must be more severe than those actually encountered. Since it is often impossible to know the total packaging cycle, extremely stringent test conditions are usually employed.

The diffusion of migratory substances into food products increases with increasing temperatures. It has been reported that extraction is fairly uniform for the first 5 days at a temperature of 140°F (60°C). After the fifth day, the rate decreases and extraction ceases by the tenth day.

The thickness of the sample used should be uniform. A 1.0 mil sample is recommended, since this is the thickness of the most commonly used plastic films and sheets. Total migration should be expressed in parts per million, and a ratio of 0.3 ml solvent/cm^2 of surface is preferable for the test.

After allowing the sample to be exposed to the solvents chosen at the specified time and temperature cycles, quantitative identification of migratory compounds is made by gas chromatography or conventional analytical methods.

Subpart G relates to radiation and radiation sources in the handling of food.

Subpart F is the only subpart which contains regulations which require compliance. If a material can be added directly to food, has GRAS clearance, or has been prior sanctioned, it is not regulated as an indirect food additive. Accordingly, the food-packaging industry is concerned with those violations of FDA regulations covered primarily within these Subpart F regulations for indirect food additives.

FOOD ADDITIVE DEFINITION

The definition of a food additive is of great value in fully understanding the law. The 1958 amendment provides that a food shall be deemed to be adulterated if it contains a food additive. Once a food contains a food additive or if a food packaging material is capable of giving rise to a food additive, penalty provisions of the act apply. The definition of a food additive is most important. A food additive is not just any substance added to food directly or indirectly.

The Food Additives Amendment as passed by Congress included a special definition for a food additive. The legal definition of a food additive is:

> . . . any substance the intended use of which results or may reasonably be expected to result, directly or indirectly, in its becoming a component or otherwise affecting the characteristics of any food (including any substance intended for use in producing, manufacturing, packing, processing, preparing, treating, packaging, transporting, or holding food; and including any source of radiation intended for such use), if such substance is not generally recognized, among experts qualified by scientific training and experience to evaluate its safety, as having been adequately shown through scientific procedures (or in the case of a substance used in food prior to January 1, 1958, through either scientific procedures or experience based on common use in food) to be safe under the conditions of its intended use.

This term (food additive) does not include pesticide chemicals or any substance used in accordance with a sanction or approval granted before the enactment of this act.

A food additive is actually any substance that can become a component of food by direct or indirect means unless that additive is known to be safe or is prior sanctioned. If that additive is known to be safe, such as GRAS materials, or is prior sanctioned, then that substance is not a food additive. The FDA has declared some substances to be nonfood additives by listing them on the GRAS list and has declared other substances to be nonfood additives by prior sanctioning. These substances appear in subparts B and E, respectively.

All other materials of the food industry which are used both in food directly and in packaging materials for foods are treated by the FDA in a separate manner. The FDA requires a petition to be filed by the chemical manufacturer requesting a permit to use a product in food or in a food-packaging material. After the FDA reviews the toxicological information in the petition and finds it satisfactory, the FDA will issue a regulation for the safe use of the chemical in question. This then regulates the use of the material. Regulations which cover direct additives for food appear in Subpart D, and regulations which cover indirect additives appear in Subpart F. The regulations appearing in Subpart F are most relevant.

By issuing a regulation for a particular substance, the FDA has declared that substance safe and, therefore, not a food additive so long as the use of the substance complies with provisions of the issued regulation. A regulation states that migration of certain substances from a coating shall not exceed 0.5 mg/in.^{2} of contact surface in the presence of a certain solvent under particular extraction conditions. If, in fact, the extraction level from this regulated coating under these conditions is higher than this amount, then the packaging material in question is capable of introducing a food additive into the food supply. That is, the packaging material components could enter the food supply at a level beyond limits known to be safe by the FDA. In such a case, violation of the 1958 Food Additives Amendment exists and all the provisions of the Food, Drug, and Cosmetic Act apply, including the seizure provision.

NEW AMENDMENT TO FOOD ADDITIVE LEGISLATION (1983)

The first attempt to modify the 1958 Food Additives Amendment began on June 1983 when the Senate Committee on Labor and Human Resources held hearings on the need for food safety reform. The Committee heard testimony from various FDA officials, the U. S. Department of Agriculture, scientists, consumer groups, and industry representatives.

Legislative momentum continued with the introduction of the Food Safety Modernization Act (FSMA) of 1983 on October 6, 1983. Identical bills were introduced in the House by Rep. Edward Madigan and in the Senate by Sen. Orrin Hatch. The bills include provisions addressing indirect additives; the definition of "safe"; modification of the Delany Clause; consideration of an additive's health benefits; scientific peer review; phaseout authority; informal rule making for food contaminant tolerances, and conforming amendments for the Federal Meat, Poultry and Egg Inspection Act. The bills do not provide for a premarket notification system as a possible alternative to the currently used additive petition process.

MANUFACTURING REGULATIONS

Good manufacturing practice is carefully outlined in federal regulations. Regulation 121.2500, "General Provisions Applicable to Subpart F," states that the use of any and all of the additive substances regulated in the 96 other regulations of Subpart F is predicated upon the use of good manufacturing practice. Paragraph (a)(1) states:

> The quantity of any food additive substances that may be added to food as a result of use in articles that contact food shall not exceed, where no limits are specified, that which results from use of the substance in an amount not more than reasonably required to accomplish the intended physical or technical effect in the food contact article, etc.

This means that a food additive can be contained within a food packaging material, but that such additive must be contained at the absolute minimum required to accomplish the intended technical effect.

There are special situations where this provision would apply in packaging operations. For example, in the production of a packaging material where a lacquer coating is applied and the coating is not

TABLE 9.2. Recent Developments in Regulatory Food Packaging

Subject	Regulation	Description
FDA Carcinogen Policy	46 Fed. Regis. *14*, 464–*14*, 470 (April 2, 1982)	1. Clarification of "Food Additive" 2. Delaney Clause only applicable when additive as a whole causes cancer 3. Risk assessment as one of the tools for determinations of carcinogens
Polyvinyl chloride	FDA recognition that 38 Fed. Regist. *12*, 931 and 40 Fed. Regist. *40*, 529 are invalid	FDA acceptance of PVC as prior-sanctioned material provided it is made with GMP[a] of low RVCM[b] content. Acknowledgment that 1975 regulations invalid.
Acrylonitrile (AN)	Monsanto Co. *v.* Kennedy 1979	Monsanto petition with data showing *no* extraction of acrylonitrile from bottle to carbonated beverages
Aseptic packaging	21 CFR 184, 1366 21 CFR 186, 1(a)	Clarification of concentration of hydrogen peroxide to be used
Irradiation	46 Fed. Regist. *18*, 992 (1981)	Deals with level of radiation used for foods and labeling required.

[a]GMP, good manufacturing practice
[b]RVCM, residual vinyl chloride monomer

properly dried, the use of the material as a food wrap with the coating in direct contact with food could result in the solvent retained in the coating migrating to the food, especially if aluminum foil is part of the construction.

Any time that a packaging material gives rise to a food additive in excess of the minimal amount allowed by this provision under good manufacturing practices, industry must face application of the Food, Drug, and Cosmetic Act penalty provisions.

In other Subpart F regulations, reference is made to good manufacturing practice. In the Adhesives Regulation 121.2520, the quantity of adhesive that contacts packaged dry food must not exceed the limits of good manufacturing practice. The quantity of adhesive that contacts packaged fatty or aqueous foods must not exceed the trace amount at the seams and at the exposed edge between packaging laminates that may occur within the limits of good manufacturing practice. In the case of coatings, this means using only the amount of adhesive required to obtain firm bonding of the seams and laminates.

On December 15, 1967, a proposed good manufacturing practice regulation for the sanitary processing and packaging of human foods was published in the *Federal Register*. The rules state that the packaging processes and materials shall not transmit contaminants or objectionable substances to the food product, shall conform to any applicable food additive regulation, and shall provide adequate protection from adulteration.

It is apparent from provisions of all good manufacturing practice regulations that food-packaging materials are being considered separately more and more in connection with food handling and food adulteration. Greater emphasis is expected on food-packaging materials. With this greater emphasis could come a greater likelihood of court action as a result of alleged violations of packaging regulations.

FUTURE ACTIVITY

Current FDA activity is mainly concerned with clarifying the status of (1) additives considered to be safe under Section 402 (a)(1) of the Federal Food, Drug, and Cosmetic Act; (2) GRAS additives, (3) prior sanctioned or approved additives, and (4) components not regarded by FDA as additives under conditions of intended use. It will probably take the FDA years to check the above materials. After this is done, the federal authorities will review all food additives in terms of safety.

There is a growing opposition among packaging material manufacturers to the 1958 amendment (particularly the Delaney Clause). No differentiation is made between direct addition of ingredients to food and minute quantities of additives migrating from the packaging material. A proposal has been made that all components of food-packaging materials of 0.2% or less (based on package weight) be exempt from the law.

THE FDA AND DRUGS

The packaging of drugs is different from the packaging of food or cosmetics because of the 1962 New Drug Act. The amendment states in effect that the drug and the package are one unit. When the FDA evaluates a drug, it thus evaluates the package. This requires that drug and packaging people work closely together in the development of new packaging for drugs.

In the specific example of a semirigid foil container or a flexible pouch, a plastic is in contact with the drug. Although plastics are often considered to be fairly inert materials, the Packaging Institute has pointed out that plastics may contain unreacted monomers, catalysts, residual solvents, antioxidants, plasticizers, and release agents. These materials are extracted by liquid drugs, which very often contain alcohol. Another problem encountered is that of permeation and sorption. When a drug and a plastic are in contact with one another, there is permeation in both directions. Sorption can reduce the efficacy of liquid drugs stored in plastic-lined packages. It was recently reported by Ciba that in one drug mixture, up to 85% of the total potency of specific ingredients was lost due to sorption by nylon and polyethylene within a short period of time. Since drugs are administered in small doses, sorption can raise havoc with drug effectiveness.

The specific FDA regulation relative to drugs states that "drug packages to be suitable for their intended use must not be reactive, additive, or absorptive to an extent that the identity, strength, quality, or purity of the drug will be affected."

The FDA has also stated that the "unit dose drug packages . . . present a new hazard . . . chronic toxicity." The FDA says that this develops when foreign substances are leached from the package into the drug.

The question now is what to do to resolve this situation. Cross-linking and polymer density can be increased. This will reduce permeation and sorption. But is this enough?

It might also be possible to reduce the amount of plastic. Normally, a container is lined with a 1 mil film or with 1–6 lb of a solution coating per ream. New techniques are needed to get coverage without using so much plastic. With more developmental activity, the FDA status of unit-dose drugs may be resolved.

The FDA is moving toward a change in its drug regulations that would let it approve the marketing of some new drugs solely on the basis of clinical tests abroad. This is based on an attempt to respond to a longstanding complaint of pharmaceutical companies—and of some doctors—that drugs with proven success in Europe may be denied to Americans unnecessarily (Table 9.2).

The Tylenol Incident

The Tylenol tragedy of 1982 in which several deaths occurred from the ingestion of poisoned Tylenol capsules led the FDA to issue tamper-resistant packaging regulations for certain over-the-counter drugs, cosmetics, contact lens solutions, and tablets in November 1982 (47 Fed. Reg. *50*, 442). The FDA regulations do not apply to food, and there is every indication that the regulations will not be extended to cover food. On October 13, 1983, President Reagan signed into law P.L. 98–127, which makes tampering with foods, drugs, cosmetics, and other consumer products a felony under federal law. The legislation would make it a federal crime to tamper with or attempt to tamper with any consumer product, or its labeling or container with "reckless disregard for the risk that another person will be placed in danger of death or bodily injury and under circumstances manifesting extreme indifference to such risks." It also is illegal to communicate false information that tainting has occurred if such tainting, had it occurred, would have created a high risk of death or bodily injury to another person. Conspiracies to commit a tampering offense and threats of tampering are also covered. Maximum punishments range from $10,000 to $100,000 in fines and from 1 year to life imprisonment, depending on the severity of the injury that occurs as a consequence of the crime.

TABLE 9.3 Changes in Drug Regulation Since 1938

Date	Regulation of Act
1962	New Drug Act (Kefauver-Harris Amendment)
1965	Drug Abuse Control Act
1967	Computerized data bank organized
1970	Comprehensive Drug Abuse Control Act
1971	NDA and IND guidelines organized with cooperation of Pharmaceutical Manufacturer's Association (PMA)
1972	Biologicals covered by FDA
1976	Medical Device Amendment
1978	GMPs for Medical Devices

Package tampering has not yet been a widespread occurrence in either Europe or Asia. There have been sporadic reports of tampering in the United Kingdom, Italy, and Brazil. In Japan, the Ezaki Glico Co., Japan's second largest confectioner, was the target of a Tylenol-style poisoning scare.

THE COLOR ADDITIVE AMENDMENT

In 1906, the original U.S. Food and Drug Act provided for a voluntary system of certification of the purity and safety of dyes used in or on foods. Under the 1938 revision, limits were established on the minute amounts of lead, arsenic, and other heavy metals that could be formulated in the colors. It also added a provision setting down the exact specifications of each individual color. Manufacturers submitted samples to the FDA for certification. Covered under this act were the coal tar colors, but not the nonaniline colors, such as iron oxide, titanium dioxide, and naturally occurring colors.

The Color Additive Amendment of 1960 embraces all colors, both natural and synthetic, used in foods or in packaging materials contacting foods. It was designed to replace the outmoded provisions which governed the use of color in foods under the 1938 act. Tolerance limitations were established when required as well as other conditions for safe use. The system of official certification was continued.

In essence, three groups of colors are permitted for use in coloring of packaging materials for food. Certified FD&C dyes and alumina lakes are acceptable. FDA-approved pigments and FDA-sanctioned colors are also permitted. Any colorant that has an impermeable barrier between it and the food is not subject to the Color Additive Amendment. It should be noted that when the colorant is not approved, the burden of proof relative to migration, offset, bleeding, etc., rests with the user.

An ink should be formulated by using certified or FDA-approved colors. It should also use ingredients that are generally recognized as safe or approved materials under the Food Additives Amendment. Lists of acceptable ingredients are available in the various periodical editions of the *Federal Register*.

THE "FAIR PACKAGING AND LABELING" BILL

In 1966 the Federal Packaging and Labeling Act (FPL Act) was enacted. This act became effective July 1, 1967. Specific extensions have been authorized by the FDA, but otherwise all food products must comply with the laws. The regulations state that the following information is required to appear on the label of food packages.

First, the food must be identified. This information must appear on the principal display panel and be in bold type. It should be in lines generally parallel to the package base when the unit is displayed. The common name of the food should be used whenever possible. If the food is offered as "whole," "sliced," or "chopped," this must be clearly visible to the consumer in either a statement or a vignette.

Second, the name and address of the manufacturer, packer, or distributor must be identifiable. It should be conspicuous and denote the type of firm supplying the product.

Third, the net quantity of contents must be stated. The use of "Jumbo Pound," etc., is forbidden. This information must be placed in the bottom 30% of the area of the display panel and appear as a distinct item. If the food is liquid, fluid measure is used. If it is solid, weight is used. For fresh fruits and vegetables, the bushel pack system is necessary. The net contents statement must be accurate and may not include the weight of the packaging material. For packages containing between 1 and 4 lb or between 1 pt and 1 gal., a dual declaration of net weight or net volume is required. The first statement must show total ounces or total fluid ounces. The second statement, in parentheses, must express the total in pounds and ounces or in pounds and fractions or decimal fractions of a pound for solids, or in quarts, pints, fluid ounces, and fractions thereof for liquids. A separate declaration of net contents in terms of metric units may also be added. The net contents statement must show the total amount of food in the package. When part of the contents is not used as food but thrown away, as with the brine in a bottle of olives, only the amount of solid food should be stated.

Fourth, a statement of the net contents of a serving is required. This may appear in units of weight, liquid, or dry measure or any other commonly used term—i.e., tablespoons. If the contents are noted in terms of servings, the serving size must be given.

Fifth, all ingredients must be mentioned. This statement must be legible and appear on a single panel of the label. Ingredients must appear in decreasing order according to their percentage of the total product weight. If a special ingredient is advertised as being exceptionally valuable to the food, a declaration of percentage content is required.

In the converting industry, the FPL act is not applicable. The manufacture and sale of converted forms of paper, film, and foil do not subject the manufacturer to the provisions of the act. The law is aimed at firms that actually package or label the products or distribute them in interstate commerce.

Voluntary assumption of the responsibility to manufacture labels that conform to the FPL Act is an individual matter for each convertor to decide.

OTHER GOVERNMENTAL REGULATIONS

Poison Prevention Packaging Act (1970)

The Poison Prevention Packaging Act (PPPA) of 1970 was introduced by Senator Moss of Utah in 1969. It was proposed as an amendment to the Federal Hazardous Substances Act (FHSA) of 1960 and was restructured in 1970 prior to passage. The bill introduced was designed to make it impossible for children under 5 yr of age "to obtain a toxic amount of product in a reasonable time." While a large number of children under 5 are poisoned every year, it is doubtful whether safety closures and/or child-resistant flexible packaging are the total answer. There is certainly no substitute for an inaccessible storage place for dangerous containers.

Assisted by a technical advisory committee and a study conducted by Dr. Wilton Krogman, a testing procedure (protocol) for special packaging was established for the PPPA. This includes a panel of 200 children used to test the effectiveness of various safety closure designs. The children are between 3½ and 4¼ yr old. This group, which is three times larger than the minimum number needed to produce significant statistical results, is given 10 min (two 5-min periods) to open the closures. The children are even paired off to include the possibility of a "team attack" by the youngsters.

In addition, a panel of 100 adults (70% female) is used, ages 18 to 45 yr, inclusive.

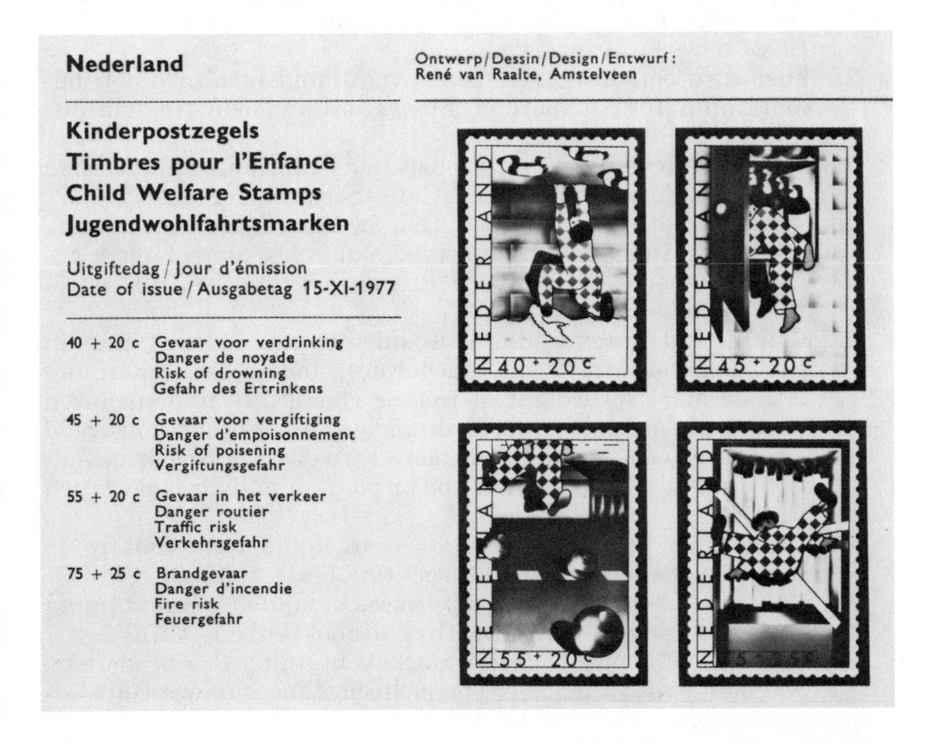

FIG. 9.4. Child Welfare Stamps from the Netherlands show accidents occurring to children. The stamp in the upper right features child poisoning.
Courtesy of the government of the Netherlands.

Dr. Robert G. Scherz, chief of pediatrics service for Madigan General Hospital, proposed an alternative testing protocol for safety packaging. Alternative Protocol, Dr. Scherz's procedure, reduces the number of children in the test panel from 200 to 100 but requires all 100 to be male. The adult panel is reduced from 100 (70 females and 30 males) to 70, all of whom are females.

The technical advisory committee reviewed these procedures and selected the Press Protocol. This test method was published in the *Federal Register*. The Consumer Product Safety Commission (CPSC) is now emphasizing adult resecuring of child-resistant closures so as to maintain their safety feature. In a recent case against Miles Laboratories, the closure on its vitamins failed the test in adult resecuring ability. A cap change ensued and all were satisfied.

Substances that require safety closures to protect children from serious personal injury or serious illness are:

1. Aspirin: any aspirin-containing preparation for human use in a dosage form intended for oral administration (August 14, 1972)[1]
2. Furniture polish: nonemulsion type liquid furniture polishes containing 10% or more of mineral oil and/or petroleum distillates (September 13, 1972)
3. Methyl salicylate: liquid preparations containing more than 5% by weight of methyl salicylate (September 21, 1972)
4. Controlled drugs: any preparation for human use that consists in whole or in part of any substance subject to control under the Comprehensive Drug Abuse Prevention and Control Act of 1970 (October 24, 1972
5. Sodium and/or potassium hydroxide: household substances in dry forms, such as granules, powder, and flakes, containing 10% or more by weight of free or chemically unneutralized sodium and/or potassium hydroxide and all other household substances containing 2% or more by weight of free or chemically unneutralized sodium and/or potassium hydroxide (April 11, 1973)
6. Turpentine: household substances in liquid form containing 10% or more by weight of turpentine (July 1, 1973)
7. Methyl alcohol: household substances in liquid form containing 4% or more by weight of methyl alcohol (July 1, 1973)
8. Sulfuric acid: household substances containing 10% or more by weight of sulfuric acid, except such substances in wet-cell storage batteries (August 14, 1973)
9. Kindling and/or illuminating preparations: pre-packaged liquid kindling and/or illuminating preparations which contain

[1]The date given is the effective date or date proposed for packaging standards as published in the *Federal Register*.

10% or more by weight of petroleum distillates (October 29, 1973)

10. Prescription drugs: any drug for human use that is in a dosage form intended for oral administration and that is required by federal law to be dispensed only by or upon oral or written prescription of a practitioner licensed by law to administer such a drug (April 16, 1972)
11. Economic poisons: household substances packaged in liquid forms in less-than-1-gal. containers or in nonliquid forms in packages of 5 lb or less that have been found to be highly toxic by human experience (September 14, 1972)
12. Ethylene glycol: household substances in liquid form containing 10% or more by weight of ethylene glycol; includes automotive antifreeze and certain automotive brake fluids (December 28, 1972)
13. Paint solvents: solvents for paint or other similar surface-coating material solvents (such as thinners, removers, and brush cleaners) in liquid form which contain 10% or more by weight of petroleum distillates, benzene, toluene, xylene, or combinations thereof (February 9, 1973)
14. Promotionally distributed samples: promotional samples of any household substance distributed directly to the household but not presented in person to a responsible adult member thereof and for which any cautionary labeling is either required or recommended, except for substances in pressurized spray containers for which the only hazard is that the contents are under pressure (February 9, 1973)
15. Acetaminophen-containing products (August 1978)
16. Iron-containing drugs and dietary supplements containing 250 mg of iron or more (August 1978)
17. Over-the-counter (OTC) antihistamines

The manufacturer or packer of any of the substances listed must provide the CPSC with a sample of each type of safety packaging used as well as the labeling for each size of the product and any noncomplying package.

The FDA has also set rules for noncomplying packages permitted under the PPPA. Manufacturers must meet five conditions to market products in noncomplying packages intended for the elderly and handicapped. The conditions are (1) the regular closures must be difficult for these people to use, (2) child-resistant packaging must be used with popular sizes of the product, (3) packages with safety closures must be adequately distributed and advertised, (4) the noncomplying size must be the one most likely to be used by the elderly, and (5) the exempt package must bear required label warnings.

In addition to the FDA, several other governmental organizations regulate packages. The Meat Inspection Act of 1906 established all

Meat Inspection Division (MID) laws under the U.S. Department of Agriculture (USDA). Labels and the overall visibility of all meat packages must be approved by the MID (now the Technical Services Division of Consumer and Marketing Services). Poultry and poultry products are also subject to Department of Agriculture approval. The CPSC regulates consumer safety in relation to various products.

The Treasury Department supervises the packaging of alcoholic beverages and tobacco. Narcotics are supervised by the Bureau of Narcotics. Imports are subject to the Bureau of Customs, and postal shipments are enforced by the Post Office Department.

Overlapping jurisdiction often occurs, whereby a product must meet both FDA and USDA regulations. In such cases, approval is obtained by each agency. It is important to note that rules and regulations differ considerably. The most stringent regulations are FDA laws, and if a product meets these regulations, additional governmental approval is fairly easy.

The differences between FDA regulations and USDA regulations revolve around two areas: packaging colorants and packaging approval procedure.

New Tamper-resistant Regulations

With "copy-cat" contaminations occurring in the aftermath of Chicago's Tylenol poisonings, the Government and the pharmaceutical industry announced new packaging regulations. Tamper-resistant protection was required for almost every nonprescription shelf drug within 15 months after November 8, 1982.

Within 90 days of that date, all newly manufactured capsules, liquids, and sprays (excluding skin products) had to have tamper-resistant containers. Within 180 days, tablets and suppositories had to comply, new packages had to have a distinctive design around their seal to show meddling, and a warning had to be evident advising consumers not to buy the item if the seal looked broken. After 15 months, products left over from before the regulations could no longer be sold. The rules let manufacturers use different kinds of barriers, such as bubble packs, shrink wrap, and vacuum seals. The packaging costs consumers up to 2 cents more per item.

FDA VERSUS USDA

Both the FDA and the USDA are involved in food-packaging regulations. An understanding of their rules are of great importance to the packaging industry.

Packaging Colorants

The FDA permits the use of heavy-metal-content inks in flexible food packaging where the inks are separated from the food by a barrier.

The USDA does not permit the use of heavy-metal-content inks in flexible food packaging even where the inks are separated from the food by a barrier. The usual barrier materials, such as films, coatings, and aluminum foil, are considered by the USDA to be inadequate to protect the food from these inks because of mistakes they feel may occur in manufacturing whereby the barrier metal might be inadvertently omitted from the package. Another situation of concern to the USDA is the handling of flexible packaging in the home, wherein meat may be placed on top of a package having heavy-metal inks. None of the problems are considered to relate to metal cans and jars. Therefore, heavy-metal-content inks could and are being used on these packages.

The ruling by the USDA places a particular hardship on flexible heat-in packaging, since heavy-metal-content inks such as chromium yellow are normally needed to withstand high temperatures.

In plastics, the FDA has approved few colorants, whereas the USDA has an accepted list for these materials.

Packaging Compliance Procedure

The FDA issues food-packaging regulations which say that it is the responsibility of the food-packaging material manufacturer to produce materials that comply with these regulations. No direct contact with the FDA is required.

The USDA, on the other hand, requires the packaging material manufacturer specifically to clear beforehand all material to be used to package meat and poultry.

The USDA has published proposed regulations that are being promulgated under the Federal Meat Inspection Act. Part 317 of these proposed regulations deals with packaging. It is entitled "Labeling, Marking Devices and Containers." It gives instructions on how to get USDA approval for meat or poultry packaging and labeling.

In comments provided to the USDA with regard to the proposed regulation, Sections 317.3, 317.4, and 317.5 were attacked by such associations as the National Flexible Packaging Association and the Society of Plastics Industries. They contend that an undue hardship was worked on the packaging industry by establishment of an overdetailed procedure and the requirement for a records keeping and composition disclosure.

It would seem that matters would be made easier if the USDA were to rely on the indirect food additive regulations issued by the FDA. The present need to treat meat and poultry packaging differently from other food packaging seems to create unnecessarily heavy compliance problems for the packaging industry. In addition, the FDA has been reconsidering its overall policy on the toxicological insignificance of food-packaging migrants. There were some discussions on permitting 0.05 ppm migration to food of components that are not heavy metals, carcinogens, or pesticides. If industry could propose a regulatory position acceptable to both the FDA and the USDA that would eliminate the need to treat meat and poultry packaging differently than other food packaging, a great deal would be simplified.

Other Government Regulatory Agencies

Although the FDA is the biggest and most viable regulating agency in the field of food and drug packaging, U.S. local and state laws and the laws governing foreign nations are also of concern to the industry. When exporting, the internal regulations of specific nations must be considered in shipping packaged products. For example, all canned meats shipped from abroad to Egypt and other Islamic nations must carry a statement on the label denoting that the animal was killed according to Islamic ritual (similar to Judaic principles, this involves a *shohet* cutting through the windpipe and gullet with a sharp implement).

Both state and local laws are significant to food and drug packaging operations where the product is sold only within the specific locality. Cities may outlaw specific packages, and yet the same packages may be perfectly legal in other parts of the state.

Foreign laws are more complex than U.S. local laws; this discussion will focus on international regulations prior to local laws. Shipment qualification tests will also be covered.

General Foreign Guidelines

Two words are of prime concern to each country having food-packaging regulations. These are toxicity and solubility. The governments want to know how toxic the questionable component is and how much of it can get into the food supply. Specific regulations differ, depending on the general sophistication of the country and its volume of packaged foods. Regulations are sometimes vague or not even enforced. To review foreign legislation is a most difficult task, since language and geographical factors often tend to confuse the issue. Some general guidelines for obtaining foreign approval follow. If the subsequent five statements are followed, approval by foreign nations should not be a problem.

1. Know the ingredients.
2. Know the legal status of those ingredients in the country where they are manufactured and in the countries where they are sold.
3. Obtain letters from suppliers regarding this status.
4. File the letters in a safe place.
5. Make only provable statements to customers, and say nothing in writing that cannot be verified.

State Regulations

Federal laws apply only to interstate traffic. If the food or package is produced and consumed within one state, it cannot be regulated by federal legislation.

Food laws vary considerably from state to state. In instances where federal regulations cannot be applied, there are no provisions to prevent a manufacturer from using any additive desired. Additive use is usually covered by a general phrase such as the use of any "nonpoisonous material necessary to effect good manufacturing practice." State laws also reflect state's special interests. California has stringent water laws. Historically, state control has been limited to weights and measures, intrastate shipments of foods and drugs, and health law regulations for dairy products and fresh seafoods.

To provide examples, the laws of several states relative to food and drugs follow:

South Carolina. Although there are no regulations concerning cosmetics, South Carolina's drug and narcotic laws follow federal regulations. Food laws pertain only to sanitation.

New Jersey. The 1955 New Jersey Food, Drug, and Cosmetic Law states that certain chemicals are prohibited in food manufacture. Labels on food packages must follow specifications, and there are quite-detailed provisions regarding carbonated and nonalcoholic beverages.

Virginia. The state laws of Virginia closely follow federal laws. Some FDA standards of identity have been adopted, and a special standard has been provided for sausage ingredients.

Hawaii. The meat and fish laws of Hawaii are quite detailed and specific. Color may be used only in casings for sausage products, and there must be no penetration of dye into the product. The label must also indicate the presence of dye. Cold-stored fish and seafoods that have been kept in storage for 30 days or more must be so labeled. In general, Hawaii follows all federal laws closely and issues special laws that control locally consumed foods, such as poi and Oriental food items.

Alabama. Alabama follows USDA regulations rather than FDA rules, although FDA labeling requirements have been used.

New York. Although food additive regulations in New York closely follow federal law, additives must be approved by state officials. New York issues GRAS lists and also adheres to prior sanction exceptions. Its standards of identity are similar to federal standards.

It should be noted that all state laws must conform to the basic provisions of the FDA rules. Interpretations generally differ. Extraction and migration data are standard in all states. Variances occur in restrictions on package design, labeling, size of lettering, and color employed.

City Regulations

Large metropolitan areas also often have regulations regarding deceptive packaging and product visibility. The laws passed in New York City and Chicago requiring prepackaged meats to be displayed in transparent plastic trays are examples of city laws. No meats can be sold in foam or molded pulp trays in supermarkets in these cities. The Tylenol incident in 1982 caused Chicago to issue a city regulation mandating the use of tamper-resistant closures on all drug packages.

Laws such as these come from local health departments, which focus their attention on firms that, because of their size or area of business, do not fall under federal or state inspection. Counties also have health departments. A good example is Los Angeles County, which has an agency that covers the city of Los Angeles as well as other municipalities in Los Angeles County.

Attempts have been made to coordinate the efforts of state and local officials to bring some degree of order to the many differing laws. The Association of Food and Drug Officials of the United States (AFDOUS)

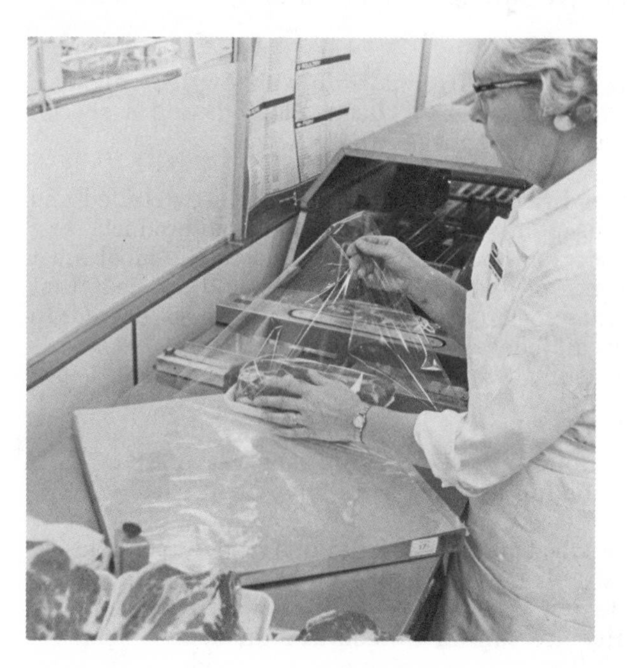

FIG. 9.5. Meat packaged in nontransparent trays have been outlawed in New York City and Chicago by local ordinances because the underside of the meat is not visible.
Courtesy of Reynolds Metals Co.

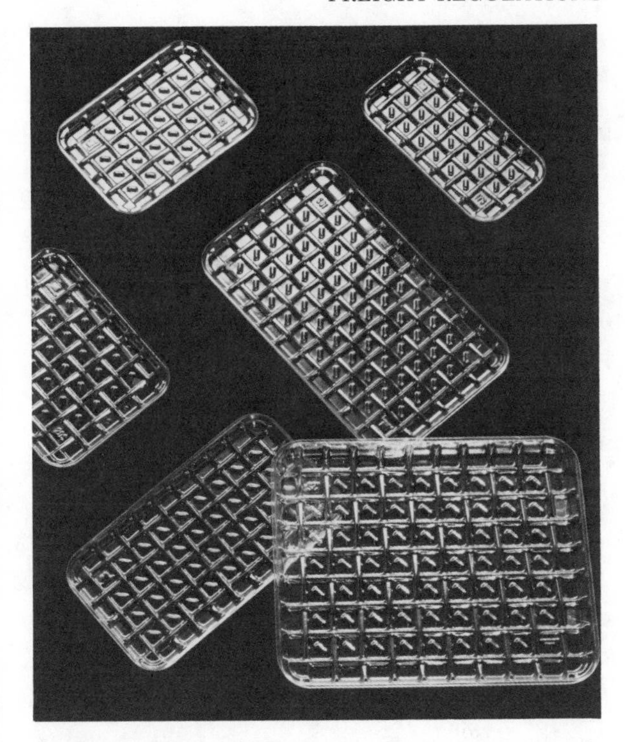

FIG. 9.6. Transparent plastic trays are widely used to package fresh red meats.
Courtesy of Mansanto Co.

consists of groups responsible for the enforcement of food and drug laws. This organization meets yearly, with regional conferences set in alternate months. Such meetings do much to reduce the degree of confusion in state and local regulations.

FREIGHT REGULATIONS

A new package is the product of many hours of costly research, development, and testing. Developing a new package and obtaining classification approval for transportation is not a simple, clear-cut project; 2 to 3 years is not an unusual length of time from inception to market for many items.

After a new package has been perfected and is ready for marketing, it must be presented to the National Classification Board (for motor freight) and Uniform Classification Committee (for rail freight) for appraisal and issuance of a test permit for shipping purposes in order to determine the durability of the package in actual transportation (Table 9.4). Regardless of how good a package is, it cannot be shipped legally in interstate commerce via common carriers without proper "Classification" approval and the Classification Tariffs properly referenced for collection.

TABLE 9.4. Shipment of Hazardous Materials—Other Methods

Method	Regulation/Organization	Regulation
Mail	U.S. Code, Chapter 83, U.S. Postal Service	Title 39, Parts 123 and 124 of Title 49
Private carriers	Interstate Commerce Act	49 CFR 177
Contract carriers		
Bus	National Bus Traffic Association	Rule 16, Section B, Northeastern Tariff No. A-603-C
United Parcel Service	United Parcel Service	Guide (in text)

The procedure for obtaining Classification approval begins with issuance of a test permit to the party that will actually ship the commodities packaged. On each shipment made, the bill of lading must be referenced with the test permit number, inspections must be made at the destination, and reports must be prepared on each shipment to submit to the classification committees.

After a sufficient number of shipments have been made, application is made to the respective classification committees, supported by the test reports, for Classification approval of the new package. If the committees are satisfied with the reports and feel a sufficient number of test shipments have been made, they will then place the new package, as specified, on their docket for hearing. At the hearing, further evidence must be presented in favor of the package . . . opposition is usually expected, even on a new innovation in packaging.

For example, when the Reynolds Metals Co. introduced a new package called "Shrinkase," it was strongly opposed by the Paper Box Manufacturer's Association. When compared, the quantity of paperboard used in the Shrinkase package was considerably less than the amount used for a conventional paperboard box. Despite this opposition, Classification approval was obtained for shipping all canned vegetables, meats, fruits, and juices via motor carrier, and this container is now listed as Package 500 in the National Motor Freight Classification. Green Giant Co., with Reynolds Metals Co. assisting, obtained approval for rail shipments of canned vegetables in this package; it is listed as Package 793 in the Uniform Classification.

BIBLIOGRAPHY

ANON. 1972A. Guiding principles for responsible packaging and labelling. (Rep. No. COM-74-10990-SET). National Business Council for Consumer Affairs, Sub-Council on Packaging and Labeling, Washington, D.C.

ANON. 1972B. New Noise Standard. Grav. Environmental OSHA News *1* (1), 17–18.

ANON. 1972C. Legislation. Mod. Packaging *45* (12), 55–58.

ANON. 1973A. California laws push metal coatings supplies to water based coatings. Packag. Eng. *18* (3), 28.

ANON. 1973B. "What's Minnesota up to?" Food Drug Packag. 29 (5), 1–24.

ANON. 1973C. Legislative Up-Date. Mod. Packaging *46* (8), 28–29.

ANON. 1974. Aerosol hazard to atmospheric ozone. New Sci. *63* (916), 781.

ANON. 1975A. Health and safety guide for paperboard container industry. U.S. Dep. of Health, Education and Welfare. National Institute for Occupational Safety and Health. Washington, D.C.

ANON. 1975B. FDA finally finalizes labelling regulations. Mod. Packag. *48* (4), 68–71.

ANON. 1975C. N.Y. environmental council recommends mandatory Oregon-type container laws. Food Drug Packag. *23* (8), 4, 24.

ANON. 1975D. FDA allowing PVC testing. Food Drug Packag. *33* (1), 9.

ANON. 1975E. Legislation gets into the act and today's concerns are just a prologue. Mod. Plastics Int. *5* (10), 56–59.

ANON. 1975F. USDA regulation on jar closures extended to 1977 as industrial reacts. Package Eng. *20* (11), 25.

ANON. 1976A. Overpackaging target of bill. Food Drug Packag. *34* (6), 7.
ANON. 1976B. A guide to government packaging regulations—state and city legislation. Pack. Print. Diecutt. *22* (8), 6–7.
ANON. 1976C. Selected statistical data. Containers Packag. Q. Ind. Rep. *29* (2), 5–12.
ANON. 1976D. A guide to packaging regulations. Tin Int. *49* (12), 441.
ANON. 1977A. A guide to packaging regulations. Tin Int. *50* (2), 66–67.
ANON. 1977B. FDA man predicts package reviews. Food Drug Packag. *36* (5), 3, 9.
ANON. 1977C. Code of federal regulations, 21. Food and Drugs Parts 100 to 199, Rev. April 1, 1977. U.S. Office of the Federal Register, Washington, D.C.
ANON. 1977D. A guide to packaging regulations. Tin Int. *50* (5), 188–190.
ANON. 1978. The 1978 packaging and labelling regulations. Environ. Data Services Rep. No. 12, pp. 11–13, 16.
ANON. 1980. Evaluating risk vs. benefit. Aerosol Age *25* (9), 46–48, 50–51.
BELOIAN, A. 1973. Nutrition Labels: A great leap forward. FDA Consumer, pp. 10–16.
GITTELMANN, D. H. 1976. Environment, health and safety and plastic additive selection. Society of Plastics Engineers 34th Am. Tech. Conf., Atlantic City, New Jersey, pp. 237–238.
GOERTH, C. R. 1981. The legal impact—declare your right to that package design. Package Dig. (U.S.) *18* (13), 28, 30.
HARVEY, B. 1979. Some legislative problems in the control of toxic substances. Ann. Occup. Hyg. *19* (2), 135–138.
HECKMANN, G. 1974. Laws and regulations affecting plastic packaging. Food Product Dev. *8* (7), 55–56.
HOLT, P. 1982. Packaging legislation—why the government needs to be involved. Austr. Packag. *30* (2), 37, 38, 40, 42.
HUTT, P. B. 1973. Safety regulation in the real world. Food Cosmet. Toxicol. *11* (5), 877–884.
INGS, J. 1977. Restrictive Packaging Legislation. Austr. Packag. *25* (8), 23–25.
MACARTHUR, M. D. 1979. The clean air act. How will the law affect flexible packaging. Flexog. Tech. J. *4* (4), 4–6.
MACARTHUR, M. D. 1982. Federal product labelling law proposals. Paper, Film Foil Convert. *56* (6), 80–82.
MAGRAM, S. H. 1974. Labelling Regulations. Aerosol Age *19* (1), 28–30, 38.
MARDER, H. 1975. Washington forecast a breather on regulations. Mod. Packag. *48* (1), 26–27.
OSHA. 1974. Vinyl Chloride. Fed. Regist. *39* (92), 16896–16900.
PAINE, F. A. (editor) 1973. Packaging and the Law. Butterworth & Co. Ltd., London.
SACHAROW, S. 1976A. A guide to government packaging regulations. Pack. Print. Diecutt. *22* (2), 6–7.
SACHAROW, S. 1976B. A guide to government packaging regulations. Part 3. Federal trade commission and department of commerce. Pack. Print Diecutt. *22* (4), 10, 13, 68–69.
SACHAROW, S. 1976C. Labelling regulations in the U.S. Aust. Packag. *24* (4), 27–28.
SACHAROW, S. 1979. Packaging Regulations. AVI Publishing. Co., Westport, Connecticut.
SUCKLING, J. 1970. Legal aspect of packaging. Packag. Technol. *16* (114), Suppl. 14, liii-lvi.

10

Strategic Planning in Packaging

Life can only be understood backwards; but it must be lived forwards.

Kierkegard

In past generations, strategy was a concept reserved for generals and chess masters. In the "Pepsi generation," strategy has been elevated to inclusion in children's games and business strategy has become gamesmanship. The era of the MBA has generated a prime channel called strategy through which those graduates who do not go into "plastics" pass on their way to the executive offices. Strategy and strategic planning, once the instinctive forte of the entrepreneur, have become a precise quantitative program within many large corporations and are permeating rapidly outward to envelope all American organizations in the private and public sector.

Regardless of the definition applied, and much confusion does reign as a consequence of the exact meaning of words, the concept of strategy is sound, meaningful, and beneficial to organization members. Thus, whether it is called planning, strategy, strategic planning, tactics, forward thinking, or something else, the notion of making decisions today on the basis of expectations is a powerful tool in conducting professional, social, and personal business.

In the relatively superficial worlds of the nation's business periodicals, strategy deals with total corporate efforts—and it is the great corporations, such as General Electric and IBM, that recognized the need for strategy and pioneered its use in business. If many observers have difficulty translating from the multibillion dollar General Electric to the Yellow Ribbon Button and Bow Co. or the Buggy Whip Corp. of Eastern Pennsyltucky, how can a rational person even consider applying strategy to packaging? It is not suggested that strategic planning is advisable for the entire packaging community, with its thousands of diverse companies, although all of us know well that the industry might greatly benefit from some thoughtful strategic planning. Neither is a concept specific to a packaging system, materials, or machinery firm proposed in this discussion, although, of course, the subject is wholly applicable to such organizations.

Rather, the issues addressed would directly merit serious consideration by packaging users. One immediate reaction should be, "How can this author be so arrogant as to suppose he knows my problems and could suggest solutions?" This response is quite similar in nature to the opposition to any strategic planning in any organization. A second reaction might be, "How could strategic planning be applied to my one-, two-, or five-person laboratory, shop, studio, or office?" Once again, this reaction is virtually the same as is encountered within corporate business units. Another common response is "Since I am already planning, why introduce a formalized procedure?" Another is that the computer can generate the requisite plans. Still another is that the packaging manager is only taking direction from marketing, R&D, or production, and so is only a cog in a much larger plan. In addition to the universal human fears of change and organizational obstacles to altering the structural fabric, a long list of rationalizations on why strategy is not for the packaging department may be conjured up more easily than an explanation of how strategic planning benefits the individual and the organization.

Had the concepts of strategic planning been implemented a generation or more ago, the counterproductive abuses of the "hale fellow well-met" salesman and the cost-is-everything tunnel vision of the old-time purchasing agent would never have been, and the present residual consequences of fully amortized inflexible equipment and systems in-place would not be limiting profitable progress. The world of packaging in those days was very clearly "Put the product into a can, bottle, box, or pouch and put the pouch into a box, etc." Choices were limited to determining materials, structural design, and suppliers. And suppliers were very price competitive. In those good old days, Coca Cola came in one size, in one material, in one multiple, and was distributed to one class of retail outlet. Today, the 6½-oz, returnable, wasp-waisted Coca Cola bottle has been replaced by containers that are 10 oz, 12 oz, 16 oz, 32 oz, 0.5 liter, and 2 liters in size; of glass, metal and plastic, aluminum, and two- and three-piece steel in construction; in 6, 8, 12, and 24 packs; returnable and nonreturnable; and so forth. Retailer groups dictate the type and count, delivery time, shelving, and pricing of product. The federal government oversees the materials and labeling, and state governments tell the bottler what size and type container may be used. If carbonated beverage packaging does not fit the image of packaging complexity, consider breakfast cereals, with box sizes, Federal Trade Commission, PCB's, sugar-sweetened, non-sugar sweetened, promotions, co-extrusions, warehouse distribution, and so forth.

Examining packaging from the perspective of the packager shows influences from marketing, production, engineering, finance, the legal department, and so forth, to the extent that absolutely no agreement exists as to where the packaging department belongs within the organization.

Packaging is at least as complex as any other function within a consumer or industrial product manufacturing firm. Packaging is probably more difficult than other functions because its work is visible and when it does not function well or economically, the message is communicated instantly. Unlike the accounting or legal departments, there is no way to hide the mistakes of either omission or commission.

Justification for introducing strategic planning has required a disproportionate part of this discussion, which should be largely devoted to the process itself. It must be noted, however, that books on strategic planning dedicate entire sections to the reasons for such planning and other sections to the means of overcoming objections. Within firms committed to strategic planning, the process is often a nuisance to business unit managers, an exercise in futility to operating people, and a tedious paperwork control task to the planners. If the implementers and beneficiaries are doubters, then outsiders must be totally puzzled in the never-never land between intrigue and skepticism.

Suffice it to say that if one does not have the time to do the job right the first time, one certainly does not have the time to correct it again and again. And think of the numerous occasions in which errors have to be rectified again and again! In the great American game of Criticize the Company, it should be remembered that the organization today is the result of whatever strategic planning was conducted in the past. To paraphrase a not-too-ancient slogan, give strategic planning a chance.

The foregoing "pre-ramble," intended to justify the use of strategic planning, has purposely omitted the prerequisite of all strategic planning: its definition. Although the definition has been alluded to throughout the foregoing pages, it has probably floated past without recognition because it was not pointed out as such. We could arrive at

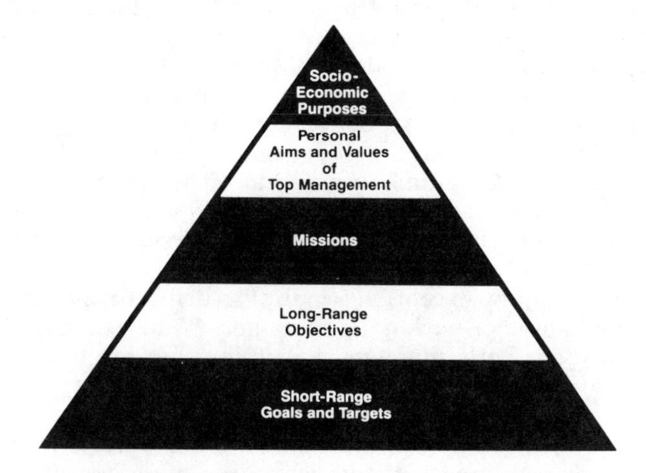

FIG. 10.1. Corporate hierarchy.
From Steiner, 1979.

one of a number of definitions by several different routes, which lends credence to a theory that it is an abstract concept, totally lacking in substance.

Quite the contrary. Strategic planning might be one of the most substantial of the many disciplines within business management.

Strategic planning is the process that establishes the objectives, defines the policies, generates the philosophies, develops the implementation details, enumerates the resources required and the benefits to be derived, and links today's issues to actions to be taken next week, next month, next year, and even into the next decade. In essence, strategic planning is the "futurity of current decisions," making sure that the future, whether tomorrow morning or next year, is realistically included in every business decision and action. If strategic planning were simplistic, it would have been made an integral component of every business long ago. Strategic planning is not simplistic, but rather simple: the concept is easy to comprehend and put into practice, but the process of reaching this point requires intense thought and effort. The simplest of all structures have been based on the best engineering.

An instant analysis of the definition is that the future is management's responsibility and does not belong to the operating folks. Indeed, at all the various levels within the organization, the scope of the future belongs to management. The top executive should worry about very long-term implications for the entire organization; his decisions can affect billions of dollars, tens of thousands of people, and the very existence of the organization.

Marketing management should be concerned with the organization's products and their definition, positioning, pricing, selling, shelving, use, and, of course, packaging—all in the future, of course. Marketing sets the objectives and, for many, the restrictions, and then expects every manager of production, design, engineering, sales, etc., to perform in his or her specialty. Each functional portion of the company is assigned the duty today to produce something tomorrow or next week. And that assignment usually includes packaging. Packaging does not happen by itself; packaging is not conjured up by magic; and it is certainly not a function that can be performed with any credibility by anyone hired in from the street, although of course, some organizations try anything.

In general, with few exceptions, manufacturing firms have packaging departments and/or packaging managers. And a manager, by title or responsibility or both, manages a function that might or might not include people, factories, machines, etc. Thus the packaging manager manages packaging—its structure, design, use, synergy with distribution, equipment, retailers and consumers, cost, safety, legality, and even environmental impact. If the packaging manager does not manage these elements of packaging, then who will? And if these elements are not managed, then what is the result?

The next question obviously is "What is management?"—a question asked repeatedly by students and practitioners alike. Perhaps one could answer by listing what managers do.

1. *Establish goals and objectives.* What do the consumer, the retailer, and marketing management expect the package to do? What is the manager trying to achieve?
2. *Plan the strategy for achieving the goals and objectives.* You must decide whether to call in suppliers, do it yourself, imitate the competition, self-manufacture, use a package that is already in the line, etc.
3. *Generate a philosophy of operation.* This determines how strategy is to be implemented. Decide whether to emphasize design, cost, or engineering, quantitative or qualitative considerations, people or things, a scientific or an instinctive approach, the pursuit of several options or in-depth focusing on one alternative, etc.
4. *Set the policy through action plans.* Detail the objective, the strategy and the nature of the approach, and the specifics of who does what and when.
5. *Plan the structure of the organization that must execute the policy,* even if this means coordination of resources outside of the control of the packaging manager. If the packaging manager has a complete staff of designers, engineers, chemists, etc, plus the required laboratories and pilot plants, then he must deploy these resources for maximum efficiency. If not, then the required resources must be tapped and integrated structurally to achieve the objectives within the prescribed time frame.
6. *Provide the staff.* After all the management work is done, staff people must do the actual work required.
7. *Actuate the staff.* The people assigned or requested to perform the task must be motivated to produce effectively and in a timely fashion.
8. *Provide the physical resources.* The paperboard, plastic, metal, glass, drop testers, cutters, rub testers, permeation testers, consumer focus groups, etc., must be made available when and where they are required; the staff people must have the necessary tools or they will either expend needless effort improvising or perform suboptimally or even erroneously.
9. *Establish procedures.* What time does the staff start in the morning? When is coffee break? Where is information recorded? Who prepares the report? To whom are reports sent? What mechanisms are used to communicate with suppliers?

10. *Budget.* Within organization financial guidelines, the budget must be drafted, negotiated, built, approved, and followed. To achieve an objective, each individual project has an associated cost. What happens if the budget runs out and the project is incomplete?
11. *Set standards of performance.* By word and deed, by dictate or by example, a key managerial function is to demonstrate the type of output that is tolerated, expected, or appreciated.
12. *Control.* No one in the manager's span of control should have greater knowledge of the overall scope of programs and progress than the manager. Others might have deeper knowledge on a specific aspect, and still others might have broader information on highlights. The manager is wholly responsible for knowing the progress of each of the component parts and of using that information to improve, accelerate, retard, or modify the action to be taken in all parts. Feedback on progress must be comprehensive, timely and based on the facts. Which football team today does not have a game plan and scouts high above the field of play observing, analyzing, and reporting back to the field? No football team worth its coach will go without two-way telephonic communication to control the game.

There are a dozen or more key operating functions, not including public relations, compliance with government rules, completing expense reports, reporting upward to the boss, smoothing the perpetual irritants among the staff members, and a thousand other things we all encounter each year. And for each of the dozen or more projects the packaging manager juggles simultaneously, the same 12 functions apply, but with different players and priorities. As described, and as it is in the real world, management function is like a dozen drops of mercury on a teeter-totter, undulating and flowing faster and slower—and too quick for the mind to grasp in its proper perspective.

Is it any wonder that management cannot be by the tried and true past intuitive, reactive, or even opportunistic approach? A packaging manager cannot muddle through on the strength of his own wits and perform. A packaging manager's responsibility to protect his company's products throughout their distribution is not an adaptation of some instinctive response to a stimulus. It is evident that shooting from the hip has no place in responsible packaging management.

Today's and tomorrow's packaging is the result of formalized and systematic strategic planning: the process that sets objectives, defines the means with which to attain those objectives, and details the steps required to follow those means, linking today's decisions to the future result and optimizing the resources and outcomes. Even comprehending the definition requires some attention, although once understood, the meaning of strategic planning is simply common sense uncommonly applied. It is important to point out here that strategic planning does not include making future decisions. It provides the framework within which today's and tomorrow's decisions are and will be made by the persons responsible.

Strategic planning has a higher probability of providing effective benefits than the alternatives because the process forces the packaging manager to ask and answer critical questions relevant to the issue at hand. Too many packaging people in the past built, sold, installed, and delivered materials and equipment and then learned from the production floor whether or not their ideas had any merit. The warehouses and boneyards of food and packaging companies are filled with the debris of packaging systems that were dreams rather than planned fulfillments of needs. Accountants have long since written off the financial costs of certain ideas that were proposed by silver-tongued orators.

Strategic planning is the parent of computer simulation of the future. Through compelling questions and answers, strategic planning introduces forces into the decision-making process that are indispensable to the process.

1. *It simulates the future in an objective manner.* How many times does the salesman demanding a new package development guarantee a million or a hundred million dollars in sales? If all these sure things were laid end to end, Americans would be consuming 15,000 pounds of food annually, ten times their actual per capita consumption. By establishing a scenario of the future in context, a model that is the basis for packaging development is generated. What is the most probable number of cups of coffee, cost of polyethylene resin, availability of tin-free steel, number of supermarkets using universal product code (UPC), etc.
2. *It applies a systematic approach*—a uniform, detached methodology for information accumulation and analysis, for projecting requirements, etc. This is an organized mechanism for appraisal and for structuring an approach to achieving an objective. System implies a beginning and an end and an orderly path between, coherence and logic to all the components.
3. *It compels the establishment of an objective.* If a tourist neither knows nor cares where he is going, any road from William Penn's statue in Philadelphia will do. If, however, he plans to visit Valley Forge, he should travel east on the Schuylkill Expressway. Having a precisely defined objective is a key beginning for strategic planning. So simple is this element that it is overlooked in favor of the simplistic "make a profit" or "increase sales" or "reduce costs." One result of reaching an objective could be profit or productivity, but the objective is something concrete, such as unit-portion packaging of coffee at 140°F to retain its initial temperature and flavor through 3 months of distribution.

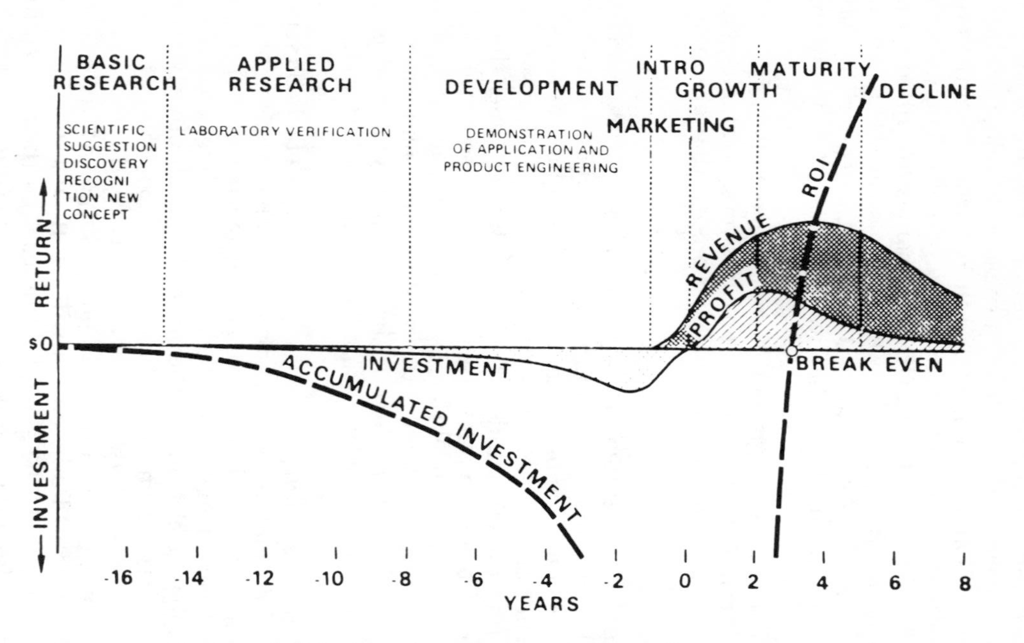

FIG. 10.2. The classical S-shaped curve of product growth.
From Bright, 1978.

4. *It identifies and aids comprehension of further opportunities and threats.* By forcing the manager to examine the future in a pragmatic rather than a visionary fashion, strategic planning allows realistic scenarios of what will probably occur to be prepared. In general, of course, the future is a series of incremental steps from the present. Catastrophic events such as the 1973–1974 oil embargo, fortunately rare, are not predictable. Abrupt changes in consumer behavior without long warning are improbable. Thus development of a new machine capable of producing 1000 retortable pouches a minute has a low probability. A reduction in aluminum weight per can is probable. As a result of identifying the realistic probabilities, opportunities are revealed, such as the need for internal can pressurization with thin-walled aluminum cans. By showing the future as it will probably be, threats may be seen, such as the introduction of aseptic packaging in paperboard laminates, which represents a threat to use of metal cans for the products affected.
5. *Strategic planning offers a basis for other decisions.* A foundation of sound information, analyzed in an orderly fashion, provides a solid rationale for decisions at the core of and peripheral to packaging itself. The notion of using inert gas to preserve rancidity-prone snacks offered the possibility of long-term distribution through store warehouses, thus bypassing weekly store shelf delivery. Did the marketing objective precede the packaging development or was it the other way around? However, aseptic packaging in paperboard composites is generating a whole new family of hard questions—e.g., on which shelf should the new package appear? Is the product quality credible? In what count multiple should the package appear? If these questions are not answered prior to national introduction, the result could be confusion—just what were the results of BrikPak and BlocPak packaged foods in Canada? Does anyone really know?
6. *It gives a measure of performance.* By establishing objectives, the packaging manager knows at the start what he must do, in what time frame, and for how much. Thus he has established the mileposts by which the project, his staff, and he himself are measured—no more post facto analysis, Monday morning quarterbacking, finger pointing in a Utopian strategically planned world.

It should be clear by this point that any packaging project, from the shortest fix in the plant to the longest, most complex total system development for a new product concept, is amenable to strategic planning in its broadest sense. Mitigating a problem with a wrap that does not seal properly in the plant requires setting an objective based on production and market needs and making a plan. Can the machine be removed temporarily from production while the problem is resolved?

What if the problem is the material, or that the machine has a heater burn out, or that someone turned down the thermostat, or that a batch of material was uncoated? Is a mechanic, a machine service person, a seal tester, or a scheduler needed? Will the fix require an hour or a day? Obviously, production down-time due to seal problems is so common that a formal written program is usually (but not always) not required.

In contrast, a project to package down-home rhubarb meringue pie for hotel, restaurant, and institutional distribution in the Southeast demands a fairly well-structured program considering the several alternatives in materials, structures, packaging machinery, secondary and tertiary packaging modes, time, cost, design, manning, samples for consumer test, use testing, simulated shipping tests, flavor evaluations, etc. The number of combinations and permutations possible in this relatively simple example approaches infinity, and skillful experimental design and planning are required to eliminate the need for every variable to be specifically evaluated in comparison with every other variable.

In the much more complex hypothetical example of packaging liquid coffee for a 3-month shelf-life, a comprehensive long-term program is almost mandatory. Clearly stated objectives and problem definitions are necessary even just to communicate the problem to the many and varied specialists.

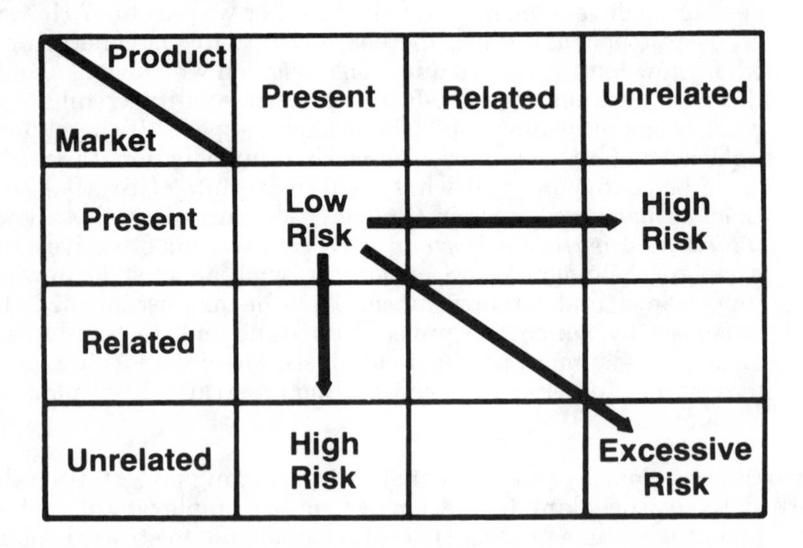

FIG. 10.3. A product market matrix depicting the risks involved in development related to orientation of either product or market to current business.
From Steiner, 1979.

Some purists might argue that defining the objective and how to get there is strategy and that all the support activities are tactics. By any name, the process of definition of the objective and the means to reach it is planned, whether strategically or tactically. Strategy generally circumscribes the entirety of a program, while tactics are the details of implementing the strategy. Nomenclature is of less importance than the concept.

One key feature of strategic planning for packaging managers that differs from that usually performed by and for upper and/or marketing management is that the packaging manager formulates, implements, and is *responsible* for the strategic planning in packaging. A corporate business planner or marketing strategist who holds this position for more than a 1 year cycle is a rarity. Consider the differences that exist between persons who merely prepare plans as if they were another business school case study and those who prepare plans as if they were accountable for the plans' outcome.

Particularly in the 1980s, nothing is perfect. Strategic planning is so new as a general discipline that it is hardly fully developed or studied. Strategic planning has its limitations, the most important of which is that it provides no guarantee of a positive result, only the notion that the probability of positive results is significantly greater with a properly formulated and implemented strategy than without. Among the several limitations to strategic planning are the following.

1. The social, political, economic, market, or regulatory environment might prove to be different than was planned. Imagine the surprise in Monsanto that day in 1977 when the FDA announced it was about to ban polyacrylonitrile as a carbonated beverage-bottle plastic. After 11 years of careful planning and execution, the environment had unpredictably changed. Conversely, the ban on polyvinyl chloride (PVC) could be lifted, after 11 years, and the environment would be quite different from that planned for packaging processed meats, vegetable oils, etc.
2. Planning is initially expensive, requiring the attention and effort of numerous persons, including operating managers and a full-time coordinator or planner. Beyond the planning process will inevitably be the changes dictated by implementation of plans. Planning might be viewed as a beginning cost or as an investment. Either way, at the outset, monetary outlays are required.
3. Internal resistance: in no human endeavor has there ever been an absence of resistance to change, to more efficient operations, or to the perceived threat to job or security. Obstacles to planning can range from subtle failure to understand the simplest assignments, to refusal to cooperate, to overt actions designed to undermine the process. Most disconcerting of all, however, are those who go through the planning process and concur with the

paperwork, but then ignore the plan and proceed along their normal paths.

4. Day-to-day operations: breathes there a planner who has never encountered managers who do not have time for such front office fads as planning? It seems that every manager is more concerned with solving today's immediate problems than with tomorrow on the grounds that if he does not survive today, there will be no tomorrow. The less managerially inclined the manager is, the greater the inclination will be to use today's crisis as a rationale to defer or ultimately avoid planning.
5. Difficulty: strategic planning is not an easy task. Strategic planning requires combinations of pragmatism and vision, reasoning and imagination, courage and patience, qualitative and quantitative skills, and communications both up and down that are really the domain of the experienced and disciplined. Strategic planning is neither simple nor simplistic; it is a tough, scientific process of analyzing and projecting how business functions in its environment.

Strategic planning offers no guarantees that rigid adherence to the rules laid down in any given cycle will provide a trouble-free ride. No manager can become a bureaucratic drone (with his inviolable sacred rule book) with his plan for the year and be assured that the objectives will be achieved. The world is not yet a sufficiently rigid environment for a given input in one spot to always provide predicted outputs elsewhere. The use of strategic planning will, however, significantly increase the probability that the decisions made and actions taken during the planning cycle will be correct in the context of the function and the organization's objectives.

Strategic planning forces the manager to introduce organized thinking, to anticipate what might happen, to assign probabilities to future events, to establish systems through which functions may be performed, to forecast and be ready for contingencies, and, above all, to plot the courses of action to be taken. The development of a retort pouch, for example, would require a program not unlike that undertaken by U.S. Army Natick Laboratories in the 1960s. Involved were product formulations, microbiologists, flavor-testing experts, biochemists, economists, flexible packaging suppliers, paperboard carton suppliers, potential users, and so forth. Had the project not been comprehended in its entirety by the Natick Laboratories' people, the requisite tasks would have been performed at random or sequentially, or not at all, and the package would not exist in this country today.

How, then, does a packaging manager who will invariably not have the luxury of a planning assistant perform strategic planning? As indicated above, the principles must be relatively simple or else they will be neither comprehended nor implemented. Planning is analogous to polymer chemistry or to the universe according to Einstein—

frightening from a distance, but so orderly when fully understood that it is almost elementary.

The four basic models are the premises, formulation, implementation, and information flow. Establish what you must do and how to do it, do it, and make certain that you know you have done it or know how to correct what went wrong.

Expanding the basic conceptual model into a typical strategic planning model that might be used by a packaging manager reveals sequential but overlapping steps:

1. Formulate the tasks that must be accomplished over the long term for the organization. Determine the sources, the information and analyses of the specific problem.
2. Obtain inputs on the past and present environments—e.g., what the marketing department and production want. What resources are available, and which are informative and which require reinforcement?
3. Enumerate and evaluate alternatives for use in the strategic plans—for example, add to the staff, use more suppliers, hire contract engineers, reduce the work accepted, transfer some tasks to other departments, and so forth.
4. Define the principal objectives: packages to be developed, specifications to be prepared, tasks to be performed, materials to be evaluated, suppliers to be graded, etc.
5. Formulate the major strategies, such as schedules, methodologies, number of staff members, budgets, etc.
6. Develop the short- and medium-range plans required to meet the long-term objectives using the major strategies.
7. Decide precisely what is needed immediately to begin implementing the plans.
8. Monitor progress and performance in accordance with preestablished standards.
9. Recycle the process on an annual basis, reviewing and improving the process and implementing it more in line with the strategic plan.

This process of developing a strategic plan, whether formal or informal, imposes a discipline on both the process and its implementation that improves chances that the departmental functions will be more profitable to all other productive departments and to the organization as a whole.

Although the sequence described is typical, it is neither the ideal one nor the only one. There is neither a perfect strategic plan nor perfect means to produce a perfect plan. Just as the strategic plan is a plan to obtain an end, producing a good strategic plan requires planning. Budget preparation is, of course, only one small part of planning and is really an arithmetic product of the planning process. The packaging

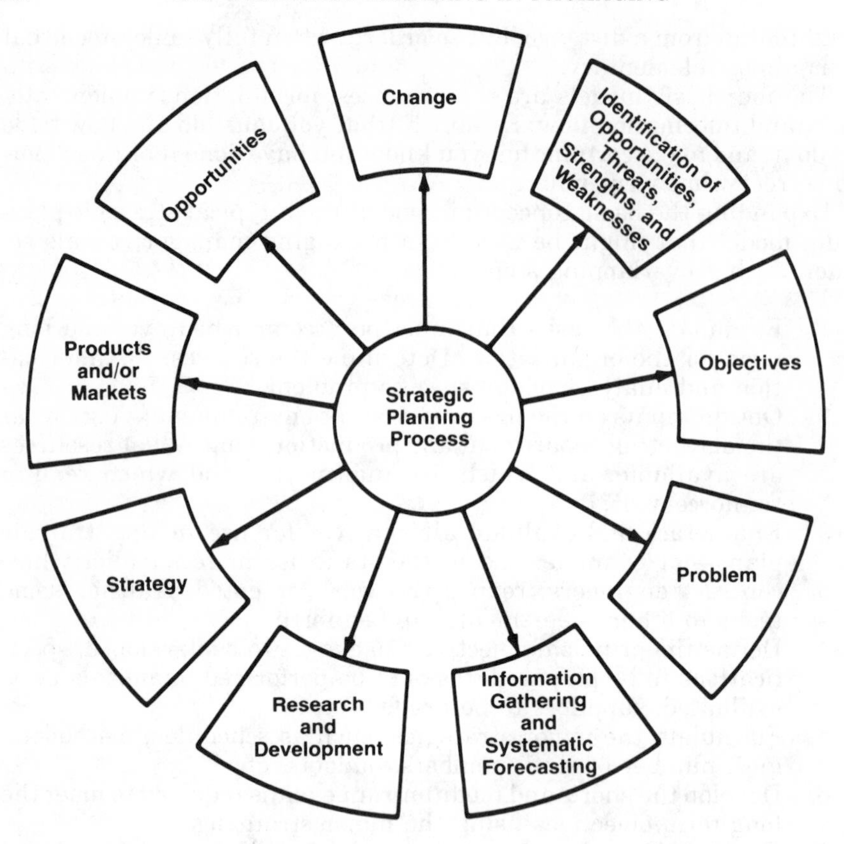

FIG. 10.4. The strategic planning process in its entirety.
From Steiner, 1979.

manager should have a plan. It could be a do-it-yourself plan, especially if the company does not impose a series of planning steps on the department. If the organization has a designated planner, the packaging manager could abrogate his responsibility to that person, who of course has almost no understanding of either the packaging function or the packaging process. The packaging manager can delegate the planning to one of his staff members and in that way possibly discover a rising star. Usually, however, delegation to an administrative assistant leads to incomplete data. If planning is to be performed, it should be performed well. Otherwise the exercise is counterproductive.

On the other hand, a plan prepared in a vacuum can be wasteful. Although in many organizations the planner and the executives are charged with making sure that the plans of the packaging department coincide and coordinate with those of the company and of other departments, the packaging manager cannot depend on this skillful interlocking.

The packaging manager should anticipate that designated planners will be looking at broader scenarios and at financial issues and thus may brush over functional plans. Thus, the packaging manager should undertake the task and responsibility of making sure that his plans indeed coincide with those of marketing, R&D, engineering, and production, at the very least. He should also encourage the participation of his own staff in the planning process, since his staff is to be a tool in the execution of the plan. With these actions, the packaging manager at least alerts those who are part of the plan to what his ideas of his function are. He will want to know how many new products, packaging changes, etc., marketing plans to introduce during the plan cycle. He will want to know whether engineering is designing a new line that will have a new filler and thus require a change in bottle specification or whether production is considering mechanization of tertiary packaging, which would necessitate a new design of corrugated fiberboard shipping cases. He will want to know whether one of his staff members is planning to leave to enter college or is being considered for promotion to purchasing manager, or whether personnel has recommended a reduction in the packaging department because its financial benefit to the organization has not been adequately presented.

In searching for information to develop and support his plan, the packaging manager will, of course, encounter some opposition and the inevitable absence of support. The process of strategic planning, by its very existence, alters relationships between departments, changes the information flow, brings new ways of decision making, and highlights differences. The process also, of course, heightens the perceived risk and fear of failure and security makes new demands upon people and in some persons, generates greater rather than less uncertainty than the process is designed to do.

When the packaging manager engages in dialogue with the people involved, a greater understanding can be achieved and the threats can be alleviated, if not eliminated.

It is important to point out that, in general, excellence in planning is the basis for excellence in implementation, and the best planning takes place in organizations with the best management.

To this point, strategic planning has been defined, justified, circumscribed, qualified and planned. The process itself must now be developed.

Strategic Planning. This usually begins with a situation analysis or audit. By whatever name, the exercise involves an in-depth examination of the present and recent past. What did you do last year and the year before? How many primary, secondary, and tertiary packages were developed? How many specifications were prepared? How many total packaging audits were performed? How much does your organization spend for bottles, cans, cartons, cases, etc.? Is the trend up or down? How many new products are packaged as line extensions? Which regulations influenced the packaging last year? What consumer complaints dictate packaging changes? What retailer suggestions or demands have been revealed by surveys or field observations? Which suppliers are anticipating major changes? And, of course, what do the salesmen insist be done to prevent collapse of the entire company? A situation analysis is a representation of past performance and more important, a data base. It is crucial to recognize that change occurs incrementally, as a result of which a 1985 plan will look very much like a 1984 "actual." Thus, if the 1984 performance is accurately depicted, a good 1985 and forward plan can be prepared.

Current Resources. An inventory should be made of who, what, and how much is available, both internally and externally. How many people and of what type? What kind of equipment? What kinds of consultants? Can talents, instrumentation, equipment, production time, etc., be borrowed from other departments, from suppliers, etc.?

Competitive Environment. What kinds of packaging are being used by the direct competitors? What kinds of equipment do they use? What packaging development resources do they have? How much do they

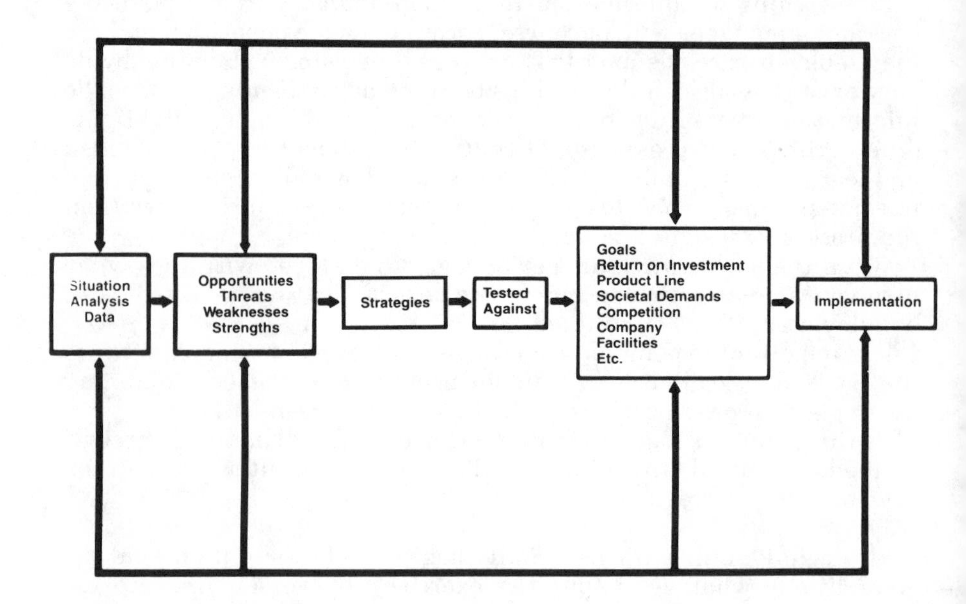

FIG. 10.5. The planning process begins with an analysis of the present situation and proceeds to implementation.

pay for packaging? What packages do they have in test market? What type of packaging development do they have underway, according to your intelligence?

An audit of the packaging used by direct competitors on an organized periodic basis is invaluable.

The sporadic, reactive examination of one or two competitive packages does not provide the meaningful perspective that comprehensive study does. There should be an exchange of packaging specifications among competitive users in the same way that game films are exchanged among football teams. It would save so much time and money in purchasing and analyzing samples from the field.

Audits should also be performed on indirectly competitive items to identify trends that might eventually have an impact.

Supplier Audit. Unique to packaging should be an updated appraisal of supplier performance, plants, activities, and trends. Each supplier should be queried on the outlook for his material—the costs, availability, and any plans that will directly affect packaging. For example, one might ask: What are the trends in polyethylene pricing? When might we expect linear low-density polyethylene film to be available? Are you installing injection molding equipment in a plant adjacent to our factory? What is your prognosis on the license you have taken out on the new European package?

Almost all good suppliers will be happy to conduct a complete presentation on their industry trends and issues of interest to their customers. It is vital for the packaging manager to be an active participant in these presentations, and to solicit them if by chance they are not offered gratuitously or if they are delivered only to the purchasing departments.

Analysis. Acronyms of various types have been generated for the self-analysis process, such as SOFT (strengths, opportunities, weaknesses, threats) and WOTS (weaknesses, opportunities, threats, strengths). Examine the packaging department for its strong and weak points. Examine the packaging problems presented and projected for the threats and opportunities. Each of the key issues, whether internal department or packaging matters, should be analyzed in terms of the four key areas: strengths, threats, opportunities, and weaknesses.

Develop Objectives. "The Past in Prologue." A data base has been established. The environment has been described. Everything has been analyzed in the harsh light of what can or cannot be done.

Now (or earlier) set that all aside and decide where you want to go or what you want to do. If you are very fortunate, the marketing or production departments will have already provided considerable input on this topic. And, as every packaging manager knows, with an infinite amount of resources, these imposed objectives could be achieved in an infinite time. In the perspective provided by as complete a knowledge as is available on the subjects, a meaningful mission statement must be formulated. Is the packaging department a leader, a structural design group, a technical service group, a testing agency, a specification writer, an identifier of future trends, or a vestige of past organizations? Regardless of what it is, what will it be in the long- and medium-term future?

The indispensable long-range objective must be suitable, measureable, feasible, understandable, flexible, acceptable to all, and, above all, limited. A problem defined is a problem half solved. An objective clearly formulated defines the roles of the packaging department and its associated organizational units. The objective can be multifold—realistic and "stretch"—if everything goes very well.

Objectives must be realistic and not idealistic. Too many mission statements are wants and not pragmatic assertions of what the organization is about. Some are very narrow (e.g., "the packaging department will perform shelf-life tests") or too broad (e.g., "perform all the research required to develop packaging"). Does the latter mission include consumer testing, polymer formulation, product alteration, etc.?

The group's purpose should be sufficiently flexible that, should the environment change, its mission can be altered to accommodate the change. The introduction of the plastic carbonated beverage bottle certainly must have created a whole new posture in carbonated beverage packaging departments, for many reasons.

A general discussion such as this cannot provide a simple, all encompassing objective for a packaging organization. Rather, this discussion provides parameters and cautions that make sure that the process is disciplined and not merely a semantic exercise.

Formulate Strategy. With the baseline environment and objectives established and detailed, the time has come to develop the strategies and tactics that enable the department to achieve those objectives in the situation at hand.

Various methods have been proposed that, in one or another situation, work well: matrix, life-cycle, niche, adaptation, imitation, intuition, synergy, and serendipity or chance.

Since luck favors those who are prepared for it, serendipity is the only acceptable form of chance. But serendipity is really an extension of planning that short-circuits the excruciating process of development. If it happens, great; if it does not, then the plan is still in place to be followed. But you cannot plan for serendipity.

Imitation of competitive packaging is an acceptable strategy if it fits the marketing strategy. Imitation is, of course, following the leader, which reduces the risk and is often most successful. On the other hand, it implies that some opportunities will be bypassed. The Anderson Clayton Co. could not have succeeded in introducing tubs of soft margarine had a rigid imitative strategy been employed.

Adaptation of existing or competitive packaging is a common packaging department strategy which can yield imitation and, at the same time take advantage of the competitor's good and bad points. Since development is usually incremental, anyway, adaptation in its broadest sense can be a good strategy.

Niche strategy implies positioning the packaging in between other packaged products through packaging graphics or structures. When the Franklin Nut Co. introduced dry-roasted nuts in glass packaging rather than traditional metal cans, it was establishing a niche. Unfortunately for Franklin Nuts, the niche was perceived as so large that others imitated the concept.

A leadership strategy is not at all unusual. Oscar Mayer, Tropicana, Coors, and General Foods are often in the lead in developing packaging that is innovative, fits a perceived need, and is consumer-oriented. General Foods stretches its suppliers and the others stretch their own internal developmental resources to the leading edge of packaging development. The U.S. Army Natick Laboratories stretched both themselves and numerous suppliers in a high-quality innovation process in their development of the retort pouch. Innovation requires time, money, and people and carries with it high risk.

Once again, the strategy for any single organization could not possibly be stated here. Strategy depends on objectives, resources, and the environment, and every organization has its own matrix of these given variables. Developing a coherent and meaningful strategy depends on knowing and analyzing these parameters and deciding on the strategy that fits best at a specified time. Corollary to the basic strategy are subsidiary strategies and tactics. Although a stand-pat, follow-the-leader approach might be used for one product line, an innovative strategy might be used for new products or for products in the downswing of their life cycle.

Above all, no strategy in packaging can be static. All packaging strategies must be dynamic, ready to alter direction when the situation warrants, but not abruptly or reactively.

The basic strategy must be thoroughly evaluated, as the choice among the alternatives suggested above might to some degree be dictated by the marketing or production departments. Regardless, those functions directly or indirectly affected by the packaging strategy should be informed so that their imports are included. A marketing department might object to an imitative strategy, whereas a production department always resists the change that would be necessitated by an innovative strategy.

Internally, the strategy rotates about such issues as use of suppliers, other departments, technical orientation, and testing protocols. The basic objective will have considerable influence on the internal strategy. Often, however, strategy is based on the philosophy, training, or experience of the manager. Other packaging strategies are an integral part of the company strategy. Thus, one packaging manager might be heavily oriented toward a sound, in-depth technical research effort prior to any field or market testing. Another might opt for very rapid development, drawing heavily on suppliers, with the bulk of the development taking place as rectification after entering field testing. Some programs divide the activity, with the packaging department performing only the technical product-protection testing while marketing handles structural and graphic design and traffic oversees the physical-distribution testing. In many companies, the division of effort is in terms of appearance, marketing, equipment, engineering, economics, purchasing, materials, packaging, etc. The problem with such a strategy is evident: no one is responsible for the total, and, in packaging, the whole is much greater than the sum of the parts.

Medium-range Plan Formulation. This step in the process might be described as functional planning or tactics. Having established the objective and strategy or direction, it is necessary to develop direction for the various pieces that must come together to create a cohesive program. If the department has several groups, plans should be made for each of the groups for the entire time period. If other departments are involved, then their roles relative to packaging should be developed.

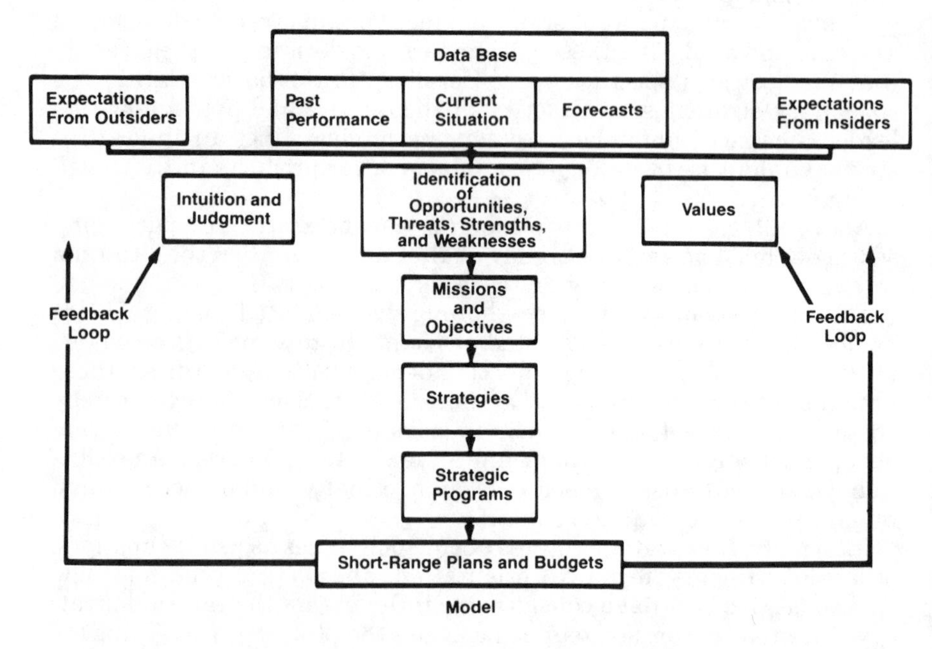

FIG. 10.6. The process of strategic planning encompassing the basic concepts.
From Steiner, 1979.

Each known project should be listed and scheduled, with the various predictable elements incorporated. For anticipated projects, tentative scheduling and required resources should be penciled in.

In essence, this step means enumerating projects initiated by packaging, marketing, production, engineering, etc., and incorporating all into a schedule and allocating resources on this basis. Obviously, the known projects are firm and the unknown but inevitable must be planned in phantom form.

Translation into Current Decisions. Besides planning, what must be done today to initiate the plan? Describe in systematic detail who is going to do what, when, where, etc. Which suppliers, consultants, contractors, staff members, etc., are expected to perform which tasks for which periods? From this scheduling, the budget can be prepared. Linkage to other programs and departments structurally and in terms of the budget must be ensured.

Contingency Plans. What happens if everything does not proceed according to plan? List all the possible events, both positive and negative, that might occur that are not already in the plan. What would be the scenario if these events were to take place? At what point would the counteraction be triggered by an unplanned event? For example, what action would be taken if either the government or Dupont withdrew ionomer from the market? Would there be a preliminary announcement to permit an orderly transition? Several items have been removed precipitously. In contingency plans, only those events that have a finite probability of occurring should be included. Contingency plans are critical because they will include thoughtful plans and not reactions to unexpected events.

The ten steps described are neither definitive nor all-inclusive. Rather, they are intended as a structure that could be used to prepare a comprehensive strategic plan. They could be somewhat truncated or greatly expanded. They could be prepared in a narrative text form, within the constraints of a set of rigid forms, or anything between. It is important to note that these steps or their equivalent serve to impose a discipline on the process that is vital for obtaining a meaningful result.

The steps may be analyzed and charted qualitatively, semiquantitatively, quantitatively, or even, in many more formalized organizations, by using computer modeling techniques. Since all results are measured in some manner, quantitative methods usually have more meaning to the manager and to the executive staff. In essence, quantification means reducing the substance of the plan to mathematical data and incorporating these data into chart form. Time, manning, and financial data are the references most commonly employed, although, of course, percentages, value added, productivity, etc., can be used if these data express the concept more definitively.

Thus a plan for packaging might incorporate such data as the number of packaging systems to be developed, the expenditure per development, manpower requirements, the probability of success projected, the return per unit of investment expressed in the present value system, etc. By reducing the operation to relatively simplified mathematics, the sensitivities of varying inputs can be readily computed. Quantitative techniques might be traditional, involving an income statement, a budget, or cash flow analysis, or they might employ sophisticated modeling and forecasting tools that are commonly used in high-technology development programs, such as aircraft design.

In contemporary business, since profit return on investment and cash flow are simultaneously the driving forces and the principal measures, the packaging manager or the planner is compelled to convert strategic planning into fiscal terms. In this manner, the executive staff can readily comprehend the returns that will be derived from the investment in the department.

Even if the department is treated as a cost center, its benefit to the organization may ultimately be converted into quantitative terms. To make sure that the investment in the packaging department is not subverted to other alternative investments, the packaging manager is forced to portray the ultimate benefit to be derived from his departmental activity.

Among the methods of measuring benefit might be pointing out the investment in packaging materials made by the organization in any given year—usually in excess of 10% of the factory cost of the product being manufactured. Cost-reduction programs are perpetual functions of packaging departments, and so highlighting of these cost savings represents a first approximation of financial benefits. Estimates of the value of packaging to new product introductions, of the merit of packaging innovation applied to mature products, of the role of monitoring packaging suppliers, etc., represent areas in which quantitative techniques might be readily employed.

In preparing the strategic plan, it is instructive to grasp the importance of the information that is used and that is communicated. The basic plan should be realistic, with some stretch or optimistic objectives, because invariably, someone upstream will find a rationale to reduce the investment. Remember that information will never be perfect. The packaging manager has one information bank on a subject which is different than the facts; he may request further information, which will invariably be different from both his own and from what the informant will perceive. Just as information is imperfect, so also is communication. If the information base can be incomplete and distorted, it will be, generally to a greater extent than the packaging manager realizes. It is therefore vital that the information used in the strategic plan be as accurate and complete as possible—and, if possible, slightly inflated on the positive side in the expectation that the plan that is submitted will be contracted.

Some managers view the planning process as among the most difficult of their responsibilities because it requires so much rigorous thought and analysis. Others view it as a restful exercise, with the implementation of the plan relatively simple because all the steps are clearly described and scheduled. A strategic plan has been likened to an instruction book. Open the book and follow the instructions and *voila!* instant success.

A strategic plan is basically a road map or a flight plan. It gives a lot of direction, but it does not tell you how to fly or how to compensate for the minor deviations that are inevitable.

Thus, when a strategic plan is followed, it acts as an excellent guide but not as a rigid rule book. The manager's talent, skill, and action are indispensable for implementation.

Incorporated into this process of implementation must be the mechanisms to measure the progress and results and to control the process through feedback. The manager can never lose sight of his staff members, all of whom are human and each of whom is different. People require direction, control, and feedback in order to optimize their production. And the key to implementing any strategic plan is to have people who can perform and who respond to the effective decisions and actions of their manager.

No manager can perform alone, and his interactions with his people, using the guidelines of their strategic plan, heighten the probability that the results will benefit all.

The dangers in strategic planning are multifold and can never be minimized as a threat. Strategic planning for the packaging department is ultimately the responsibility of the department manager and as such basically cannot be delegated. Today's problems are important, but tomorrow's problems are also important because tomorrow will eventually come. Many, if not most, of today's problems are the result of yesterday's paucity of effective strategic planning. The packaging manager must be familiar with his organization's goals and must have clear objectives for his own department. He must involve his staff and the relevant department heads around him. Above all, he must not dwell on past failure of planning to work as well as expected. With experience and repetition of the process, the mistakes become fewer, the projections become sharper, and the strategic plan becomes a much more refined tool that is an integral part of the everyday operations.

Those persons who have been exposed to strategic planning in their education and experience will, of course, recognize that the principles of strategic planning are the same, whether for the giant corporation, the federal government, the individual worrying about next week's grocery bills, or the packaging department.

We have grown a complex economy and society with a broad array of specialists. Division of labor is a requisite because the hard professions, such as medicine and engineering, demand so much education and experience as a result of a mushrooming body of knowledge that each hard science is subdivided; chemists are no longer just chemists, they are analytical, instrumental, quantitative organic, polymer, or nuclear chemists. And, as everyone knows, the soft disciplines have followed with their jargon and specialization.

Although the universe is ultimately very orderly, it is now extraordinarily difficult to fathom the order because our span of specific knowledge and ability to comprehend in a finite time is limited. We have resolved this dilemma by designating a leader who manages by delegating each of the individual steps and has the responsibility of interpreting the pieces into a coherent whole.

Subject ______

Objective	Actions Required	Resources Required	Time Required	Responsibility	Contingency	Return
XX00XXX	XXXXX	XXX	3 mo	Joe	Return to R&D	1982 $1 million
XX00XX	XXXXX	XXXX	6 mo	Phil		$1 million $2 million
XXXXX	XXXXX	$XXX,000	10 days	Alice		$5 million total
XXXX	XXXX	$XX,000	1 yr	Carol	Punt	1983 $2 million $5 million $10 million
XXX	XXXXX	XXXXXX		Ted		$17 million total
	XXXX	XXXXXX			Push Panic Button	1984
	XXXXX	XXXXXX				$10 million $50 million $100 million
	XXXXX					$160 million total

FIG. 10.7. A representative plan of action within a strategic plan.

So it is with the packaging manager. A generation ago, the packaging manager, if he existed, could be the marketing and technical manager and handle the packaging job on alternate Tuesdays during coffee break. In those allegedly simpler times, the packaging manager called in his own supplier and was offered a standard-size three-piece can or a standard-style folding carton in one of three board types and weights and put the packaged product into an A-flute corrugated fiberboard case. If his product could be packaged in a pouch, he had a choice between cellophane and glassine. Someone else worried about the machines, and few were available.

Some organizations still operate in a similar manner: give the product to the packaging manager and have him throw it in a box. Those who have no appreciation of marketing, costs, efficiency, and protection receive the results that that method delivers.

The professional packaging manager, inundated with information on new materials and machines on one side, with new demands from the marketplace on another side, and with pressures relative to economics, performance, compliance, etc., knows very well that his task is demanding and taxing. We in this country have essentially passed the era of simple materials and single machines. We are well into a period of multipurpose multiple materials used in special equipment to produce packages that must perform on the production floor, in distribution, on the retail shelf, and in the home. And, in the event that one system does not perform well, a viable alternative exists in a totally different material system.

We did not invent complexity as a scheme to thwart the consumer or the government. Just the reverse—complexity is a product of an expanding knowledge of what to do, how to do it better, and what each of us as individual consumers wants.

This environment clearly demands managerial talent, which in turn needs tools that enable us to not just cope, but rather to effectively use the knowledge available to us.

Strategic planning has been demonstrated to be a powerful weapon in extracting the greatest amount from the vast body of information and requirements that face the packaging manager. Strategic planning is not a panacea and does not compensate for a lack of skill,

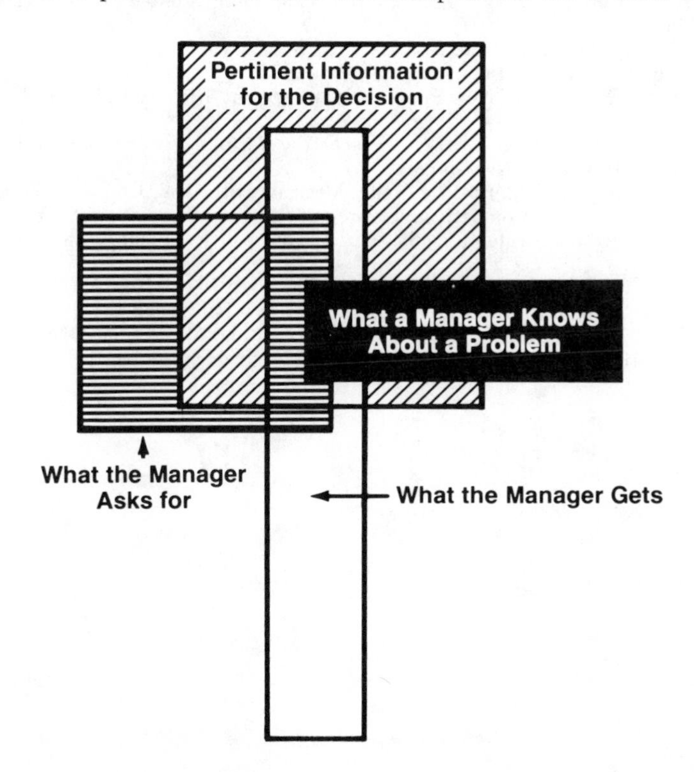

Fig. 10.8. The communication process in organization: the information needed for planning versus that wanted versus that obtained.

knowledge, experience, and diligence. Neither is it an academic exercise to soothe a turbulent operation. Packaging managers know that no single material, machine, or design can satisify all product and market requirements and that the perfect material is elusive. Similarly, strategic planning, when treated as an effective tool, given proper attention and used to think through the future consequences of today's decisions, will return rich rewards to the manager who really uses it.

Every packaging manager knows the benefits of using every material and test available to him. Now he has a managerial tool that enables him to take full advantage of his own resources.

Tomorrow is the result of the strategic planning you begin today.

BIBLIOGRAPHY

ANON. 1972. Perspectives on Experience, Boston Consulting Group, Boston, Massachusetts.

BRIGHT, J. R. 1978. Practical Technology Forecasting. Industrial Management Center, Austin, Texas.

BRODY, A.L. 1981. Strategic planning in packaging, Presented before Pack Info '81, Philadelphia.

BUZZEL, R. D., GALE, B. T. and SULTAN, R. G. M. 1975 Market share—A key to profitability, Harvard Bus. Rev., Jan-Feb 1975.

CRAWFORD, C. M. 1983. New Products Management. Richard D. Irwin, Homewood, Il.

JONES, H. and TWISS, B. C. 1978. Forecasting Technology for Planning Decisions, Petrocelli, Princeton, New Jersey.

LEVITT, T. 1969. The Marketing Mode, McGraw-Hill, Book Co., New York.

PORTER, M. E. 1980. Competitive Strategy, The Free Press, New York.

QUINN, J. B. 1977. Strategic goals: Process and politics, Sloan Management Rev. Fall, 1977.

STEINER, G. A. 1979. Strategic Planning. The Free Press, New York.

STERN, M. E. 1966. Marketing Planning, McGraw-Hill, Book Co., New York.

URBAN, G. L., and HANSEN, J. R. 1980. Design and Marketing of New Products. Prentice-Hall, Englewood Cliffs, NJ.

11

Technological Forecasting

My interest is in the future—because I'm going to spend the rest of my life there.

Charles Kettering

An annual rite in business is the preparation of the 5-year plan. The federal government produces a five-year budget forecast each spring. At year's end, many newspapers and magazines create predictions for the coming 12 months.

Many persons view these prognostications with wonder and ponder how the authors generate their picture of the future. Few persons look back on predictions to verify their accuracy.

Why do so many persons invest in an exercise that can be so unrewarding when examined in hindsight? If not for a need to plan for the future, few persons would have a desire to know what will probably occur next year, 5 yr out, or in the next decade. Current decisions must be based on best estimates of tomorrow, and so planning and forecasting are intimately allied.

Since packaging is increasingly entering the matrices of long range planning, the need to comprehend the future is becoming an imperative for packaging professionals. Packaging soon will no longer be an immediate response to a request from a customer or from marketing. Rather, packaging professionals will employ the tools of strategic planning to establish courses of action.

All planning demands insights into the future not attainable by emotion, hunch, or guesswork. Inputs are required from the economy, raw material supply forecasts, consumer demand patterns, labor rates, equipment costs, government regulatory expectations, and a variety of other sources.

In no way does emphasis on a single crucial element of forecasting, technology, minimize the need to integrate all variables into forecasting for planning. Rather, the impact of technology on packaging in the past has been dramatic. And the effect of technology on packaging in the future will continue to expand. The intent of this chapter is to stimulate thought and action that will nurture a discipline that can be meaningfully applied by all of us in packaging.

This chapter is based on the literature of technological forecasting and the teachings of Dr. James Bright and his disciples, who have been pioneering this new pathway. An introduction of present principles in the context of packaging should provide a framework for further study and application by both forecasting and packaging professionals.

Thus, this chapter defines technology and its offspring, technological forecasting; the process of technological innovation and development in perspective of packaging will be reviewed. The chapter further enumerates and details several of the currently employed methods of technological forecasting, including those which are qualitative and those which are quantitative. For example,

- Qualitative/quantitative
  - Intuitive forecasting
    - Delphi
  - Trend extrapolation
  - Morphological analysis
  - Monitoring
  - Dynamic modeling
- Probability assessments
  - Scenarios
  - Cross-impact analysis
- Time methods
  - Learning curves
  - Time series curves
  - Time series analyses

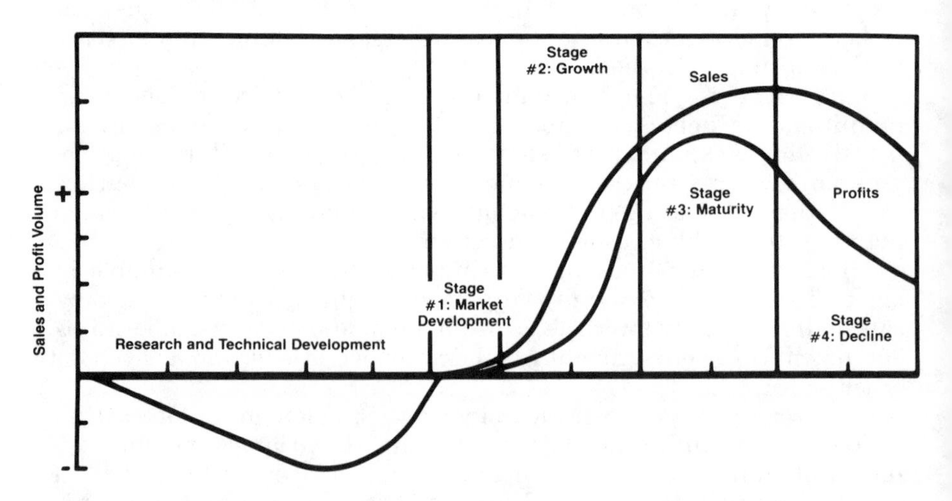

Fig. 11.1. The product life cycle.

Each of these discussions of technique contains principles, procedures, advantages and disadvantages, examples extracted from packaging, and some proposals on practical applications. This vehicle of exposing principles and probing into contemporary methodology is not suggested to generate expertise. Rather, introductory concepts can spark an expanding interest—or the conclusion that perhaps insight into the future of technology has limited value to some.

TECHNOLOGY

Technology has been the foundation of packaging since the notion of applying scientific principles to practical problems was born. In inventing canning, Nicholas Appert was not employing technology, but rather preceding the Edisonian approach: try, try again—and keep trying until you find a solution. Not until Pasteur deduced the propagation of microorganisms did a scientific premise for canning arise. Almost two decades were to elapse before Prescott and Underwood were to wed Appert's empirical solution to Pasteur's discoveries to form a technology that could rightfully be designated canning.

Today's packaging does not derive from Appert's and Edison's trial and error. Little of the packaging to be used in the future will be the result of constructing a three-dimensional structure from an artist's pristine materials. Packaging is the result of knowledge of the skillful integration objective of product protection and its corollaries; comprehension of the physical, chemical, and biological forces affecting the product; and intimate understanding of the scientific rationale for each of the components of the packaging system.

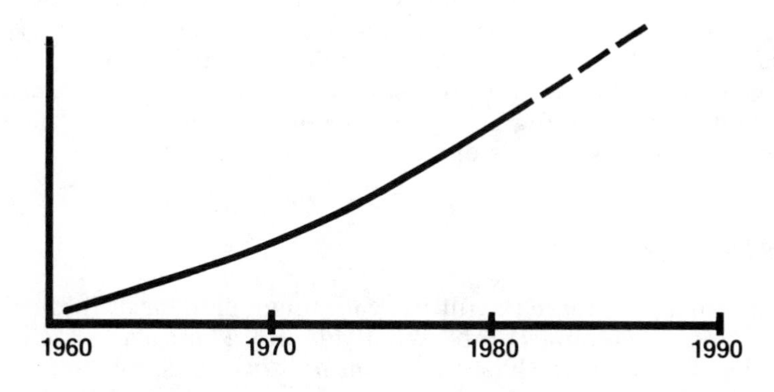

Fig. 11.2. Technological forecasting: a simplified trend extrapolation curve.

Consider such disparate packaging examples as solid-phase pressure forming (SPPF) of polypropylene and internal gas pressurization (LN_2) of aluminum cans containing noncarbonated liquids. Both reflect the application of a broad diversity of principles—of physics and its subdisciplines, thermodynamics and mechanics, of chemistry and biochemistry, and of mathematics—and the introduction of rigid predictive techniques before the initial bench model was constructed. Both appear to have demanded extensive, directed experimental effort to expand and refine them from the primitive physical models. Both could have been predicted before any investment had been made in physical models. SPPF and LN_2 can pressurization systems are simultaneously the culmination of a century of building on knowledge and a point on the continuum of packaging development. The fact that packaging is in perpetual change implies that those involved in it must plan for the change. Planning itself is an emerging science in which all the known elements, including technology, are forecast. Thus, a good technological forecast is supportive of a good plan but does not make a good plan.

WHY PLAN?

Since everyone in the packaging community must live in the future, and the future will certainly be different from the present, it is desirable—or even necessary—to plan for that future. Planning is the introduction of the future into current decisions, the systematic introduction of expectations of the future into present management to make sure that when tomorrow inevitably comes, action will be organized and responsive, rather than reactionary. Planning is an effort to comprehend all of the future in the context of today. Forecasting is an effort to predict the future on the basis of today.

When a lightbulb engineer predicts a bulb which can be made from plastic sheet with double the output and life and half the cost of present glass bulbs, the packaging technologist can begin to plan for lower volume packaging and the supplier of single-faced corrugated sheet should start looking elsewhere for an outlet for his commodity. Technological forecasting is one component of planning.

WHY FORECAST?

If you do not anticipate the future, you cannot plan for it. In the past, forecasting has been inaccurate. Wholly accurate forecasts are close to impossible. Critics, negativists, and conservatives anchored to the past can find fault with forecasts—of the weather, of stock market performance, and, in packaging, of the retortable pouch and ionizing radiation.

Forecasting may not provide a detailed and accurate portrayal of the future and will certainly not divine the totally unexpected, such as an eruption of Mount St. Helens or a war between an old, established European power and a Latin American country. Forecasting will, however, provide a dynamic picture of what could, should, and will happen in the future. Without forecasts, people cannot plan, and without planning, managers need not persist in business functions.

WHAT IS TECHNOLOGY?

It is easy to dismiss technological forecasting with an exhortation on the difficulties of political, economic, sociological, and marketing forecasting—since all of these interact with technology, why the concern about technology?

The classical definition of technology is the application of scientific principles to the physical world, or, more simply, the application of science. In the technical worlds, science is the comprehension of the vastness of the unknowns of space, molecules, and cells and is really the pure chemistry, physics, biology, and mathematics.

Technology is the disciplined application of knowledge gained by science to some man-oriented product or service. The gas laws of Boyle and Fick defined the rate of passage of gaseous molecules in space. These principles, applied in conjunction with those of polymeric chemistry, provided the scientific basis for the technology of plastic bottles.

Those familiar with the scientific basis know what is possible in the practical world—that the laws of thermodynamics, gravity, Arrenhius, etc., are immutable, but that man's laws are arbitrary and thus changeable.

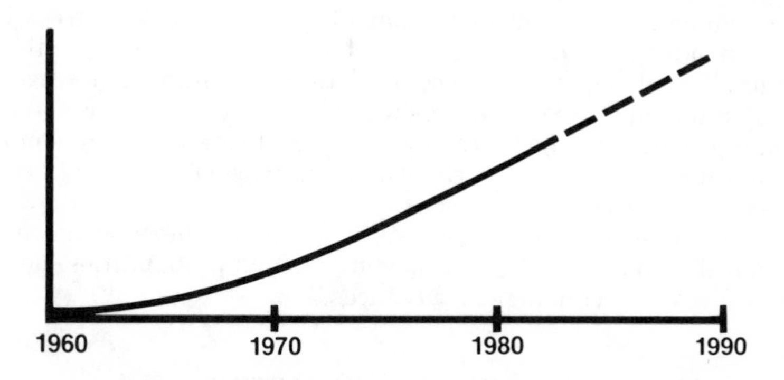

Fig. 11.3. Technological forecasting: trend extrapolation with only one variable.

In the context of technological forecasting, technology may be defined by three contemporary terms: physical, software, and definitional. Physical technology concerns the machine, the tool, or the material used—the plastic film extruder; the horizontal form, fill, seal machine; the laminated label; etc. Software technology concerns the processes and procedures for marrying the physical entities—extrusion, adhesion, filling, etc. And definitional technology concerns standards or references—e.g., the caliper shall be $\pm 10\%$ from the mean.

WHY IS FORECASTING POSSIBLE?

If events occurred in a totally random fashion, then, of course, forecasting of anything would be virtually impossible. However, occasions such as 1982 drug-tampering incidents are rare in history. Progress, and especially technological progress, is measured in small, logical, and generally quite predictable increments: metal cans move from 200 lb tinplate to lighter-weight tinplate to chrome/chrome-oxide–coated steel to black plate, etc.; paperboard moved down in caliper, and now the well-publicized thrust of development is toward board additives to increase strength and thus reduce the quantity of fiber material content—always in the direction of functional economics.

Incremental progress is based on the fact that each step must be preceded by another step and is itself the precedent for the next step. Further, society in the form of the converting or packaging operation is usually not capable of absorbing very large chunks of change. Thus a breakthrough is a rarity and, when viewed in detailed historic perspective, turns out to be a series of incremental moves that simply were not previously visible to the reporter.

On the other hand, just because all technological progress occurs in logical sequence does not mean that all technological progress is an extension of past trends. Technological advance is not just a reflection of technological logic; it is a combination of inputs of a variety of societal, economic, and political as well as technical influences. Aseptic packaging using chemical sterilization was technologically sound in 1972, but it did not appear on the American market until 1981 because of the well-publicized FDA issues.

Thus, forecasting technology requires a comprehension not just of the scientific and technological movements and probabilities, but also of the causes of technological advances.

ORIGINS OF TECHNOLOGICAL FORECASTING

Since the dawn of man, forecasting has been a serious matter for farmers, astronomers, religious leaders, militarists, and a host of other people. Endless volumes and films of stories have been generated on early attempts at prophecy. In the overwhelming majority of cases, these efforts were opinions, viewpoints, utopian dreams, self-motivated pronouncements, or self-fulfilling exercises. All of these ancient rituals appear to be in the annual announcements of packaging predictions of things to come. At every assembly of packaging professionals, spokesmen for packaging suppliers pontificate on how next year their organizations will introduce unbreakable glass at the cost of raw sand, transparent steel, waterproof paperboard, impermeable plastic, or another impossible dream. And how, within 30 days—never 4 weeks or 1 month—a major packager— almost always hinted to be one of the major blue chippers—will be introducing its new product in the supplier's new packaging breakthrough.

These predictions might be described as the modern extension of medieval witchcraft—using a few known facts to weave a credible pattern—but they are wishful thinking, not forecasts.

One of the great problems is that these are not a harmless form of recreation, such as the weekly football pools; they are believed, and thus draw investments in time and money. Many packaging professionals probably chase one of these rainbows several times a year.

In the 1940s, the federal government observed the chaos in efforts to develop complex weapons systems. Recognizing the need, staffers conceived the notion of organized, systematic technological forecasting as one basis for development. Thus, the federal government bureaucracy is actually the source and/or major user and proponent of technological forecasting—as a consequence of the need to anticipate the defenses required against any enemy's offensive weapons.

Of course, Appert's invention of canning was in response to the need of Napoleon's army for rations on which to wage war.

WHAT IS TECHNOLOGICAL FORECASTING?

A variety of words and phrases have been used almost interchangeably to describe looking into the future, but to be precise, each should be examined for its true meaning. *Propaganda* is infrequently used to depict the future and is an obvious attempt to sway rather than a realistic probe. On the other hand, *prophecy* is an often-used term which means wish or fantasy, often self-serving and often counterproductive. *Speculation* is contemplative reflection on what might be, with no definitive basis that it could be real—it has value as a tool for thinking. *Prediction* is a word that implies that some thought has gone into the statements regarding the future, but not a lot. *Forecasting* implies a systematic, logical process, bringing to bear all of the known inputs and probabilities. The much maligned weather forecast is much superior today to what it once was because of satellite observations, computer analysis, and greater knowledge of cause and effect; today's weather forecast is truly quite accurate.

Definitions are important because words are used to communicate, and actions are based on these words. Thus, in dealing in the future, it is desirable to be as precise as possible. The three components of contemporary forecasting are the objective, the qualifications of the predictor, and the methodology employed in preparing the forecast. This brief diversion demonstrates the characteristics that should be employed by the analytical forecaster or planner to assess the work of their others—and their own.

OBJECTIVE OF THE FORECAST

1. Entertainment: an amusement for stage effect.
2. Concerned interest: statements made that are based on honest fear, e.g., plastics will wrap the world.
3. Alarm: the extension of observations to a potential doomsday, e.g., packaging waste will suffocate our entire environment.
4. Anticipation of problems or benefits: e.g., by 1999, we will be nutritionally deficient because of packaged foods and so. . . .
5. Utopia: e.g., the world will be a better place with the energy savings arising from using returnable bottles.
6. Decisions-making: e.g., forecasting as a tool to support planning.
7. Self-satisfaction: e.g., what would happen if a package with an oxygen barrier of 10^{-6} cc/day were developed.
8. Propaganda for goals: e.g., if my bill passes, packaging costs will go down.
9. Commercialization: the development of future portrayals that permit the introduction or expansion of products or services.

The purpose of the forecast should be known so as to place a weighting on the information. Journals and panel discussions often provide views on the future. Each looks at it in the perspective of one narrow field with no regard for other fields— e.g., the consumerist sees the consumer reacting to (all) packaging, the polyester converting expert sees polyester making further penetration into the market, and the glass bottle R&D manager sees new vigor in glass as a result of the new glass hot-end coating.

QUALIFICATIONS OF THE PERSON OR INSTITUTION MAKING THE STATEMENT ABOUT THE FUTURE

1. Expert: Technical competence in the discipline about which the statement is made.
2. Chemist or Scientist: Competence in a related area, but not in the area about which the statement is made: e.g., a polymer chemist making predictions about radiation sterilization.
3. Psychologist or Accountant: Competence in a wholly unrelated area: e.g., a psychologist or an accountant discussing the en-

vironmental effects of returnable glass bottles for carbonated beverages.

4. Writer or novelist: a person who was trained as or who works exclusively at writing one of the many volumes for the public, the subject of which is often a planetary doomsday that will come if the hero does not stop somebody or something.
5. Essayist: not far removed from the writer of books, with pieces prepared with much thought on the form and rhetoric and appearing in lengthy version in sophisticated media.
6. Advocate: advocates come in many styles, ranging from professional lobbyists supporting packaging to those opposing the industry. In form, the advocate is almost the antithesis of the essayist because his action is really brief and emotional reaction, as when the consumerists and students of the 1970s vigorously campaigned against nonreturnable packaging.
7. Journalists: a writer with a deadline to meet, seeking to write a piece for print or electronic media that will fit a space and capture attention; ranging from the members of the trade press known to the packaging community to the local newswriter seeking facts to the person whose mind was set before the item was conceived.
8. Spokesman: an advocate, an apologist, a person whose sole task is to provide data that shade the future in the most favorable light for the establishment which employs him or her— e.g., the public relations director of a trade association.
9. Analyst: a person or persons trained in projective analysis or mathematical models who know nothing about the subject of packaging itself, but who apply the facts provided by a questionnaire to a methodology to produce a result. Persons in this category may predict the future of packaging today and the future of subway flower pot decorations next week.

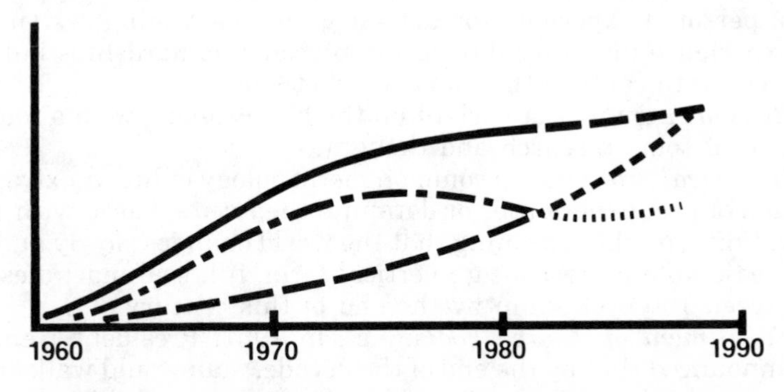

Fig. 11.4. Technological forecasting: simplified trend extrapolation of multiple variables, with each variable extrapolated individually.

With this probably incomplete list of persons and their types of competences and motivations offering insights into the future, several conclusions may be drawn:

1. Portrayals of the future are as fragile as the person delivering them, which accounts for many inaccuracies.
2. If the future is conceived as depicted today, the identity of the forecaster and his or her motives must be known.
3. Few are called to account for the consequences of their scenarios of the future when the future arrives.
4. Looking into the future requires direction and a consciousness of the biases of the abundance of personnel in predictive fields.

Compounding the problem of determining who is looking into the future and what background and/or experience he or she has in conducting such an exercise is the issue of the method employed. Just as most people automatically assume a certain level of competence in those who deliver forecasts, so do most people believe that the forecast is derived from some soundly based technique that has been validated by experience.

A brief diversion into some of the more commonly employed methods used should squelch that notion.

METHODOLOGIES USED TO PROVIDE A PICTURE OF THE FUTURE

1. *Speculation or fantasy*: pure science fiction; the idea that science, whatever that is, will discover a new polymer that is as strong as steel, as transparent as cellophane, as flexible as paper, and as inert and impermeable as glass.
2. *Opinion*: probably the most common technique; an extrapolation of personal experience or bias—e.g., my packaging system will experience difficulty during these economic hardships but will emerge to capture the major market share.
3. *Reasoned opinion*: a variant on the basic theme, with a foundation of some research and thought.
4. *Historical projection*: a common methodology of linear extrapolation of past experiences or data into the future, based with some validity, on the reasoning that the world changes slowly and in a predictable pattern—e.g., vertical form, fill, seal machines will exceed 100 cycles/min by the end of this century.
5. *Fulfillment of desired goals*: e.g., in 1961, President Kennedy announced that by the end of the decade, man would walk on the moon—and surely, in 1969, Neil Armstrong took his one giant step.

6. *Modeling*: employing quantitative analytical techniques based on experience to construct a structure, including the future.
7. *Repeating predictions of others*: a technique of parroting what is read in the Sunday supplement, heard on television, or heard at a meeting; embellishing what someone else has said or written—e.g., the perpetual misperception that plastics packaging is petrochemically unsound.
8. *Projection of existing trends to extremes*: a distortion of linear extrapolation to an extinction point—the decline of use of wooden crates for packaging would lead to the inescapable conclusion that no wooden crates will be used by 1990.

It should be noted that the level of qualification of the predictor and the methodology employed are not related. Good methods may be used by unqualified persons, and poor methods may be used by highly qualified and motivated individuals. With so many variables, it is easier to generate an authoritative and wholly inept technological forecast than to skillfully fashion a sound scenario using the best available people and techniques. Thus, a second essential building block is that competence be reinforced by technique.

A third basic and independent variable in forecasting, and especially in technological forecasting, is the level of effect predicted. A prediction stipulating that people will eat food in the twenty-first century has far less impact on current decisions than one which asserts that, for example, the price of aluminum for cans might rise 40% by the end of the year. The following series of examples is intended to demonstrate the effect of the level of impact of the forecast.

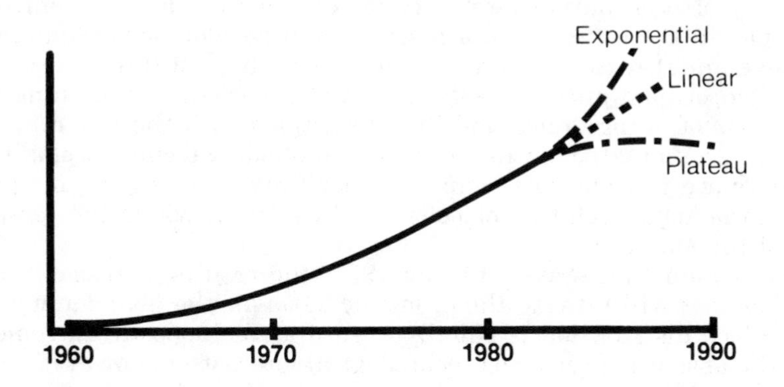

Fig. 11.5. Trend extrapolation: simplified result trend using each of three extrapolation procedures to demonstrate the possibilities.

1. *Acquisition of knowledge*: e.g., by the mid-1980s, science will discover a new principle for surface sterilization.
2. *Availability of a new technical capability based on the new knowledge*: e.g., it will be possible to sterilize the surfaces of packaging materials and food or medical products by the end of the 1980s.
3. *Identification of specific means of applying the technology*: e.g., by 1990, a prototype surface sterilization unit will be used to preserve packaged food.
4. *Initial commercial trials*: e.g., by the early 1990s, stores will be selling surface-sterilized packaged foods.
5. *Affect of technology on other elements of the economy*: e.g., factories will be built, capital will be invested, and old refrigerated distribution channels will fade away as the new shelf-stable packaging technology replaces frozen packaging technology.
6. *Affect of technology on the behavior of individuals or society*: e.g., food costs will decline by $n\%$ by 1995 as a result of the decline in food waste and thus less than 1% of the population will be involved in food distribution.
7. *New problems from societal adoption of the new technology*: e.g., because food distribution will be so simplified, millions of supermarket workers will have no jobs by the end of this century, truncating the average effective career lives of the population to 17 years.

Using the best techniques, the forecaster must also consider the level of the forecast—a single level, such as the specific increment of scientific information expected to be developed from a specific test for the body of information necessary for commercial development. For example, in the continuum of research and development required to achieve indefinite ambient-temperature shelf stability for concentrated orange juice, a vast amount of knowledge concerning the properties of orange juice and its interactions with the environment must be accumulated, assimilated, and applied. A technical plan that states that a test for water vapor permeability of packaging material will be the major technical objective is obviously trivial in the perspective of the objective.

And a plan that states that by 1990, the results of research and development will provide the economic basis for the abandonment of frozen food distribution is equally absurd at the opposite extreme.

In the absence of reasoned technological forecasting, however, management can be over-awed by the image of technical instrumentation and accomplishment, as embellished by the popular media, which are equated with an ability to meet business objectives. A procedure to measure the gas permeation of one flat sheet may be likened to a single bit on a floppy disk. The level of emergence or impact must be delineated before the effect of a technology or technological resource can be forecasted.

The levels of forecasting may be simplified as follows.

1. *Acquisition of knowledge*: a long and painstaking process in which comprehensiveness and completeness are not mutually exclusive.
2. *Analysis and application of the knowledge* towards the technical objective: an apparently obvious process that is often overlooked in the search for a quick and easy route using the resources available; in the final analysis, the reason for failure of most packaging innovations is oversimplification.
3. *Incorporation into a business enterprise* to fulfill a commercial objective: the translation of forecasting into meeting a customer or consumer need or desire, if and when the technical accomplishment indeed fits some need—e.g., with the development of microprocessor-controlled, energy-efficient, continuous-motion equipment, it must be asked whether need exists for new high-speed carton overwrapping machines.
4. *Interactions* with the perimeter disciplines of the economy and society: e.g., what happens to the many three-piece, soldered-side-seam can making plants when a new integral sterilize, fill, seal system renders canning archaic?

Thus, technological forecasting is not as simple as "Tomorrow's test results will give the answer" or "In 1 year, the breakthrough [whatever that is] will occur." As those who pioneered aseptic packaging and the retortable pouch will attest, technological forecasting is a complex process reaching far beyond the apparent.

INTERACTIONS EXTERNAL TO TECHNOLOGY

In the past, technologies and developments were fairly independent of each other—metal can makers could make cans with the confidence that only metal cans could contain certain foods, and glass bottle makers knew that only glass could be used to package beer. Today, nations, people, companies, and technologies interact, support, reinforce, and even counter each other. Imagine building a technical resource around a single raw material packaging process today and expecting it to provide for the coming decade or even year.

Communication has moved so rapidly that any of us can talk to anyone else in the world or reach into past knowledge instantly. We can receive immediate responses to questions and problems—if a technical development is reported in France, people know about it in the United States when they arise the next morning. The converse is that we often react to the raw data emotionally rather than contemplating the full nature of the technical impact. Technological forecasting in an environment of correct information is complex; in the perspective of instantly communicated theatrics, it becomes a venture for the bold.

Technological forecasting may be employed to determine the future in terms of:

1. *Future attributes of a system, material, or process*: e.g., types of oxygen barrier thermoplastic available and new techniques for hermetically sealing thermoplastics.
2. *Future processes*: e.g., reduction of net energy input to convert fibers into paperboard.
3. *Future operations*: e.g., the use of gas-barrier thermoplastics to package ambient-temperature, shelf-stable concentrated juice products.
4. *Second-order effects*: e.g., the impact of shelf-stable juice packaging technology on composite cans and on frozen food distribution.

Since technological forecasting can be rewarding when performed intelligently, it may be effectively applied in a broad variety of meaningful exercises. Future products and processes may be identified, as well as the planning and research and development required for their attainment. Planning personnel should request the results of all forecasts available as a basis for their activities, particularly in predicting the market impact of competitive technologies. In the broader sense, technological forecasting can be an invaluable aid in depicting the overall impact of the introduction or expansion of a technology on an industry sector or on an industry. For example, the net results of the low-pressure, linear, low-density polyethylene resin production processes introduced a few years ago have hardly been felt yet, even though there has been major production for some time. What will be the ultimate impact of aseptic packaging technology on the metal can making industry or on paperboard multiple packaging?

Having introduced the concepts of technological forecasting in the contexts of uses and abuses, shortcomings, popular notions, and related but obviously counterproductive futurism devices, this chapter now focuses on some validated methodologies. Since every discipline develops or creates its own jargon, it is essential that these nomenclatures be comprehended before the methodologies are described.

Intuition: an organized assimilation and evaluation of the informed judgments of persons knowledgeable in the field.

Trend extrapolation: since the technical world moves in an incremental fashion, building fact on fact, and not on the wings of dramatic breakthroughs, technically logical extension of current technology is a valid means of projection.

Monitoring: because alminformed judgments of persons knowledgeable in the field.

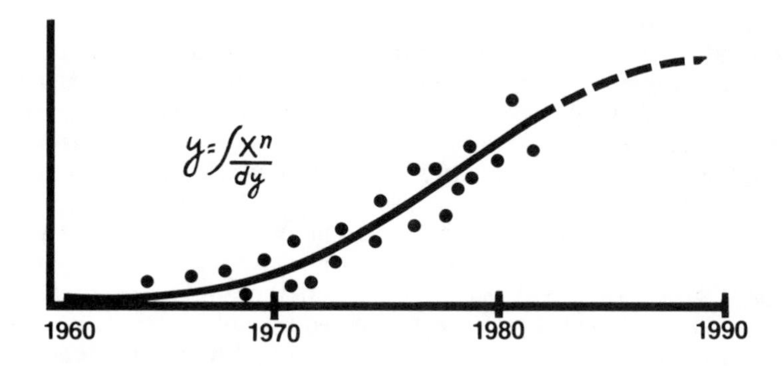

Fig. 11.6. A mathematical curve, fitting as a method of trend extrapolation.

Trend extrapolation: since the technical world moves in an incremental fashion, building fact on fact, and not on the wings of dramatic breakthroughs, technically logiechanism for foretelling the technological future.

Structural analysis: goal-oriented forecasting develops the relative importance of the technology and projects the probability of its emergence on the basis of the benefit to the organization, industry, economy, or society.

Modeling: a process of measuring the dynamics of technological developments and mathematically projecting the future; in essence, an extension of the role of mathematical modeling into technological forecasting.

Cross-impact analysis: determination of the many interacting elements affecting the technology, placing them in matrices, and determining the probabilities of tracking through these matrices.

Scenarios: exploration of a full range of the future by persons knowledgeable in the field that projects the possible results with judgmental probabilities.

PROCESS OF TECHNOLOGICAL INNOVATION

In order to comprehend technological forecasting, it is necessary to understand technology and the process of technological development. Although it is obvious that the varied portrayals of science and technology in the popular media are fictional fantasy, the many differing inputs from industry, research institutions, business managers, and consultants lead to a confusing picture. Technological development is not inspirational breakthrough, and is not an infinite number of scientists conducting an infinite number of laboratory experiments to produce an infinitesimal bit of knowledge. Neither is technological development a mighty edifice equipped with the latest instruments and peopled by bustling, white-coated thinkers incapable of social interchange. Conversely, technological development does not arise from a mass of small, unrelated laboratory tests performed in a chatter of immediate technical service responses. Technological development also is not the Edisonian "Try anything and everything until you succeed."

Since the process of technological development does not fit the popular view, what is it? Technological innovation is the sequence of thoughtful actions by which scientific and technical knowledge and information are systematically transformed into physical realities that can be applied to realistic needs and desires of a business enterprise or to an economic or societal need. The key terms are a body of knowledge, intelligence, and realistic needs—and effective application of resources keyed and controlled by thought and directed to a man-oriented objective.

The process of technological innovation is often regarded as encompassing eight phases: (1) origination, (2) concept proposal, (3) concept verification, (4) laboratory demonstration of the concept, (5) field trial, (6) commercial introduction, (7) widespread adoption, and (8) proliferation.

Again employing a hypothetical example, the innovative process may be expanded to facilitate understanding.

Origination. Technological innovation begins with an idea derived from a scientific suggestion; with a discovery obtained by observation in the laboratory or in nature, by serendipidity, or by intentional searching; or with recognition of need or opportunity and deliberate targeting of a solution. For example, several scientists separately publish observations that ionizing radiation can disrupt nuclear structures within biological cells and that microorganisms on the surfaces of packaging materials are responsible for deterioration of food products having water activities in excess of 0.85.

Concept Proposal. This is a proposal of a hypothesis or design that focuses the scientific knowledge on specific problems or applications. For example, ionizing radiation such as gamma rays can destroy the microbial flora on packaging material surfaces. Searching the scientific literature would provide information of the known effects of various ionizing radiations, such as electromagnetic radiations from decay or particulates such as alpha or beta rays, on biological cellular nuclei and on packaging substrates such as cellulosic or polyolefinic polymers.

Verification. Under controlled conditions, with appropriate standards and references, small-scale laboratory experiments or field observations are conducted to ascertain the validity of the theory—a physical demonstration. For example, a known quantity of identified microorganisms is applied to the surface of a packaging material which is exposed to a measured quantity of ionizing radiation, and contaminated packaging is retained without exposure as a control. Measurements are made of the numbers of organisms surviving the exposure and the changes in physical properties of the packaging substrate.

Laboratory Demonstration. In the verification phase, only the theory is corroborated, using any valid experimental techniques—for example, a square centimeter of yeast cells on a 4 in. × 4 in. plaque might be exposed to an x-ray beam. In the laboratory demonstration, actual pieces of packaging material are quantitatively contaminated and exposed to a cobalt-60 source for specified time periods. Quantitative measures are made of rates of death of various microorganisms exposed to various radiations, and physical measurements of the packaging are made.

Field Trial. The process or product is translated from a laboratory bench experiment into a prototype in which copies of the original in practical form are initially laboratory- or pilot-plant tested and subsequently moved through real or simulated channels. These trials are designed to show how to produce commercial quantities and to provide feedback that will help to rectify the inevitable flaws. In the example, designated packaging is conducted through a radiation source and sterile product is filled and sealed aseptically. Statistically valid quantities of packaged product then undergo controlled-temperature shelf-life tests and go through distribution channels. Safety evaluations are conducted and regulatory data are gathered for submission to government agencies. Economic analyses are conducted.

Commercial Introduction. This is the initial marketing or operational use in a profit-making facility. In the continuing example, a pilot plant to radiation-sterilize packaging materials and to aseptically package product is engineered, constructed, debugged, and ultimately put into operation to produce sterile, shelf-stable, packaged food product which is launched in retail markets. Economic analyses, market plans, and business plans emerge from this phase.

Widespread Adoption. The product or process is recognized as a significant innovation and so is introduced into full-size commercial factories, modified and imitated by competitors, and, in effect, reported in financial and trade journals. In our example, the pilot plant reverts to experimental operations, and one or more full-scale, continuous radiation-sterilization aseptic packaging systems for foods become operational. Competition uses alternate forms of ionizing radiation, of packaging, and of filling/sealing equipment.

Proliferation. This involves adaptation of the original technology to other applications and functions. Having been proven scientifically and technologically viable, the principles of the process/product are applied to new products or processes—for example, aseptic packaging of medical instruments, radiation sterilization of bulk packaging materials, and introduction of the process to packaging machines.

Technological innovation follows the classical S-shaped curve (Fig. 11.1), for which long time periods are required, from conception through evaluation and into the logarithmic growth commercialization/proliferation stage. The phase portrait of technological innovation also clearly demonstrates the multiple exploratory trials (the word *exploratory* meaning determination of technical feasibility) and low investment at the outset as a result of the relatively low probability of any single concept proving sufficiently attractive to progress into the succeeding stage. Thus, most of the initial concepts are shelved for reevaluation at a later time. With no initial inputs, however, no entry into the rapid growth stage is possible: nothing in, nothing out. Possibly most important in the long-range scheme is that the peak of each exponential growth stage should represent the initiation state for a subsequent technological innovation.

At no stage in the process of technological innovation are any short-circuits visible; the process of developing a hypothetical ionizing radiation aseptic packaging system or a real high-oxygen-barrier plastic is much like the process of placing a man on the moon:

Explorer: unmanned
Vanguard, Discoverer, Pioneer: unmanned
Mercury: 2 manned suborbital launches, 4 manned orbital launches, 2 unmanned launches
Gemini: 10 manned launches
Apollo: 11 launches with Apollo, with number 9 orbiting the moon and number 11 making an actual landing—more than 27 launches

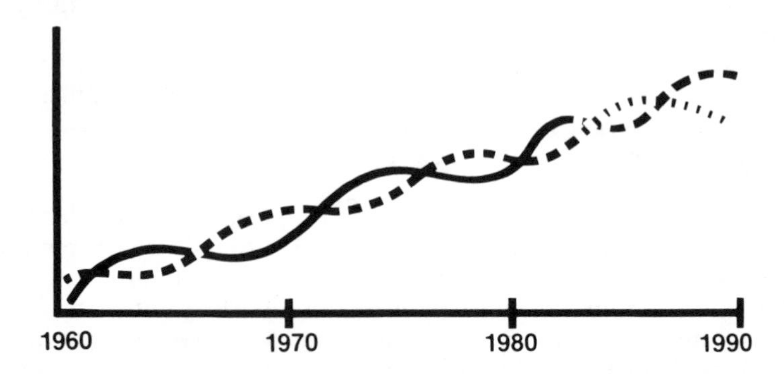

Fig. 11.7. Tracking leading indicators as a method of forecasting; the solid line represents the leading indicator and the dashed line the trends being forecasted.

These launches represent a sequence in development ranging from shooting an inert object into the outer atmosphere, to firing a capsule down range, to firing a manned vehicle down range, to a one-man orbit, to a two-man orbit, to orbiting the moon in a three-man vehicle, to final landing—a typical well-planned build-up from the easiest step to the climactic event. In contrast to a frequently expressed and published program of "attempt the moon landing first and everything else is easy," sound technology dictates developing a firm foundation and building from it.

The process of technological innovation is characterized by a number of significant basic tenets:

1. Time is generally required. Vanguard was launched in 1958; Neil Armstrong walked on the moon in 1969.
2. Many variables influence its nature and direction. Technological innovation is not a unilateral or one-dimensional process.
3. Government is more than ever directly or indirectly an influence—by encouraging in the case of military orientation or delaying in the case of regulation.
4. Technological capability grows in an exponential manner. This is a most significant assertion based on experience. Technology builds on itself, as the past 20 yr of plastic packaging clearly demonstrate.
5. Technological progress is not necessarily equated with sales dollars or profits, and the converse is not necessarily true, either.
6. Interactions of technologies frequently stimulate new technologies, especially now.
7. The method of paying for the technological innovation is often critical.
8. Business, society, and economics often alter, mask, or deter technological innovation.
9. Technological progress is inevitably delayed if the investment must demonstrate an immediate profit.
10. Technological innovation moves most swiftly when demand pulls the innovation through the system.

INTUITIVE FORECASTING

Since no man known can actually observe the future, opinion, judgment, and intuition must be a basis whenever man attempts to predict the future. A single person's opinion is suspect, since no means exist to probe the inner recesses of the mind and psyche to determine the supporting information and assumptions. In the newly emerged discipline of technological forecasting, the use of the much maligned committee to overcome the deficiencies of one-man opinion becomes intuitive forecasting. In contrast to the commonly accepted definition of committee, however, in some technological forecasting, no member of the committee is known to any other, nor do the members ever communicate to each other directly.

Rand Corporation's Delphi technique, introduced more than 20 yr ago, was designed to focus a broad range of expertise, disciplines, and experience on the issues and to anonymously impose challenge of thought against thought to stimulate better reasoning and information accumulation and application.

Delphi techniques have become relatively popular means of forecasting the future, as they attempt to capture a range of substantive inputs and integrate them into a coherent whole. Further, in a business organization, Delphi techniques can be administered relatively easily and can be extended with considerable feedback. The disadvantages include the fact that the inputs are only as good as the range of knowledge and courage of the committee members; that the members cannot directly discuss the topic; that, for practical purposes, it is qualitative; and that it is in reality an averaging technique.

In a typical Delphi procedure, a study director or administrator selects the areas of interest and edits a series of questions to be communicated to a preselected panel or committee of experts. The questions deal with the issue and ask that the expert provide a list of expected events, a date when he or she expects a specific event to occur, and the probability that the event will occur by that date.

The administrator receives the data and tabulates them. All deviant data are further probed to obtain a rationale for the nonconformity with the mean or median—thus forcing the panelist to justify his prediction.

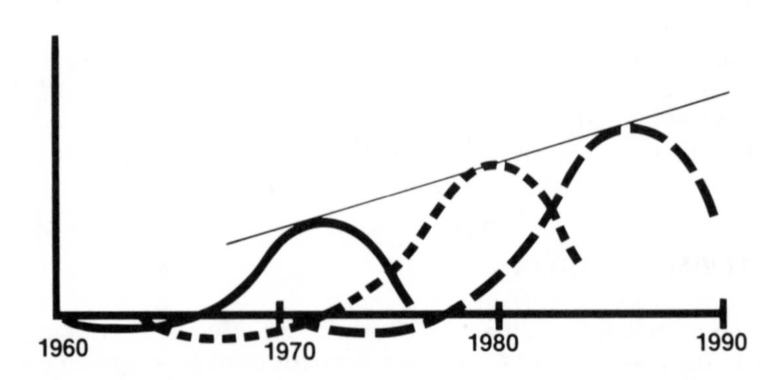

Fig. 11.8. A tangent of the peaks of the growth curves of technologies is linear over a long period of time.

The raw data are sent back to the panelists for refinement, counter-arguments, etc., and the final results become the technological forecast.

Although simple to describe, in actuality the process is time-consuming, as the questions must be pretested, the panelists must be selected for competence and willingness to actively and promptly participate, and the data must be statistically interpreted. In many cases, the administrator does not know the questions to ask and the panelist is an expert who cannot see beyond the details of today's problems; in addition, the organization occasionally dictates that the panel of experts include the accountant or salesperson, whom it is politically wise to involve, but whose inputs are uninformed opinion—the very opposite of the objective.

In a Delphi procedure for packaging, the basic question for a packaging machinery firm might be what the future technology will be for its folding carton packaging machinery. Obviously, a single packaging machinery firm would not have even a smattering of the total knowledge required by this demanding issue. Thus, dozens if not hundreds of panelists from the technical and mechanical departments of paperboard and paperboard-converting firms, from plastics-converting machinery companies, from plastic fabricators, etc., are recruited as the committee. No one is known to another, and so the inputs in total can be useful only to the querying organization.

In a 1979 Delphi operation, the FDA predicted the future of foods, drugs, and packaging. The panelists were principally FDA technical employees with little background in packaging, and so the results might not be palatable in the packaging community. The FDA study encompassed over 150 panelists describing more than 400 technologies in at least two rounds—a formidable clerical task, at least.

A representative Delphi result would be a list of expected future events, their descriptions, the median year of introduction, and the probability of the median occurrence.

TREND EXTRAPOLATION

Since technology progresses in a relatively orderly and incremental pattern as a function of time, extrapolation has a valid basis. Unfortunately, research on historical technological advances has not been performed to establish theories for technological trend analysis.

Trend extrapolation is based on the classical S-shaped growth curve of each individual technology and upon its derivative, the exponential growth of the tangents of the peaks of the arcs of sequential technologies. Although each technology grows in an S-shape, each succeeding technology grows in a similar S-shape, and the growth of the many in sequence is a line connecting the tangents—and consequently the concept of using linear extrapolation to forecast technological innovation.

The problems with trend extrapolation include the use of visual curve fitting, which, of course, can be naive; interpolating linear extension of historic trend lines which may be simplistic; the introduction of new forces which did not act to generate the previous trends; and the technologically generated value and other effects from society, the economy, business, etc., on future technology.

Since technological trends are orderly, however, eyeball curve fitting might not be all bad, particularly since mathematical curve fittings are not all good. Examination of historic trends in technology do not show discontinuities, and so little reason exists to believe that they will erupt in the forecasting time horizon.

Trend extrapolation encompasses the accumulation of historic inputs on a time abscissa and extending the ordinates to determine what is and when it is likely to occur. Plot the historical technical advances on a time base and linearly extrapolate the line; this is a highly simplified description of what is obviously a much more complex process, but a description which provides the principles. Several variations on trend extrapolation have been explored, some of which are discussed below.

The essential stages in trend extrapolation are as follows.

1. *Identify attributes.* These are the historic variables relevant to the technological innovation whose trend is being tracked—e.g., metallurgical properties of metals, organic coatings, and machine tools for fabricating metal.
2. *Develop parameters* for the attributes. Meaningfully quantify the attributes in order to be able to apply the information mathematically or graphically—e.g., the ductility of aluminum or steel alloys and the corrosion resistance of coatings.
3. *Establish and extrapolate the trend line*, reducing the data input to a progression whose future track can be visualized. The techniques are (a) *Intuitive extrapolation*, with the risk being human error; (b) *Mathematical curve fitting*, with the risk being that an equation selected might not reflect the actual trend-driving variables; (c) *Pattern identification*, a search for historic relationships, relating perceived forces to trends—e.g., as beer consumption increases, the need for lighter-weight two-piece metal cans increases; and (d) *Analogies*, use of unrelated but similar trends to establish a trend for the technology under study—e.g., the development of cold working of steel has been employed to develop directions for cold working of packaging plastics, such as ABS and XT polymer. This method of curve fitting is obviously subject to question, but it not only has some powerful merit, it has been successfully used.

Trend analysis is mechanistic and often subjective, and so provides a picture without the meaning. The forecaster should be compelled to question the trend portrayed by the data to determine the rationale, as in scenario generation, described later.

Among the variations on the basic trend extrapolation concept is trend impact analysis, which focuses not on the technological trend itself, but rather on the variables influencing the mathematical analysis. Trend impact analysis is based on a thesis that the trend expressed by the extrapolation must inevitably be altered by future events. For example, the federal government's incidental ingredients concept proposal might alter the availability of various plastic materials for food packaging.

Trend impact analysis procedures involve extrapolation of both the basic technological development of concern and relevant events (for example, FDA trends) and integrating them into a single trend. In effect, trend impact analysis is a multiple-variable mathematical correlation extrapolation—that is, the integration of all the known influences affecting a technology. To the planner, trend impact analysis would be the totality of the business or research enterprise, embracing all the parameters. As such, trend impact analysis, a refinement of trend extrapolation, approaches another technological forecasting technique, morphological analysis.

Because trend extrapolation is among the simplest and therefore most mathematically derived of the technological forecasting techniques, a not inconsequential body of methodologies has been adapted:

Single parameter: a time series based on a single technical variable—e.g., cogenerated energy input in paperboard mill operations. Obviously, this method fails to consider multiple direct variables.

Compound parameter: a time series based on multiple relevant direct variables—e.g., energy derived from waste wood, coal, oil, and natural gas; air emissions; energy for transportation of wood; and corrosion of boilers, all of which are input paperboard mill operations.

Leading indicators: not unlike the analogy in which a trend that has been a historic predictor is tracked—e.g., oil prices have preceded the search for substitution of plastics for other packaging materials. Obviously, past leading indicators are not valid for all time.

Envelope curves: formed by connecting the tangents of the S-shaped growth curves. This is based on the thesis that the innovative process over a long period follows an exponential growth pattern.

Step functions: introduction of each of the influencing variables to determine their individual impact and timing and thus develop a pattern of events. It is difficult to quantify the multiple variables in a time-series pattern; not unlike trend impact analysis.

Technological progress: as more knowledge is accumulated and applied, a greater base of information and resources is applied to a particular technology, and consequentially the technology expands—e.g., as more organizations have worked in coextrusion of plastic materials, the number and scope of developments have significantly increased.

Substitution theory: when a new technology displaces a preceding technology without major functional change, the new technology will tend to go to completion, with the amount of substitution predictable according to a hyperbolic tangent function based on the annual rate of substitution—e.g., the displacement of glass in packaging by rigid plastic is perhaps one of the classical examples of the substitution theory.

Correlation analysis: not unlike analogies or step functions; the blending of attributes that are not actually attributes of the technology itself to generate a trend—e.g., using polymer molecular packing and polarity theories to predict the oxygen permeability of plastic materials.

Although it is mathematically based and therefore apparently on sound scientific ground, it is essential that the elegance of mathematics not overshadow the thought and rigorous discipline demanded in trend extrapolation. The objective is to measure future technological progress, and trend extrapolation is a tool—not the reverse.

MONITORING

The process of technological innovation is incremental and is inevitably signaled by preceding events—e.g., Neil Armstrong's walk on the moon did not spring like the Phoenix from ashes, but rather from a series of carefully planned and executed steps. The development of the retorted steam tray is a plateau in a continuum whose origins can be traced to canning, sardine can production, aluminum drawing, and retort pouch heat penetration data development. Thus careful observation and analysis of published technical information are sound means of predicting future technology.

Monitoring encompasses observing and assessing the events in order

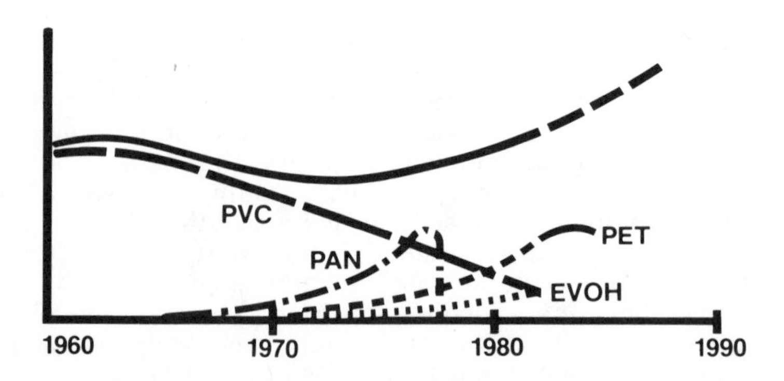

Fig. 11.9. The sum of the trend variables becomes the total technological forecast, as depicted in this hypothetical solid line showing the growth of barrier packaging plastic.

to comprehend them in the perspective of future technology. The term "monitoring" implies observation or searching, as in a library or market research function. The concept of monitoring embraces far more than the simple secondary information scan, collection of library lists or abstracts, or accumulation and tabulation of fragments of primary and/or secondary data. Monitoring is a continuing probe of the literature for data and meaningful alternate possibilities and effects. Monitoring involves selection of critical parameters (not those that receive press headlines) and synthesis of conclusions on the basis of the totality of information. For example, a review of the entire American technical and trade literature, meeting topics, known research projects, etc., for the period 1974 to 1980 shows virtually nothing on aseptic packaging in paperboard/plastic complexes using chemical sterilization, and what few references exist are trivial. Thus a tap into on-line computerized information sources would in no way reveal the explosive technological trend in aseptic packaging that began in the United States in 1981.

Those monitoring the activities, however, knew of the rapidly escalating prices of small metal cans, of government pressures to remove lead solder from cans, of the petitions to the FDA to permit chemical sterilization of packaging, of university pilot-plant aseptic packaging operations, of the body of literature, and of commercial application in Western Europe.

Monitoring involves four steps:

1. *Searching* the environment for signals that could anticipate significant technological change.
2. *Identification* of the envelope of effects if these signals are valid.
3. *Selection* of the parameters to be tracked in order to verify the velocity and direction of the technological change and the consequences of employing the technology.
4. *Reporting* the information in a timely and meaningful manner.

Monitoring is continuous search, evaluation, review, and reporting—a serious thrust into today to forecast tomorrow.

STRUCTURAL ANALYSIS

Goal-oriented or normative analysis is based on a thesis that technology expands to meet the needs of society or the economy. Thus, by identifying the needs, one can forecast the future technology required to meet those needs—the carbonated beverage industry was latently seeking a lightweight, nonbreakable package to replace costly glass, and an oxygen-barrier plastic bottle and its polymeric derivation and

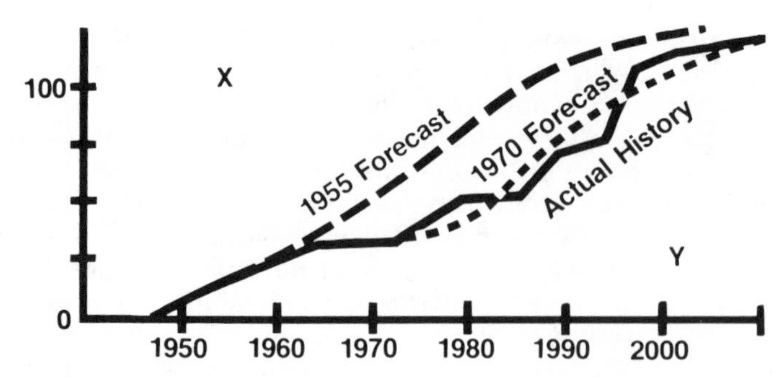

Fig. 11.10. A hypothetical example depicting substitution theory in technological forecasting.

fabrication technology were predictable. The need might generate the desire for a technological change which may be severely limited by scientific impossibility, costs, societal obstacles, etc. Thus, the many popular predictions of shelf-stable food preserved without heat by ionizing radiation or by lyophilization by the 1960s or 1970s proved invalid.

On the other hand, the confluence of need, technological feasibility, and cost justification within a prescribed parameter can be a highly effective mechanism for technological forecasting. Within a single organization in which some control might be exercised, the notion of applying structural analysis for a self-fulfilling project can be a powerful tool.

Thus structural analysis might be regarded not as forecasting but rather as planning.

Structural analysis involves definition of an objective and identification of all the parameters and variables that must be considered in reaching the objective, together with the time sequence of each, and preparation of the decision tree, mission flow analysis, or PERT chart displaying the progression. In effect, when the goal and the varying routes to the goal are known, pathways to achieve the goal can be forecasted—and when they are followed, the forecast is fulfilled.

MODELING

Possibly the most fashionable of the procedures employed today to depict the future for a managerial or business environment is modeling—a matrix of hypotheses converted to mathematical analogies which are subjected to multiple analyses to portray the infinite possibilities for the future. By holding all variables save one constant, the effect of the variable of concern can be measured—as in the popular sensitivity analysis.

Modeling thus is the expanded structural depiction of the future, embracing cause, effect, and time in the future. Modeling differs from intuitive forecasting in that it forces a discipline not necessarily always found in a single person or in a disrelated group. Modeling is unlike trend extrapolation, which deals principally with one or more variables on a time series basis. Modeling deals with a multiplicity of interactions, not all of which need be on a time scale, because it seeks the quantitative casual relationship. For example, in projecting the probabilities having to do with ambient temperature shelf-life of packaged foods, the interaction of oxygen with the food is generally a factor. Removal of some fraction of the oxygen has generally been regarded as a mechanism to achieve biochemical stability. What about the method of oxygen removal: sweep, displacement, vacuum plus pressure? What of the minute residual headspace oxygen that proved fatal in military dried food packaging? And what of the oxygen in the food product fibers? If a preprocessing step is taken, is the use of a sweep displacement sufficient to provide the requisite inert atmosphere?

Modeling does not seek the answers to the technical questions as much as it probes for the questions themselves and the interrelationships of the answers.

Models require that the assumptions and relationships be quite explicit and quantified so that they may be visualized, altered, and tested. A model, however, is not a real system, but rather a depiction, and therefore may overlook or overemphasize key variables. The use of mathematics, computers, and printouts implies a degree of precision applicable solely to the instrument being employed and not necessarily to the forecast. Good models are obviously extremely complex and therefore costly and thus may defy use in a conventional business environment, even though the alternative of no forecast or plan may be far more expensive. Since the data inputs are derived from secondary sources, such as opinion, it is vital to ensure the accuracy and precision

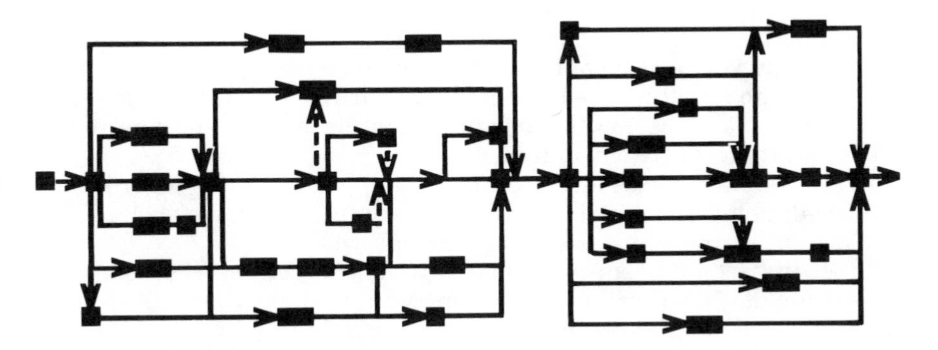

Fig. 11.11. Program evaluation and review technique—a method for forecasting by planning and implementing the plans.

of these inputs or the most elegantly conceived and implemented model can become a random, meaningless jumble. The fact that a result is a series of neatly tabulated numbers does not mean that it has meaning.

The two types of models are correlative and dynamic. Correlative models are the simpler of the two in that they are attempts to deduce the real relationship between two variables—e.g., molecular polarity is related to the ability of plastic polymers to bond to paperboard, and the imposition of a polar group to a polymer can lead to low-melt bonding. By constructing a series of correlative models and by ensuring that the correlations are not chance, the static interactions of variables can be predicted, although, of course, not necessarily on a time base. For example, an intensive study of polymer chemistry can demonstrate the probability of a zero-oxygen-permeability plastic in an economic gauge.

In contrast, dynamic models are truly causal in that they not only seek the correlation, they attempt to ensure that the relationship is true cause and effect—e.g., molecular polarity is inversely correlated by oxygen permeability because the oxygen gas molecule is electrically unbalanced.

Dynamic models are organized amplifications of the human thought process. Dynamic models foster the analysis of integral components by permitting constancy of some variables and the determination of interactions by permitting variability of some constants.

Although physical models, such as of an idealized packaging material barrier, may be constructed, the most common models are mathematical, made practical by the introduction of computers.

CROSS-IMPACT ANALYSIS

Cross-impact analysis rests on the oft-repeated thesis that many different variables affect events. Cross-impact analysis allows for the interacting forces without introducing the casual entirety through dynamic modeling. Cross-impact analysis forces explicit consideration of interactions and influences. Cross-impact analysis is a semiquantitative matrix depiction of the factors involved in a technological progression. The variables are identified and incorporated into a matrix to visually display the inputs, variables, and effects. Timing and probability of the event is included.

In an oversimplification, cross-impact analysis might be regarded as a three-dimensional version of a Delphi analysis in which each of the participant's inputs is crossed into the matrix.

Cross-impact analysis presupposes that interactions are not on a one-to-one basis, but rather in clusters, with variables such as timing, counter-influences, and reinforcements occurring.

SCENARIO GENERATION

Scenarios have been employed for many decades by military strategists to visualize the structure and outcome of a projected strategy. Scenarios have been variously defined as word descriptions of possible futures, explanations of events leading to the future, and future history.

Scenarios stimulate people to consider the range of future possibilities, and provide a basis for planning. Because this methodology is largely qualitative and verbal, it is proving relatively fashionable in managerial circles.

To develop a scenario, it is necessary to set an objective—e.g., to identify the range of possibilities, to encourage open-minded analysis, or to shift people from rigid strictness. Single scenarios are designed to focus attention on a single issue; bounding scenarios employ a central scenario and develop variations from the central theme; and significant theme scenarios are groups of alternatives with different variables on the same issue.

A scenario exercise assembles people who can contribute to constructing a vision of a future event. The procedure could follow a scheme as follows:

1. *Define the purpose* to the participants.
2. *Identify the variables* that must be considered and classify them according to the five external influencing environments: social, technical, political, economic, and ecological.
3. *Select the critical variables* from the matrix constructed with all the known influencing variables.
4. *Develop themes* for each scenario to basically establish a fabric for the future. For example, what will be the effect on packaging of a future world in which retail display space for frozen foods is restricted to half the present space? The scenario theme, which must be carefully selected, might be "Frozen Food Packaging in 1990," or "Food Packaging in a Freezerless Society," or "Food Distribution without Freezing."
5. *Develop a format* for each scenario and then prepare the scenarios or alternate sets of events.
6. *Compare the various scenarios.*
7. *Determine the impact* on the organization.
8. *Incorporate the unexpected inputs*—e.g., energy costs suddenly decline, so that freezer space is doubled, or homebuilders build 40 ft^3 walk-in freezers into every new home.
9. *Analyze the scenarios* for their probabilities and applicability to the issues being considered.

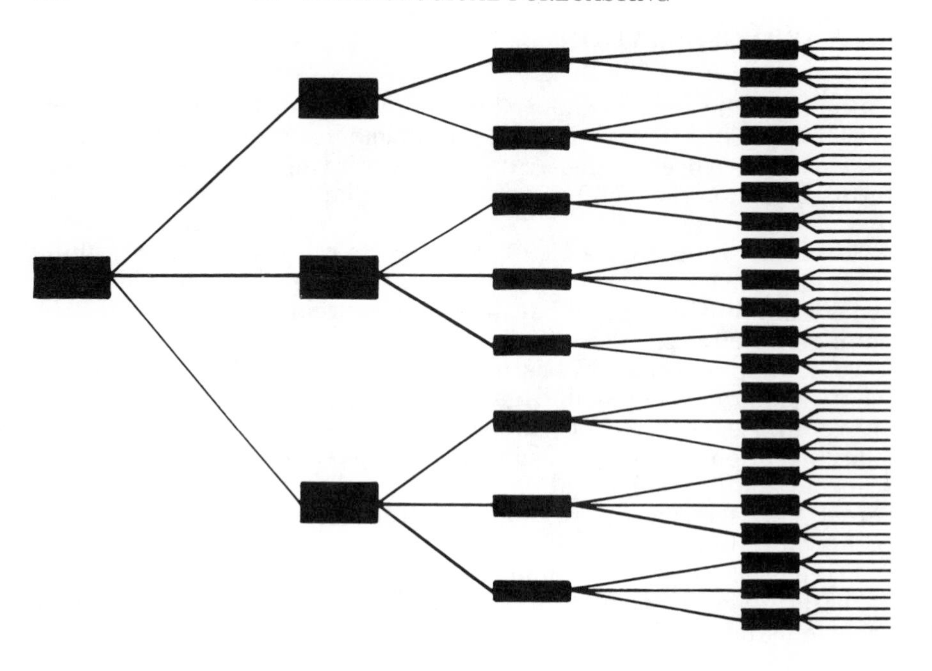

Fig. 11.12. The decision tree forecast: the probable results of alternative pathways.

Scenarios can become unwieldy exercises in imagination if they are not tightly controlled. On the other hand, well-constructed and conducted scenarios expand the vision and range of probabilities and, without hyperbole, open new directions that are likely to be searched and developed.

CONCLUSION

Technological forecasting is a newly emerging discipline that should function as an indispensable element in planning. Because packaging changes with market needs and with supply and technical forces impinging from the supply side, technological change is at the root of packaging progress. No one can be a participant in packaging without being involved as a leader or follower in its rapidly growing patterns. Inter-and intramaterial competition of the 1970s, which supplanted the one material/one purpose (and vice versa) basis of the 1950s through the mid-1960s, is itself being displaced by function without regard to material. The intermingling of materials and structures in a single package is becoming the rule, with technology being the driving force. Technological innovation is an irresistible thrust that must be recognized, understood, and forecasted if it is to be harnessed. Technological progress cannot be viewed as incomprehensible complexity, because in truth it follows fairly regular patterns.

The process of technological innovation is a sequential series of scientific observation and comprehension, conceptualization, laboratory demonstration, prototype development and evaluation, commercial development, proliferation, etc. No short cuts have been found.

The progress of technology is exponential in packaging as in other disciplines, and no single element of packaging—graphics, structural design, printing, material barrier, testing procedures, etc.—can ever be permitted to be regarded as dominant, since in the world of packaging, direction and progress originate from a range of forces—largely marketing, economic, and technical, founded in basic scientific disciplines. The conventional perceptions of prediction have been dissected to highlight their flaws—usually a total lack of objectivity and a focus on self-interest, which usually has little relationship to reality in a universe of events. We can neither rely on nor even seriously use the pronouncements of those with a product or service to sell or a deadline to meet. Art, skill in rhetoric, and the juxtaposition of fashionable buzzwords, frequently mistaken for forecasting, have no more validity than the crystal ball or tea leaves.

Technological progress is a rigorous process, and its forecasting is even more demanding because it is such a new concept to all of technology, to business, and to packaging.

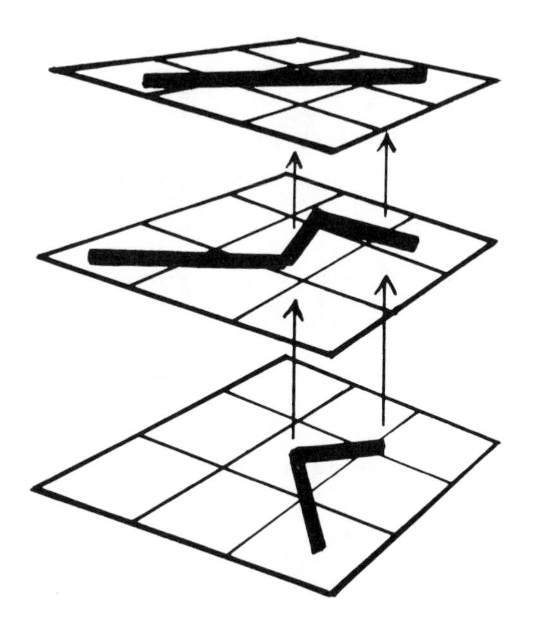

Fig. 11.13. The complex cross-impact analysis methods of technological forecasting may be simplified.

The importance of comprehending technological innovation and its forecasting has been developed. Packaging is as amenable to technological analysis as are the basic sciences or engineering disciplines—and the participants must recognize and appreciate these facts.

As significant as employing technological forecasting in perspective is the actual process, which certainly is not as well developed as it will be a decade or two from now. Nevertheless, because we are in the midst of change, we are obligated to be part of it by anticipating it, or by observing and being its inevitable victim. The processes described include monitoring, trend extrapolation, Delphi, structural analysis, scenario generation, cross-impact analysis, and modeling. Despite the paucity of exercises in application of these techniques, even in their currently primitive forms, they are significantly superior to emotion or intuition. Many organizations defer planning and consequently forecasting because of the totally mistaken notion that if you do not survive today, there is no tomorrow. True, but when you are in business tomorrow—and you have produced a set of sales and profit figures for tomorrow—you will still be in the uncomfortable position of arriving without knowing how you travelled. If an organization has a sales forecast, then it is essential to have a plan in which the technical forecast is an integral part.

Whether or not we employ the gestating processes of technological forecasting in planning our packaging futures, technological innovation will occur. Our role in that future is shaped by our actions today.

BIBLIOGRAPHY

ANON. 1981. Forecast of Emerging Technologies. US FDA National Technical Information Service, Rockville, MD.

BRIGHT, J. R. 1970. Evaluating signals of technological change Harvard Business Rev., Jan.-Feb.

BRIGHT, J. R. 1978. Practical Technological Forecasting. Industrial Management Center, Austin, Texas.

BRODY, A. L. 1983. The ins and outs of technological forecasting. Soc. Packaging Handling Eng. Tech. J. *2*, 1.

DUPREY, J. R. 1984. Future food packaging—Plastic, convenient, aseptic. Packaging *29*, 1.

JONES, H., and TWISS, B. C. 1978. Forecasting Technology for Planning Decisions. Petrocelli, Princeton, New Jersey.

MARTINO, J. P. 1972. Technological Forecasting for Decision Making. Elsevier, New York.

NORTH, H. Q., and PYKE, D.L. 1969. Probes of the technological future, Harvard Business Rev. 47, (3) 68.

QUINN, J. B. 1967. Technological forecasting, Harvard Business Rev. *45* (2), 89.

Index

N

O

P